Future Prospects for Soil Chemistry

Related Society Publications

Chemical Equilibrium and Reaction Models

Chemical Mobility and Reactivity in Soil Systems

Interactions of Soil Minerals with Natural Organics and Microbes

Rates of Soil Chemical Processes

Reactions and Movement of Organic Chemicals in Soils

Soil Chemistry and Ecosystem of Health

For information on these titles, please contact the ASA, CSSA, SSSA Headquarters Office; Attn.: Marketing; 677 South Segoe Road; Madison, WI 53711-1086. Telephone: (608) 273-8080, Fax: (608) 273-8089.

Future Prospects for Soil Chemistry

Proceedings of a symposium sponsored by Division S-2 Soil Chemistry of the Soil Science Society of America in St. Louis, Missouri, 30–31 Oct. 1995. Cosponsors include the Working Group MO Interactions of Soil Minerals with Organic Components and Microorganisms of the International Society of Soil Science; Divisions S-3 Soil Biology and Biochemistry, S-4 Soil Fertility and Plant Nutrition, S-9 Soil Mineralogy, and S-11 Soil and Environmental Quality of the Soil Science Society of America.

Editor
P.M. Huang

Co-editors
D.L. Sparks
S.A. Boyd

Organizing Committee
P.M. Huang, Chair
A.L. Page
P.S.C. Rao
M. Schnitzer
L.M. Shuman

Editor-in-Chief
Jerry M. Bigham

Managing Editor
David M. Kral

Associate Editor
Marian K. Viney

SSSA Special Publication Number 55

Soil Science Society of America
Madison, Wisconsin
1998

Copyright © 1998 by the Soil Science Society of America, Inc.

ALL RIGHTS RESERVED UNDER THE U.S. COPYRIGHT ACT OF 1978 (P.L. 94-553)

Any and all uses beyond the "fair use" provision of the law require written permission from the publishers and/or author(s); not applicable to contributions prepared by officers or employees of the U.S. Government as part of their official duties.

Soil Science Society of America, Inc.
677 South Segoe Road, Madison, Wisconsin 53711 USA

Library of Congress Catalog Card Number: 98-061821

Printed in the United States of America

CONTENTS

Foreword vii
Preface ix
Affiliation of Editor and Co-Editors xi
Affiliation of Members of the Organizing Committee xi
Contributors xiii
Conversion Factors for SI and non-SI Units xv

1   Soil Chemistry: Past, Present, and Future
    M.E. Sumner 1
2   Computational Chemistry in the Future of Soil Chemistry
    Brian J. Teppen 39
3   Thermodynamics of Soil Systems: A Personal Perspective
    P.F. Low 69
4   Kinetics of Soil Chemical Phenomena: Future Directions
    Donald L. Sparks 81
5   Elucidating Fundamental Mechanisms in Soil and Environmental Chemistry: The Role of Advanced Analytical, Spectroscopic, and Microscopic Methods
    Paul M. Bertsch and Douglas B. Hunter 103
6   Chemistry of the Soil Solution
    W.L. Lindsay and K.M. Catlett 123
7   The Chemistry of Soil Minerals
    S.B. Feldman and L.W. Zelazny 139
8   New Ideas on the Chemical Make-up of Soil Humic and Fulvic Acids
    Morris Schnitzer and H.-R. Schulten 153
9   Chemistry of Soil–Nutrient Interactions and Future Agricultural Sustainability
    Stanley A. Barber 179
10 Impact of Soil Chemical Reactions on Food Chain Contamination and Environmental Quality
    Terry J. Logan 191
11 Innovations in Curricula and Teaching Methods in Graduate Education of Soil Chemists
    Paul R. Bloom and Wayne P. Robarge 205
12 Interdisciplinary Approaches to Environmental Research: Examples from the National Science Foundation
    Margaret A. Cavanaugh and Maryellen Cameron 217
13 Role of Soil Science in the International Council of Scientific Unions: Soil Chemistry
    Winfried E.H. Blum and P.M. Huang 223

FOREWORD

Soil chemistry, as a subdiscipline of soil science, dates back to the classic ion exchange studies of H.S. Thompson and J. Thomas Way, in the 1850s. Thompson and Way's studies initiated more than 120 years of important research on macroscopic, equilibrium aspects of soil chemical reactions and processes including ion exchange, soil acidity and salinity, sorption, redox and precipitation–dissolution phenomena. These studies primarily focused on the chemistry of plant nutrients and involved the past giants of soil chemistry: Bradfield, Kelley, Mattson, Jenny, Coleman, Low, Rich, Babcock, Marshall, and Peech, to name a few.

In the 1970s the environmental quality of soils and waters became a major leitmotif in soil chemistry. Studies ensued on the chemical interactions of trace elements, radionuclides, plant nutrients, and organic chemicals in soils, sludges, and soil components. It soon became evident that in addition to understanding equilibrium processes, one must determine the kinetics and mechanisms of soil chemical reactions.

In the past two decades, major advances have occurred in soil chemistry by employing: state-of-the-art molecular scale spectroscopic–microscopic techniques that often enable one to study soil chemical processes in situ; kinetic techniques that provide information on reaction rates over time scales of milliseconds to years; and, advances in computational chemistry. Many unanswered questions remain, and arguably, the above techniques, and yet to be discovered advances, will be employed in the next millennium to help unlock the remaining mysteries of soil chemical phenomena. Such research, which is discussed in this book, is imperative to preserving our planet, feeding the world's rapidly rising population, and enhancing the quality of life for humankind.

This excellent treatise, which contains contributions from leading soil chemists, discusses past and present research achievements, but focuses on future research needs and innovations. Other important topics that are discussed include trends in graduate education and training, interdisciplinary research and funding opportunities in environmental soil chemistry, and interactions of soil chemists with colleagues in the physical and life sciences. This special publication will wisely serve scientists, students, policy makers, and funding agency personnel for years to come.

Gary W. Petersen
SSSA President

PREFACE

An impressive amount of progress has been made in soil chemistry research and education over the years. It is time to look back, to review current progress, and to identify future directions in this increasingly important area of Soil Science.

The U.S. National Academy of Sciences has recently created a new section named <u>Plant, Soil and Microbial Sciences</u> under Division 6 titled *Applied Biology and Agricultural Science*. This is the first time that Soil Science has been recognized by the Academy by listing in a name of a Section. The U.S. National Science Foundation is working on several research initiatives that will provide more opportunity to soil chemists. Dr. M.A. Cavanaugh, Program Director of the Chemistry Division, and Dr. M. Cameron, then of the Division of Earth Sciences of the National Science Foundation, accepted the invitation from SSSA Division S-2 to jointly present a paper entitled "Environmental Research at the National Science Foundation: New Opportunities for Soil Chemists" in the Symposium "Whither Soil Chemistry" at the 1995 ASA–CSSA–SSSA Annual Meetings in St. Louis, MO. Funding is indeed the key to the development of innovations in soil chemistry research, education, and placement and the impacts on the well-being of humankind. The International Society of Soil Science has recently become a member of the International Council of Scientific Unions. This is a very significant recognition of Soil Science in the international scientific community. Dr. W.E.H. Blum, Secretary-General, International Society of Soil Science is the senior author of a key paper entitled "Role of Soil Science in the International Council of Scientific Unions: Soil Chemistry" in the above-noted Symposium. The structure of the International Union of Pure and Applied Chemistry (IUPAC) was changed in 1995. The "Division of Chemistry and the Environment," which is subdivided into six Commissions, was newly founded in IUPAC; the "Commission of Fundamental Environmental Chemistry" and the "Commission of Soil and Water Chemistry" are of special interest for Soil Chemists. In view of the past achievements and current developments of Soil Science, it is indeed timely to address the issue of the future prospects of soil chemistry development.

The symposium "Whither Soil Chemistry" was sponsored by Division S-2 Soil Chemistry of SSSA and cosponsored by Working Group MO Interactions of Soil Minerals with Organic Components and Microorganisms of the International Society of Soil Science, and Division S-3 Soil Biology and Biochemistry, Division S-4 Soil Fertility and Plant Nutrition, Division S-9 Soil Mineralogy, and Division S-11 Soil and Environmental Quality of SSSA. The SSSA Executive Committee approved the proposed special publication and appointed the Editorial Committee on 16 June 1995. The objective of this special publication is to address the issue of future prospects of soil chemistry development in terms of (i) fundamentals, (ii) innovations in organization, funding, education, and placement, and (iii) impacts on global scientific community and humankind. It is hoped that this publication will: (i) provide new information on the prospects of soil chemistry development pertaining to research funding, education, placement, and the impacts on humankind, (ii) foster communications on scientific developments of soil chemists to potential users of soil science information in other scientific disciplines, and (iii) promote the soil science profession on a global scale.

P.M. Huang, Editor
D.L. Sparks, Co-editor
S.A. Boyd, Co-editor

Affiliation of Editor and Co-Editors

P.M. Huang	Editor, Department of Soil Science, University of Saskatchewan, 51 Campus Drive, Saskatoon, SK Canada S7N 5A8
D.L. Sparks	Co-editor, Department of Plant and Soil Sciences, University of Delaware, Newark, Delaware 19717-1303
S.A. Boyd	Co-editor, Department of Crop and Soil Sciences, Michigan State University, East Lansing, MI 48824

Affiliation of Members of the Organizing Committee

P.M. Huang	Chair, Department of Soil Science, University of Saskatchewan, 51 Campus Drive, Saskatoon, SK Canada S7N 5A8
A.L. Page	Department of Environmental Sciences, University of California, 2208 Geology Building, Riverside, CA 92521-0424
P.S.C. Rao	Soil and Water Science Department, University of Florida, 2169 McCarty Hall, P.O. Box 110290, Gainesville, FL 32611-0290
M. Schnitzer	Eastern Cereal and Oilseed Research Centre, Agriculture and Agri-Food Canada, Central Experimental Farm, Ottawa, ON Canada K1A 0C6
L.M. Shuman	Department of Crop and Soil Sciences, University of Georgia, Georgia Station, Griffin, GA 30223-1797

CONTRIBUTORS

Stanley A. Barber — John B. Peterson Distinguished Professor of Agronomy Emeritus, Department of Agronomy, Purdue University, West Lafayette, IN 47907-1150

Paul M. Bertsch — Director, Advanced Analytical Center for Environmental Sciences, and Professor, Department of Crop and Soil Sciences, Savannah River Ecology Laboratory, University of Georgia, Drawer E, Aiken, SC 29802

Paul R. Bloom — Professor of Soil Science, Department of Soil, Water and Climate, University of Minnesota, St. Paul, MN 55108

Winfried E.H. Blum — Professor of Soil Science, Institut für Bodenforschung, Universität für Bodenkultur, Vienna, Austria

Maryellen Cameron — Executive Officer, Office of Polar Programs, National Science Foundation, Arlington, VA 22230

K.M. Catlett — Department of Soil and Crop Sciences, Colorado State University, Fort Collins, CO 80523

Margaret A. Cavanaugh — Program Director, Division of Chemistry, National Science Foundation, Arlington, VA 22230

S.B. Feldman — American Colloid Co., 1500 W. Shore Dr., Arlington Heights, IL 60004

P.M. Huang — Professor of Soil Science, Department of Soil Science, University of Saskatchewan, 51 Campus Drive, Saskatoon, SK, Canada S7N 5A8

Douglas B. Hunter — Chemical Spectroscopist, Advanced Analytical Center for Environmental Sciences, Savannah River Ecology Laboratory, University of Georgia, Drawer E, Aiken, SC 29802

W.L. Lindsay — Emeritus University Distinguished Professor, Department of Soil and Crop Sciences, Colorado State University, Fort Collins, CO 80523

Terry J. Logan — Professor of Soil Science, School of Natural Resources, Ohio State University, Columbus, OH 43210

P.F. Low — Professor Emeritus of Soil Chemistry, Department of Agronomy, Purdue University, West Lafayette, IN 47907-1150 (deceased)

Wayne P. Robarge — Senior Researcher, Department of Soil Science, North Carolina State University, Raleigh, NC 27695

Morris Schnitzer — Emeritus Distinguished Research Scientist, Eastern Cereal and Oilseed Research Centre, Agriculture and Agri-Food Canada, Ottawa, ON, Canada K1A 0C6

H.-R. Schulten — Professor, Institute for Soil Science, University of Rostock, Rostock, Germany

Donald L. Sparks — Distinguished Professor of Soil Chemistry, Department of Plant and Soil Sciences, University of Delaware, Newark, DE 19717-1303

M.E. Sumner — Regents' Professor of Environmental Soil Science, Department of Crop and Soil Sciences, University of Georgia, Athens, GA 30602-7272

Brian J. Teppen — Assistant Professor, Department of Crop and Soil Sciences, Michigan State University, East Lansing, MI 48824-1325

L.W. Zelazny — Thomas B. Hutcheson Jr. Eminent Scholar, Department of Crop and Soil Environmental Sciences, Virginia Polytechnic Institute and State University, Blacksburg, VA 24061

Conversion Factors for SI and non-SI Units

Conversion Factors for SI and non-SI Units

To convert Column 1 into Column 2, multiply by	Column 1 SI Unit	Column 2 non-SI Units	To convert Column 2 into Column 1, multiply by
Length			
0.621	kilometer, km (10^3 m)	mile, mi	1.609
1.094	meter, m	yard, yd	0.914
3.28	meter, m	foot, ft	0.304
1.0	micrometer, µm (10^{-6} m)	micron, µ	1.0
3.94×10^{-2}	millimeter, mm (10^{-3} m)	inch, in	25.4
10	nanometer, nm (10^{-9} m)	Angstrom, Å	0.1
Area			
2.47	hectare, ha	acre	0.405
247	square kilometer, km² (10^3 m)²	acre	4.05×10^{-3}
0.386	square kilometer, km² (10^3 m)²	square mile, mi²	2.590
2.47×10^{-4}	square meter, m²	acre	4.05×10^3
10.76	square meter, m²	square foot, ft²	9.29×10^{-2}
1.55×10^{-3}	square millimeter, mm² (10^{-3} m)²	square inch, in²	645
Volume			
9.73×10^{-3}	cubic meter, m³	acre-inch	102.8
35.3	cubic meter, m³	cubic foot, ft³	2.83×10^{-2}
6.10×10^4	cubic meter, m³	cubic inch, in³	1.64×10^{-5}
2.84×10^{-2}	liter, L (10^{-3} m³)	bushel, bu	35.24
1.057	liter, L (10^{-3} m³)	quart (liquid), qt	0.946
3.53×10^{-2}	liter, L (10^{-3} m³)	cubic foot, ft³	28.3
0.265	liter, L (10^{-3} m³)	gallon	3.78
33.78	liter, L (10^{-3} m³)	ounce (fluid), oz	2.96×10^{-2}
2.11	liter, L (10^{-3} m³)	pint (fluid), pt	0.473

CONVERSION FACTORS FOR SI AND NON-SI UNITS

To convert Column 1 into Column 2, multiply by	Column 1 SI Unit	Column 2 non-SI Unit	To convert Column 2 into Column 1, multiply by
	Mass		
2.20×10^{-3}	gram, g (10^{-3} kg)	pound, lb	454
3.52×10^{-2}	gram, g (10^{-3} kg)	ounce (avdp), oz	28.4
2.205	kilogram, kg	pound, lb	0.454
0.01	kilogram, kg	quintal (metric), q	100
1.10×10^{-3}	kilogram, kg	ton (2000 lb), ton	907
1.102	megagram, Mg (tonne)	ton (U.S.), ton	0.907
1.102	tonne, t	ton (U.S.), ton	0.907
	Yield and Rate		
0.893	kilogram per hectare, kg ha^{-1}	pound per acre, lb acre^{-1}	1.12
7.77×10^{-2}	kilogram per cubic meter, kg m^{-3}	pound per bushel, lb bu^{-1}	12.87
1.49×10^{-2}	kilogram per hectare, kg ha^{-1}	bushel per acre, 60 lb	67.19
1.59×10^{-2}	kilogram per hectare, kg ha^{-1}	bushel per acre, 56 lb	62.71
1.86×10^{-2}	kilogram per hectare, kg ha^{-1}	bushel per acre, 48 lb	53.75
0.107	liter per hectare, L ha^{-1}	gallon per acre	9.35
893	tonne per hectare, t ha^{-1}	pound per acre, lb acre^{-1}	1.12×10^{-3}
893	megagram per hectare, Mg ha^{-1}	pound per acre, lb acre^{-1}	1.12×10^{-3}
0.446	megagram per hectare, Mg ha^{-1}	ton (2000 lb) per acre, ton acre^{-1}	2.24
2.24	meter per second, m s^{-1}	mile per hour	0.447
	Specific Surface		
10	square meter per kilogram, m^2 kg^{-1}	square centimeter per gram, cm^2 g^{-1}	0.1
1000	square meter per kilogram, m^2 kg^{-1}	square millimeter per gram, mm^2 g^{-1}	0.001
	Pressure		
9.90	megapascal, MPa (10^6 Pa)	atmosphere	0.101
10	megapascal, MPa (10^6 Pa)	bar	0.1
1.00	megagram per cubic meter, Mg m^{-3}	gram per cubic centimeter, g cm^{-3}	1.00
2.09×10^{-2}	pascal, Pa	pound per square foot, lb ft^{-2}	47.9
1.45×10^{-4}	pascal, Pa	pound per square inch, lb in^{-2}	6.90×10^3

(continued on next page)

Conversion Factors for SI and non-SI Units

To convert Column 1 into Column 2, multiply by	Column 1 SI Unit	Column 2 non-SI Units	To convert Column 2 into Column 1, multiply by
Temperature			
1.00 (K − 273)	kelvin, K	Celsius, °C	1.00 (°C + 273)
(9/5 °C) + 32	Celsius, °C	Fahrenheit, °F	5/9 (°F − 32)
Energy, Work, Quantity of Heat			
9.52×10^{-4}	joule, J	British thermal unit, Btu	1.05×10^{3}
0.239	joule, J	calorie, cal	4.19
10^{7}	joule, J	erg	10^{-7}
0.735	joule, J	foot-pound	1.36
2.387×10^{-5}	joule per square meter, J m^{-2}	calorie per square centimeter (langley)	4.19×10^{4}
10^{5}	newton, N	dyne	10^{-5}
1.43×10^{-3}	watt per square meter, W m^{-2}	calorie per square centimeter minute (irradiance), cal cm^{-2} min^{-1}	698
Transpiration and Photosynthesis			
3.60×10^{-2}	milligram per square meter second, mg m^{-2} s^{-1}	gram per square decimeter hour, g dm^{-2} h^{-1}	27.8
5.56×10^{-3}	milligram (H$_2$O) per square meter second, mg m^{-2} s^{-1}	micromole (H$_2$O) per square centimeter second, µmol cm^{-2} s^{-1}	180
10^{-4}	milligram per square meter second, mg m^{-2} s^{-1}	milligram per square centimeter second, mg cm^{-2} s^{-1}	10^{4}
35.97	milligram per square meter second, mg m^{-2} s^{-1}	milligram per square decimeter hour, mg dm^{-2} h^{-1}	2.78×10^{-2}
Plane Angle			
57.3	radian, rad	degrees (angle), °	1.75×10^{-2}

CONVERSION FACTORS FOR SI AND NON-SI UNITS

Electrical Conductivity, Electricity, and Magnetism

To convert Column 1 into Column 2, multiply by	Column 1 SI Unit	Column 2 non-SI Unit	To convert Column 2 into Column 1, multiply by
10	siemen per meter, S m^{-1}	millimho per centimeter, mmho cm^{-1}	0.1
10^4	tesla, T	gauss, G	10^{-4}

Water Measurement

9.73 × 10^{-3}	cubic meter, m^3	acre-inch, acre-in	102.8
9.81 × 10^{-3}	cubic meter per hour, m^3 h^{-1}	cubic foot per second, ft^3 s^{-1}	101.9
4.40	cubic meter per hour, m^3 h^{-1}	U.S. gallon per minute, gal min^{-1}	0.227
8.11	hectare meter, ha m	acre-foot, acre-ft	0.123
97.28	hectare meter, ha m	acre-inch, acre-in	1.03 × 10^{-2}
8.1 × 10^{-2}	hectare centimeter, ha cm	acre-foot, acre-ft	12.33

Concentrations

1	centimole per kilogram, cmol kg^{-1}	milliequivalent per 100 grams, meq 100 g^{-1}	1
0.1	gram per kilogram, g kg^{-1}	percent, %	10
1	milligram per kilogram, mg kg^{-1}	parts per million, ppm	1

Radioactivity

2.7 × 10^{-11}	becquerel, Bq	curie, Ci	3.7 × 10^{10}
2.7 × 10^{-2}	becquerel per kilogram, Bq kg^{-1}	picocurie per gram, pCi g^{-1}	37
100	gray, Gy (absorbed dose)	rad, rd	0.01
100	sievert, Sv (equivalent dose)	rem (roentgen equivalent man)	0.01

Plant Nutrient Conversion

	Elemental	Oxide	
2.29	P	P$_2$O$_5$	0.437
1.20	K	K$_2$O	0.830
1.39	Ca	CaO	0.715
1.66	Mg	MgO	0.602

1 Soil Chemistry: Past, Present, and Future[1]

M. E. Sumner

University of Georgia
Athens, Georgia

1–1 INTRODUCTION

I am honored to have been invited to make this presentation, but at the same time, I am intimidated by the immensity of the task. To attempt to fully cover the past and present aspects of the topic adequately would require far more time and space than I have at my disposal. To do justice to the future is impossible; indeed past attempts at doing so have usually resulted in failure, which was recognized by Winston Churchill:

> "I always avoid prophesying beforehand, because it is much better policy to prophesy after the event has already taken place."

Given the fickleness of public opinion, government directions, funding agencies, and economics, I fear to tread in this arena. Consequently, the comments to follow are subjective in nature and are based on my experiences in Soil Chemistry during the past 40 yr on the four continents on which I have worked, Africa, the Americas, Europe, and Australia. In no way, would it be possible for me to present a detailed and comprehensive account of the history of Soil Chemistry and its role in the future in such a short space. Indeed, such a task would be overwhelming given the ever-expanding frontiers of the subject. Consequently, this presentation represents my interpretation of only some aspects of the past and present, and having cursorily glanced into the crystal ball, the directions I believe to be important in the future. At the beginning of my career in Soil Chemistry under my mentors, E.R. Orchard, R.K. Schofield, and G.H. Bolt, one was required to be well versed in all aspects of Soil Chemistry, but alas today, the assimilation of such a comprehensive knowledge of the entire subject would be somewhat more difficult and, in the future, a truly daunting task. Soil Chemistry, a speciality within Soil Science, is now being divided itself with ever-increasing specialization becoming the order of the day. This is unfortunate but inevitable.

In the present age involving changes in funding sources and the organization of research throughout the world, it is appropriate to review past achievements in Soil Chemistry and to prognosticate about the future. There have been

---

[1] Contribution from the Department of Crop and Soil Sciences, University of Georgia, Athens, GA 30677.

Copyright © 1998. Soil Science Society of America, 677 S. Segoe Rd., Madison, WI 53711, USA. *Future Prospects for Soil Chemistry.* SSSA Special Publication no. 55.

many such attempts in the past that have focused primarily on Soil Science as a whole (Kellog, 1961; Cooke, 1979; Tinker, 1985; Aldrich, 1987; Nielsen, 1987; Wild, 1989; Gardner, 1991, Greenland, 1991; Greenwood, 1993; Simonson, 1991; Theng, 1991) or on particular facets of the subject such as Soil Fertility (Bradfield, 1961), Biotechnology (Läuchli, 1987), Soil Physics (Kirkham, 1961; Hillel, 1991; Philip, 1991), Environmental Aspects (Menzel, 1991), Organic Matter (Schnitzer, 1991), and Soil Microbiology (Alexander, 1991) with a few taking a careful look at Soil Chemistry (Schuffelen, 1974; Thomas, 1977; James, 1993; Sparks, 1993).

In the past, Soil Chemistry in all its aspects (teaching, research and extension) has been driven largely by the requirements for improving agricultural production. There can be little doubt that, in the future as a result of increasing environmental regulations to protect the well-being of human society, the focus of Soil Chemistry will increasingly change from Agriculture to the Environment. In a large measure, this has been under way for the past decade or more and the tempo is definitely going to increase in the next 25 yr. Fortunately, in the past, the major part of the scientific effort in Soil Chemistry has addressed the improvement of our understanding of soil chemical processes. This stands us in good stead to face the challenges that the environmental arena will present because, in most cases, the applications for our knowledge will simply change and we will not have to reinvent the wheel from scratch to tackle these new problems. We have a vast storehouse of knowledge at our disposal on which we can continue to build in the future. At the same time, we should guard against our role in this arena being usurped by those claiming to be environmentalists who have an inadequate understanding of the processes involved in the behavior of contaminants in the soil environment. We need to apply our current knowledge to the new problems posed by the environment as well as developing new technologies aimed at solving problems for which we have no current solutions.

Because of the confined nature of this presentation, I am going to take the liberty of selecting only two topics in Soil Chemistry as illustrations of how our knowledge that was developed for agricultural purposes, can be applied in the new environmental arena. In addition, I propose to develop the discussion from the accumulated knowledge on the first topic to the future demands likely to be placed on Soil Chemistry rather than spending time on reviewing past achievements in detail that have been adequately recorded in the literature. I have selected the following topics, which I believe are or are going to be important issues in the future, and at the same time, illustrate our ability to apply present knowledge to current and future problems: (i) the chemistry of soil colloids in relation to soil degradation, transport of pollutants, and contamination of water, and (ii) the sorption and desorption of contaminants by soils in relation to bioavailability and water quality. Although the present audience is largely from North America, I am going to attempt to place a global flavor on this presentation as I believe many pressing needs in Soil Chemistry research lie in areas outside the developed nations, where the burgeoning population is placing and will continue to place, increased pressure on the soil both from the agricultural and environmental perspectives. I, therefore, plan to highlight the importance of studying the chemistry of variable charge soils that have received disproportionately little attention in the past. These soils cover a wide spectrum with different morphological, mineralogical, chemical, physical, biological, and genetic characteristics but have one property in common, namely, both the magnitude and sign of their surface charge are strongly dependent on the chemistry of the equilibrium solution (pH, composition, and ionic strength). Soils falling into this category dominate the Alfisols and Ultisols with low activity clays, the highly weathered and leached Oxisols of the humid tropics, the Andisols derived from volcanic ash and the Spodosols of the cooler forested regions.

1–2 THE PAST AND PRESENT

1–2.1 Chemistry of Soil Colloids

1–2.1.1 Charge

From the earliest work by Thompson (1850) and Way (1850, 1852) that showed that soils had the power to sorb and exchange cations through that of Hissink (1924), Gedroiz (1925), and Kelley (1948) who added further concepts such as rates of reaction, replacing power, saturation, and the effect of exchangeable cations on soil properties, our knowledge of the phenomenon of cation exchange has developed. Very early on, Way (1850) found that the English soils being studied had no anion-exchange capacity (AEC):

> *"The absence of the power in the soil to unite with gypsum, or in any way to retain sulphuric acid, coupled with the certainty that sulphur is of absolute necessity to vegetation, may perhaps in part explain the advantage in applying sulphate of lime in certain cases."*

The choice of soil was understandable, but nevertheless unfortunate, as it has led us down the wrong path. Consequently, it was some considerable time before it was realized that soils could sorb (Gedroiz, 1925) and exchange anions (Mattson, 1927). While some soil colloids have been known for a long time to carry positive charges (Mattson, 1931; Mattson & Pugh, 1934; Mattson & Hou, 1937; Schofield, 1939, 1947, 1949), much of the literature on exchange chemistry and interfacial phenomena on whole soils with relatively few exceptions (Jackson, 1963; Rich, 1968) from the earliest times to the present, has concentrated on the study of dominantly negatively charged systems (Kelley, 1948; Bear, 1964; Bolt & Bruggenwert, 1976; Cresser et al., 1993). This occurred despite the fact that soil components were known to carry variable charge (Mattson, 1931; Sumner, 1963b; Kwong & Huang, 1978, 1979). The reason for this no doubt stems from the fact that the soils available in regions, where most of this research was being conducted, were all of the permanent negative charge type. The focus of the world of Soil Chemistry on these systems resulted in the transfer of the wrong technology to areas being newly developed in Africa, South America, and Asia, particularly with respect to lime. Lime rates required to bring highly weathered soils to pH 7 as was the fashion in the developed world, were excessive and then often resulted in yield depressions below control plots (Sanchez & Salinas, 1981; Farina et al., 1982). Nevertheless during the 1960s and 1970s, work on positive charge in soils continued on materials from Africa (Sumner, 1963a,b; Davidtz & Sumner, 1965; Sumner & Davidtz, 1965; Fey & le Roux, 1976; Gallez et al., 1976), Hawaii (El-Swaify & Sayegh, 1975), Australia (Gillman, 1974, 1979; Gillman & Bell, 1976), Canada (Hendershot & Lavkulich, 1978), and South America (van Raij & Peech, 1972; Espinoza et al., 1975; Morais et al., 1976) that led to a better understanding of the origins and behavior (pH and ionic strength) of the charge in variable charge systems and improved methods for measuring their properties (Gillman, 1979; Gillman & Uehara, 1980; Gillman & Sumpter, 1986a,b). It is interesting to note that even in Canada, soils occur where their point of zero charge (PZC) approaches their field pH (Hendershot & Lavkulich, 1978).

Despite the fact that Thomas (1977) refocused attention in the USA on the importance of positive charges on the adsorption and exchange of simple anions (Cl^-, NO_3^-, SO_4^{2-}) in soils in his excellent review of ion exchange, historically, most interest in anion exchange has focused on phosphate sorption from the initial investigations of Ravikovitch (1934) and Scarseth (1935). (Because of the dominantly covalent nature of P sorption, I plan to discuss this phenomenon under the section on Sorption and Desorption.) Much valuable work has continued on the charge aspects of variable charge soils in Australia (Gillman, 1979, 1981, 1984; Gillman & Sumpter, 1986a,b; Gillman & Sinclair, 1987; Gillman & Hallman, 1988), the USA (Grove et al., 1982; Gillman & Sumner, 1987), China (Zhang et al., 1989, 1991), Japan (Wada & Okamura, 1983), and Brazil (van Raij et al., 1988; Chorover & Sposito, 1995a,b). Of the simple anions, SO_4^{2-} has received the most attention (Kamprath et al., 1956; Adams & Rawajfih, 1977; Rajan, 1978), but there is more specificity in its adsorption than with the other simple anions (Alva et al., 1991b; Bolan et al., 1993). Despite this rather sizeable literature on variable charge soils and their properties, textbooks with few exceptions (Bear, 1964) have hardly made a reference to positive charge or AEC in soils until recently. In fact, the first text book devoted to the chemistry of variable charge soils first appeared in 1981 (Uehara & Gillman, 1981). This is a reflection of the dominance of contributions to Soil Chemistry from western Europe and northern and western America where, in general, the soils are young and overwhelmingly negatively charged. On the other hand, the variable charge nature of Andisols that have been extensively studied in Japan (Wada, 1977; Wada & Okamura, 1983), Chile (Espinoza et al, 1975), and New Zealand (Fieldes & Schofield, 1960), has been known for many years.

The discovery of crystalline clay minerals in soils with the advent of x-ray diffraction (Hendricks & Fry, 1930), that caused the ideas of Mattson to be discarded and consequently largely ignored, also contributed to the preoccupation of studying permanently charged systems. Thus, as a result of an advance in technology for studying the mineralogy of soils, we became diverted from a path that would have provided valuable insights. A combination of the approaches of Mattson and Hendricks and Fry would have been much more fruitful but alas we tended to place all our eggs in one basket. It is interesting that Mattson (1931) clearly demonstrated that colloids from soils common in the southeastern USA such as the Cecil and Sassafras series could have a net positive charge under the commonly occurring acid conditions, yet this information was ignored and

Table 1–1. Charge characteristics of various soil orders.

Soil order	Horizon	pH$_{Salt}$	Charge cmol$_c$ kg^{-1}	
			Negative	Positive
Oxisol (Morais et al., 1976)	A	3.0	3.9	3.2
	A	7.8	11.6	1.4
	B	3.0	2.5	5.1
	B	8.2	5.3	2.0
Ultisol (Morais et al., 1976)	A	3.0	1.0	1.1
	A	8.5	5.0	0.8
	B	3.0	2.5	2.4
	B	8.5	3.8	1.7
Alfisol (Morais et al., 1976)	A	2.9	3.8	2.0
	A	8.5	14.0	−1.6
	B	2.9	2.4	5.5
	B	8.5	9.5	1.9
Andisol (Sumner et al., 1993)	A	4.6	3.8	2.6
	B	5.1	1.9	4.2
Oxisol (Sumner, 1963b)	B	3.5	2.3	5.1
	B	8.2	8.0	1.0

appeared to go missing for >25 yr until Coleman et al. (1959), Rich (1968), and Grove et al. (1982) reestablished the fact. Despite considerable ridicule from my colleagues, the latter investigation was initiated as a result of my insistence that the soils of Georgia should behave similarly to those I had investigated in Africa. Alas, many workers, both present and future, do not or will not review what has been done in the past. Consider the difference in approach that we would all be following now had the birthplace of Soil Chemistry been on the Campo Cerrado of Brazil or the highlands of Africa where workers would have selected Oxisols and Ultisols instead of Mollisols and where ignoring the charge character of the soil and liming to near neutral pH values, would have had a disastrous effect on yield. I submit that, had this been the case, we would have initially developed the general case for the exchange behavior of soils rather than the specific case for permanently charged surfaces. Although considerable stimulus was given to the study of variable charge soils by the International Conference held in New Zealand in 1981 (Theng, 1980), relatively little notice of these developments has been taken by workers in the traditional home of Soil Chemistry. I am pleased to state that some (Sposito, 1989; McBride, 1994; Sparks, 1986, 1995) but not all (Cresser et al., 1993) recent books on Soil Chemistry have remedied this situation. Hopefully, the situation will change by the appearance in the near future of a comprehensive book on the subject (Yu, 1997).

It is fair to say that we currently have a good understanding of cation exchange phenomena in soils including the kinetics and thermodynamics of the processes involved (Sparks, 1986, 1989, 1995; Sposito, 1981, 1989). I will not dwell on this topic further except to point out that most soils considered to be of permanent charge character have a variable charge component due to the organic matter and Al and Fe coatings and interlayers present. The variable charge character of soils in some orders is illustrated in Table 1–1. One interesting aspect of these data is the indication that permanent positive charge may exist in some

Table 1–2. Soil pH and points of zero net charge (PZNC) for a range of soils from different parts of the world.

Soil type	Depth (m)	Soil pH H$_2$O	KCl	PZNC	Source
Oxisol, Australia	0–0.1	5.2	4.7	4.8	Gillman & Bell, 1976
	0.3–0.6	5.4	5.4	5.5	
	2.1–2.4	4.8	5.9	6.4	
Andisol, Japan	Bb		5.1	5.9	Wada & Okamura, 1983
Spodosol, Canada	B22ir	6.0	5.3	5.0	Hendershot & Lavkulich, 1978
Oxisol, Brazil	A	4.2	3.5	3.2	Morais et al., 1976
	B	4.9	4.4	6.1	
Ultisol, Nigeria	0.3–0.7	5.8	4.4	3.6	Gallez et al., 1976
Oxisol, China	0.7–0.9	5.0	5.1	4.8	Zhang et al., 1989
Oxisol, Jamaica	2.4–3.0		6.9	8.4	Sumner, 1994, unpublished data
Andisol, Guatemala	0.8–1.2		5.0	5.6	Sumner et al., 1993

soils as evidenced by the presence of substantial positive charge at high pH (8.5) that could arise as the result of isomorphous substitutions of Ti^{IV} for Al^{III} or Fe^{III} in kaolinite (Holdridge, 1959) or Fe oxides (Katsura et al., 1962). Thus most soils of the world with few exceptions should be considered as mixed charge systems. In the remainder of this section, I wish to present a thumbnail sketch of our current knowledge of variable charge in soils in the hope of being able to convince more people to take this into consideration.

Vast areas of the world are covered by soils with variable charge that have developed as a result of intense weathering and leaching. The resulting soils are rich in hydrous oxides and 1:1 clay minerals with only accessory amounts of 2:1 minerals if any are present at all. The interaction of the surfaces of hydrous oxides, edges of the clay minerals, and organic matter with potential determining ions (H$^+$ and OH$^-$) is responsible for the development of the variable charge. Most of these soils have Points of Zero Net Charge (PZNC) in the pH range commonly found in nature as illustrated in Table 1–2.

It should be realized that even though the pH of the soil may be above its PZNC, substantial amounts of positive charge can still be present. In addition to being dependent on pH, these charges are sensitive to changes in concentration and type of electrolyte (Fig. 1–1). Anion exchange takes place at protonated OH groups on the surfaces of Fe, Al, and Mn oxyhydroxides, amorphous aluminosilicates, and at the edges of clay minerals according to the following reaction:

$$>[Fe,Al,Mn]\text{-}OH^{-1/2} + H^+ \rightleftarrows >[Fe,Al,Mn]\text{-}OHH^{+1/2} \qquad [1]$$

Thus at low pH, these groups are positively charged, and at high pH, negatively charged. Under acid conditions at a given pH, increasing concentration increases positive charge and vice versa. In the presence of a specifically sorbed cation (Ca), the pH$_o$ (pH at which variable positive and negative charges are equal) is increased. Specific sorption of anions both inorganic and organic (Fig. 1–2) results in an increase in negative charge and a decrease in positive charge. Consequently, the net charge becomes more negative. This charge reversal caused

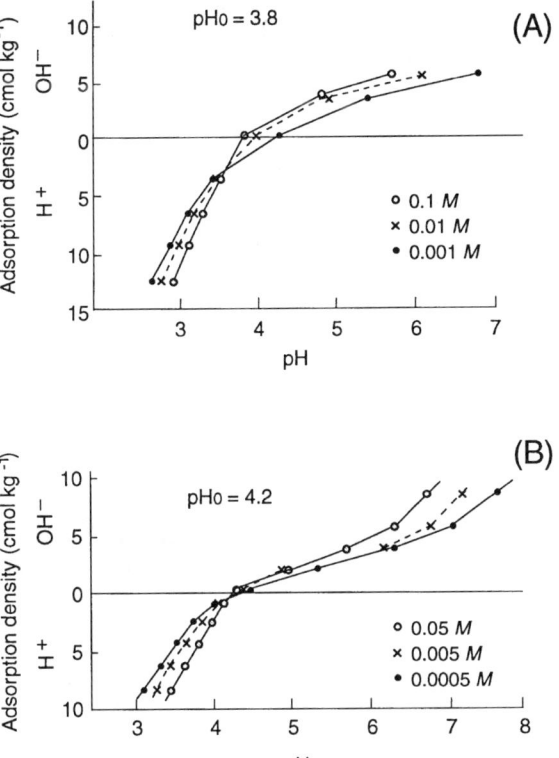

Fig. 1–1. Effect of pH and electrolyte concentration and type (A) KCl and (B) CaCl$_2$ on the adsorption density of H$^+$ and OH$^-$ by a Chinese Oxisol (Yu, 1997).

by the adsorption of such anions entirely alters the behavior of the soil colloids. Thus, unless the colloidal properties of soils are studied under conditions similar to those extant in the field, extrapolations to real world systems will produce erroneous conclusions. Under acid conditions that are common in these areas, substantial amounts of positive charge in addition to negative charge can be present, completely changing the behavior of many chemical species in the soil. A major consequence of the strong dependence of charge on ambient solution conditions is a highly variable surface charge density illustrated in Table 1–3. Although the chemical removal of Fe oxides can cause some alteration of clay surfaces, the magnitude of their contribution to positive charge can clearly be seen in Table 1–3. As a result, solute behavior in such soils is very different from that in dominantly permanent charge systems.

In addition, because organic matter decreases with depth, positive charge usually increases with depth (Gillman & Sumner, 1987) that has profound effects on solute transport (Wong et al., 1987, 1990a,b). As a result, simple anions such as Cl$^-$, Br$^-$, and NO$_3^-$ can no longer be used as conservative tracers for water movement in such soils due to their retardation by the positive charge present. The strong relationship between AEC and the retardation factor (R) for transport

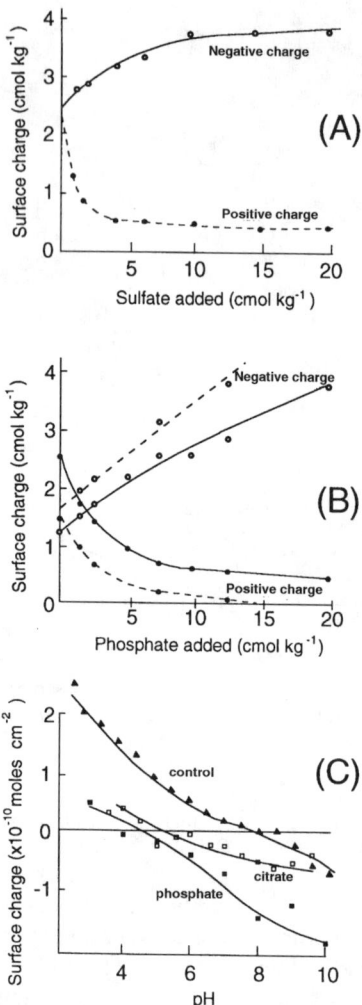

Fig. 1–2. Effect of (A) sulfate and (B) phosphate on surface charge characteristics of a Chinese Oxisol (Yu, 1997) and (C) phosphate and citrate on the net charge on goethite (Bowden et al., 1980).

Table 1–3. Effect of pH and removal of Fe oxides on the surface charge density (γ) of a Rhodic Ferrasol from China (Yu, 1997).

Whole soil		Fe oxide removed	
pH	γ µC cm^{-2}	pH	γ µC cm^{-2}
3.3	+3.6	3.2	−1.5
4.1	+0.7	4.2	−6.7
5.4	−1.4	5.8	−12.3
6.3	−6.1		
7.0	−9.0	7.1	−17.7

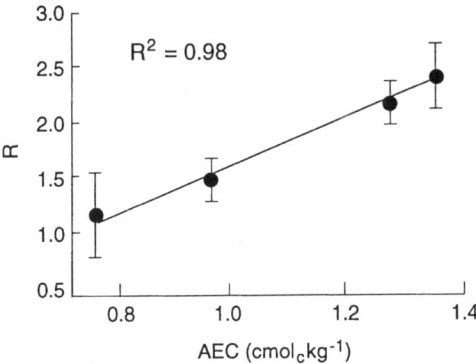

Fig. 1–3. Effect of anion exchange capacity (AEC) on the retardation factor (R) for NO_3^- in a Cecil subsoil (Bellini et al., 1996).

of NO_3^- through a Georgia Ultisol profile is illustrated in Fig. 1–3. Retardation factors for NO_3^- varying between 2 and 6 are common for such subsoils of Oxisols, Ultisols, and acid Alfisols (Wong et al., 1990a; Bellini et al., 1996), which means that between 2- and 6-fold more water is required to leach the same amount of NO_3^- as compared with a soil exhibiting no retardation (completely negatively charged). Thus, transport models developed for dominantly negatively charged soils totally lose their predictive potential unless subroutines are incorporated to account for the effects of AEC on anion movement (Bellini et al., 1996). Melamed et al. (1994) studied the effects of P additions on the transport of Br^- through columns of an Oxisol initially almost at its PZC. In the control column, a value for $R = 1.20$ was obtained and as the P treatments increased negative charge, they explained their results on the basis of anion exclusion that slightly accelerated Br^- transport ($R = 0.79$; Table 1–4).

These different data sets (Fig. 1–3 and Table 1–4) highlight the importance of knowing the charge condition of the soil when studying the transport of ions. Variable charge soils whose pH values are in the vicinity of the PZNC will exhibit both cation and anion retardation as has often been observed in the apparent salt sorption effect (Alva et al., 1991a,b; Pearce & Sumner, 1997) in which the soil appears to deionize a solution passing through it as can be seen from the initially very low electrical conductivity (EC) values of the leachate from a dry Ultisol soil column leached with a dilute electrolyte solution (Fig. 1–4). With time, the EC values increase as the potentially available variable charge sites become sat-

Table 1–4. Effect of P treatments on charge characteristics of a Brazilian Oxisol and the retardation factor (R) for Br^- in packed soil columns (Melamed et al., 1994).

P treatment	Net charge	R	Exclusion volume
mmol kg^{-1}	cmol$_c$ kg^{-1}		m^3 m^{-3}
0	+0.8	1.20	--
16.2	−1.1	0.99	0.006
32.4	−3.0	0.88	0.068
48.6	−4.8	0.79	0.119

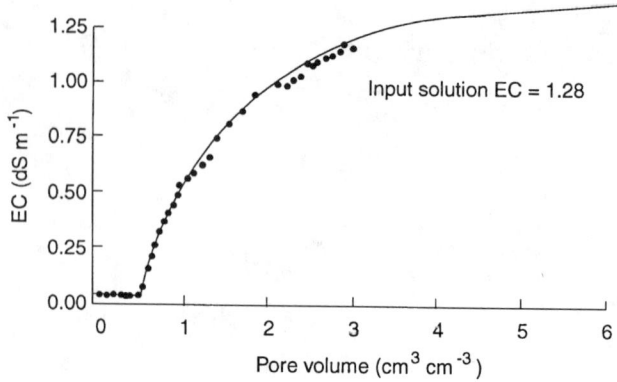

Fig. 1–4. Illustration of the apparent salt sorption effect exhibited by variable charge soils poised near their point of zero net charge (PZNC). A column of dry Ultisol was leached with a 10 mmol L^{-1} solution of $Ca(NO_3)_2$ and the electrical conductivity (EC) of the leachate monitored (Bellini et al., 1996).

urated with the cations and anions of the solute. This phenomenon is due to the uptake of cations and anions of the solute in roughly equivalent amounts by exchange for H^+ and OH^-, respectively. As the pH is increased, anion and cation retardation will decrease and increase, respectively.

The presence of positive charge also has enormous impact on the adsorption and retardation of pollutants such as 2,4-D (2,4-dichlorophenoxy acetic acid), 2,4,5-T (2,4,5 trichlorophenoxy acetic acid), dicamba (3,6-dichloro-*o*-anisic acid), MCPA (4-chloro-2- methylphenoxy acetic acid), and other anionic or weakly acidic pesticides and also promoting the movement of cationic heavy metals and pesticides such as diquat (9,10-dihydro-8a,10a-diazoniaphenanthrene), *s*-triazines [e.g., atrazine (6-chloro-*N*-ethyl-N^1-isopropyl-1,3,5-triazine-2,4-diamine)], and prometon (N^2, N^4-di-isopropyl-6-methoxy-1,3,5-triazine-2,4-diamine) down the profile to groundwater (Green & Karickoff, 1990; Wauchope et al., 1992; Green et al., 1993; Bellini et al., 1996; Naidu et al., 1997). Some examples of this behavior will be presented later. Although the fate and behavior of pesticides in variable charge soils have not been extensively studied, there is evidence that the nature and extent of pesticide interactions such as sorption–desorption, are likely to be very different in these soils relative to the dominantly permanent charge variety (Green et al., 1993).

In terms of the relative areas occupied by permanent and variable charge soils and the importance of the latter in producing a major proportion of food supplies in the third world, surprisingly little attention has been paid to the behavior of these variable charge soils. Even on the home front where such soils dominate the landscape in the southeastern USA, there is still little realization and acceptance of the variable charge nature of these soils and the implications for their management. For example, in Georgia, the study described above (Bellini et al., 1996) became necessary to convince soil physicists that the retardation of anion transport that they were measuring was indeed due to positive charges in the subsoil rather than to the partitioning of water into mobile and immobile categories.

Table 1–5. Comparison of the critical flocculation concentrations (CFC) for pure clay minerals and soil colloids containing the same minerals.

Mineral–Soil	SAR–ESP	CFC mmol$_c$ L^{-1}	Source
Georgia kaolinite (pure mineral)	5	4	Goldberg & Forster, 1990
Cecil soil clay (kaolinite, vermiculite)	5	15	Miller et al., 1990
Montmorillonite (pure mineral)	5	3	Oster et al., 1980
Altamount soil clay (montmorillonite, illite)	5	10	Goldberg & Forster, 1990
Illite (pure mineral)	5	5	Goldberg & Forster, 1990
Fallbrook soil clay (kaolinite, illite)	5	10	Goldberg & Forster, 1990
Gouldburn soil clay (kaolinite, mica, smectite)	4	3.5	Rengasamy, 1983
Gouldburn soil clay (kaolinite, mica, smectite)	4	11	Rengasamy, 1983

The extent of Soil Chemistry investigations into the properties of these soils has been inadequate up till now and this provides not only very interesting but also challenging opportunities for further highly relevant, rewarding research work. This relative lack of research on variable charge soils is a deficiency that looms into prominence in view of the environmental impact of the rapidly increasing populations in countries where such soils occur. With a better knowledge of the charge characteristics of such soils, we will be able to devise innovative technologies to manage contaminated sites.

1–2.1.2 Colloidal Stability

The behavior of soil colloidal particles has been intensively studied during the past century (Kruyt, 1949, 1952; van Olphen, 1977). While the Gouy-Chapman double layer theory and subsequent improvements (Stern, multi-layer models, and others) do not quantitatively explain soil colloidal behavior in all situations, nevertheless it is true to say that we have a fairly good qualitative knowledge, at least, of how the system works in terms of the dispersion and flocculation of soil particles (Gregory, 1989). Such understanding is the key to management strategies required to reduce nonpoint pollution of surface and groundwaters. Unfortunately, much of the work conducted in this area in the past has been focused on studying pure clay minerals as models for soils that has proven to be inappropriate (Sumner, 1993). Soil colloids, both organic and inorganic, are intimately mixed and associated. In addition, they sorb various ligands such as phosphate present in the soil that can alter their electrokinetic behavior drastically as we have already seen. Thus, they flocculate and disperse under conditions quite different from the pure varieties. The different levels of electrolyte required for flocculation of pure and soil colloidal systems are illustrated in Table 1–5. In all cases, soil colloids required two- to three-fold higher electrolyte concentrations to achieve flocculation and values as high as 10-fold have been reported

(Emerson & Dettman, 1960). This would suggest that organic matter plays a role in the stabilization of soil colloidal suspensions. On the other hand, hydroxy Al and Fe coatings that can develop positive charge, may promote flocculation. In addition, pure clay minerals and soil clays differ in their sensitivity to flocculation by Ca vs. Mg salts. Although Ca and Mg flocculate pure systems equally well, Ca is much more effective in soil clay systems. Consequently, if we want to obtain meaningful information on soils, they should be studied as they occur in nature so that a better understanding of how colloids move in soils and their potential contribution to pollutant transport can be obtained. Future studies on colloidal stability should, therefore, involve little-altered real soil materials rather than pure surrogates.

From a colloidal standpoint, we have long known that electrolyte concentration, counter ion valence, pH, and surface charge (amount and polarity) are all factors that determine the stability of soil colloidal suspensions (van Olphen, 1977). Decreasing electrolyte concentration and increasing pH and surface charge promote dispersion. This knowledge has been effectively exploited in the management of sodic soils where gypsum applications have long been effective in improving the state of flocculation of colloids within the soil (Overstreet et al., 1951; Oster, 1982, 1993). Such amelioration has permitted increased water entry and movement through these soils (Shainberg et al., 1989).

It has only recently been realized, however, that similar strategies are effective on nonsodic soils (Miller et al., 1986). A closer look at what we know about the behavior of such systems is illuminating. Unlike the case of sodic soils where Na plays a dominant role in determining clay dispersion, in these nonsodic soils, the very low electrolyte concentration is the major factor in promoting dispersion, particularly in topsoils even in the absence of Na. This behavior stems from the preferential loss of the most labile components of the organic matter present that are the primary binding agents in the formation of water-stable aggregates. This loss of organic matter facilitates dispersion when small amounts of mechanical energy are introduced in the form of raindrops (McIntyre, 1958). This can take

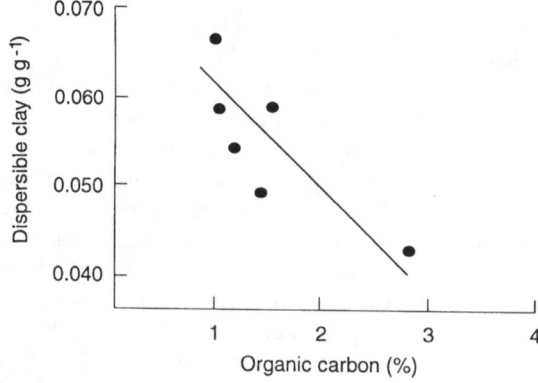

Fig. 1–5. Relationship between organic C content and mechanically dispersible clay in Queensland soil that had been continuously cultivated for different periods up to a total of 49 yr ($r^2 = 0.40**$; recalculated from the equations fitting data of Cook et al., 1992).

place at low values of exchangeable sodium percentage (ESP) provided that the electrolyte concentration is sufficiently low. The relationship between organic C content and mechanically dispersible clay obtained in an experiment in which the soil had been continuously cultivated for different periods of time up to 50 yr, is illustrated in Fig. 1–5. The effect of this dispersible clay in reducing hydraulic conductivity is illustrated in Fig. 1–6 for the soils at three such sites. Clearly, the loss of organic matter on cultivation sets in motion processes leading to severe soil degradation. In the past, it has been assumed that the organic matter remaining in such degraded agricultural soils merely adds to the negative charge that would further promote dispersion (Miller et al., 1990; Goldberg et al., 1990). Recently, Heil and Sposito (1993a,b) confirmed this view and suggested that the organic matter did not promote dispersion by increasing negative charge but rather by a steric repulsion that essentially reduced the effectiveness of van der Waals forces between mineral particles. They also drew attention to the role that this phenomenon could play in colloid mobility and contaminant transport. In a subsequent scanning force miscrocopy investigation, Heil and Sposito (1995) offered confirmatory evidence for their hypothesis.

As a result of the dispersive nature of the colloids at the soil surface (Miller & Baharuddin, 1986), such soils form seals and crusts (Miller et al., 1986; Shainberg et al., 1989) and the mobile colloids have the potential to contribute to pollutant transport to surface (Sharpley et al., 1994) and subsurface waters (Kaplan et al., 1993). While we have a reasonable understanding of the forces involved in the mobilization of colloids at the soil surface, we currently know very little about the nature of the organic constituents in the two pools mentioned above, on the one hand, for the promotion of structure and on the other, for the enhancement of dispersive character. Furthermore, their interactions with inorganic colloids while known to involve hydrous metal oxide coatings, are not well understood. With the increased disposal of organic wastes such as poultry, cattle, and swine manures as surface applications to land, all of these interactions will be fruitful and necessary arenas for future investigation. The enhanced transport

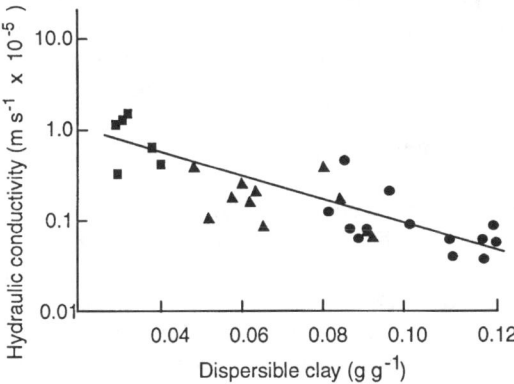

Fig. 1–6. Relationship between mechanically dispersible clay and hydraulic conductivity of three Queensland soils that had been continuously cultivated for varying lengths of time up to 49 yr (Cook et al., 1992).

of particulate P from such sources is a potential threat to surface water bodies as far as eutrophication is concerned.

While pedologists have always been cognizant of the fact that clay and organic matter can move down the profile to form textural B horizons in Ultisols and Alfisols and Bh horizons in Spodosols (Jenny, 1941), it is only recently that the importance of colloidal movement as a vehicle for contaminant transport has been recognized by geologists, hydrologists, and soil chemists (Nightingale & Bianchi, 1977; Chiou et al., 1986; Buddemeier & Hunt, 1988; Penrose et al., 1990; Kaplan et al., 1993). This is despite the pioneering work of Bloomfield (1955) and Swindale and Jackson (1956) who clearly demonstrated the mobilization and accelerated movement of Fe and Al by extracts made from the leaves of various trees during an investigation of the mechanisms involved in podzolization. Bloomfield (1955) showed that the Fe and Al organic complexes could be subsequently immobilized lower down in the profile on fresh aluminosilicate and Fe oxide surfaces. The extent of the sorption varied inversely with the efficiency of the tree species extract in mobilizing Fe and Al. The applicability of these findings to present day problems appears to have been overlooked as these investigations are seldom, if ever, cited. Among possible reasons for not considering that clay could become mobile in the profile, are the perception that textural B horizons required geological time to form and the apparent unsuitability of the conditions (dominance of divalent ions) present in most soils for clay dispersion to be possible, making the phenomenon of little interest to the soil chemist in the short term. As indicated above, colloid mobility is possible under conditions previously considered to be unsuitable largely because the models used (pure clays) were inappropriate.

In studies on the transport of contaminants within soils to groundwater, numerical simulations based on partition coefficients for the particular species used in two-phase transport models often predict that the pollutant would not reach the watertable whereas monitoring of the groundwater detected measurable quantities (Looney et al., 1987). Such lack of agreement between predicted and observed contaminant transport points to the inadequacy of current models. Now that we know that colloids can move and act as vectors for the transport of sorbed contaminants (McCarthy & Zachara, 1989; Newman et al., 1993) particularly in lighter textured soils over substantial distances and in relatively short time periods, we need to devote our energies to studying all the conditions that favor mobilization and transport of such colloids within the body of the soil. This will improve our ability to predict the fate of pollutants sorbed on the surfaces of colloids that can be mobilized. A recent study in South Carolina on colloid generation and transport in repacked columns of highly weathered subsurface material is illuminating (Seaman, 1994). The material was pretreated by leaching with dilute solutions of varying SAR (sodium absorption ratio: 7 and 14) or pure electrolytes containing OH^-, Cl^-, or SO_4^{2-} followed by deionized water. Increasing SAR did not promote dispersion as expected. Solutions of $CaCl_2$ that depressed pH, produced mobile colloids whereas NaCl produced very little at the same equivalent concentration that tended to increase pH (Fig. 1–7). In the case of Na_2SO_4, $CaSO_4$, and $Ca(OH)_2$ solutions, which all increased pH, no colloid movement was observed. The colloids produced in the leachate possessed positive electrophoretic mobilities that is consistent with the observations made with the different electrolytes. The decrease in pH caused by $CaCl_2$ promoted positive charge whereas in the presence of OH^- or SO_4^{2-}, positive charge would have been reduced as a result of the increased pH or specific sorption of the anion. These results suggest that even minor changes in the composition of groundwater can influence charge and colloid generation, and consequently, transport. This is illustrated in Fig. 1–8 for the effect of pH on the level of mechanically dispersible clay in a variable charge soil. At both low and high pH, clay dispersion is observed due to the positive and negative charges present, respectively. At intermediate pH values that are in the vicinity of the PZNC, the soil was highly flocculated. In a

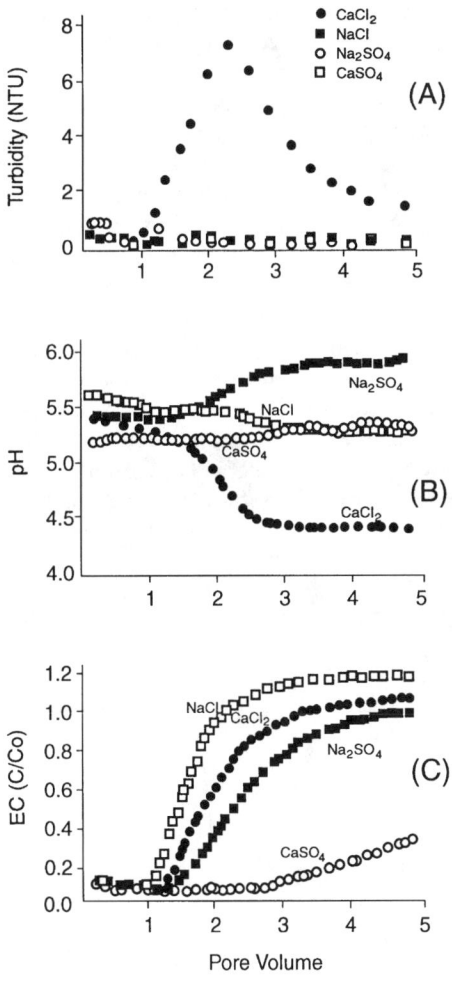

Fig. 1–7. Effect of various equivalent salt solutions (1 mmol L^{-1}) on (A) effluent turbidity, (B) pH, and (C) electrical conductivity (EC) of the leachate from columns of highly weathered Barnwell Formation (Seaman, 1994).

study on reconstituted Ultisol soil profiles in lysimeters, Kaplan et al. (1996) demonstrated the importance of organic matter in promoting the production of mobile colloids at near neutral pH values. The colloids emanating from the columns were enriched in smaller particles of Fe and Al oxyhydroxides and kaolinite at the expense of quartz and hydroxy-interlayered vermiculite relative to the soil horizons (Fig. 1–9); however, under more acid conditions that decreased negative charge, virtually no mobile colloids were observed. With the exception of one observation, the electrophoretic mobilities of the mobile colloids were related to their organic C contents (Fig. 1–10; Kaplan et al., 1993). Thus, the colloids that in pure form would normally have been positively charged (sesquioxides), are mobile as a result of the charge reversal brought about by the presence of the organic matter. This effect of organic matter also was demonstrated by Kretzchmar et al. (1993, 1995), but on the other hand, in contrast to the results of Kaplan et al. (1996), Kretzchmar et al. (1994) found no preferential filtering of mineral phases in saprolites from the southeastern Piedmont. The nature of these interactions between organic and inorganic colloids is not well understood. Gu et al. (1994) studied the adsorption–desorption of natural organic matter on Fe oxide and found that ligand exchange between carboxyl–hydroxyl groups on the organic matter and the surfaces of the oxide were largely responsible but pointed out that a better mechanistic understanding was necessary to improve our ability to predict cotransport of colloids and associated contaminants. Kretzschmar et al. (1993) suggested that a combination of electrostatic and steric stabilization may be responsible for the dispersion of mineral colloids with the latter being the more important. This field involving the mobilization of colloids within profiles is a branch of Soil Chemistry that is still in its infancy. Much more attention must be devoted to investigating the mechanisms by which mobile colloids are generated, and how their mobility can be managed while taking the variable charge nature of soils and the effect of the ambient solution composition on soil properties into account. Once these mechanisms are clearly understood, better predictions of colloid assisted transport of contaminants will be possible.

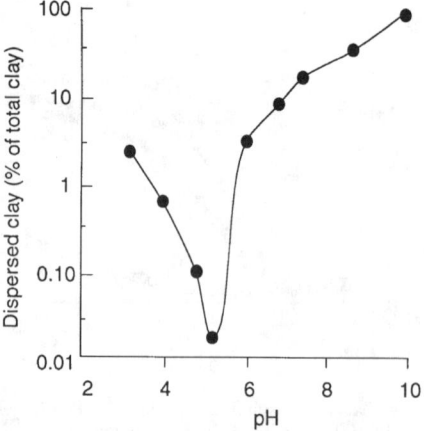

Fig. 1–8. Effect of pH on clay dispersion from an Oxisol (Sumner, 1995, unpublished data).

1–2.2 Sorption and Desorption

During the period that the focus of Soil Chemistry was largely on increasing agricultural production, the topic to which most effort was devoted was the

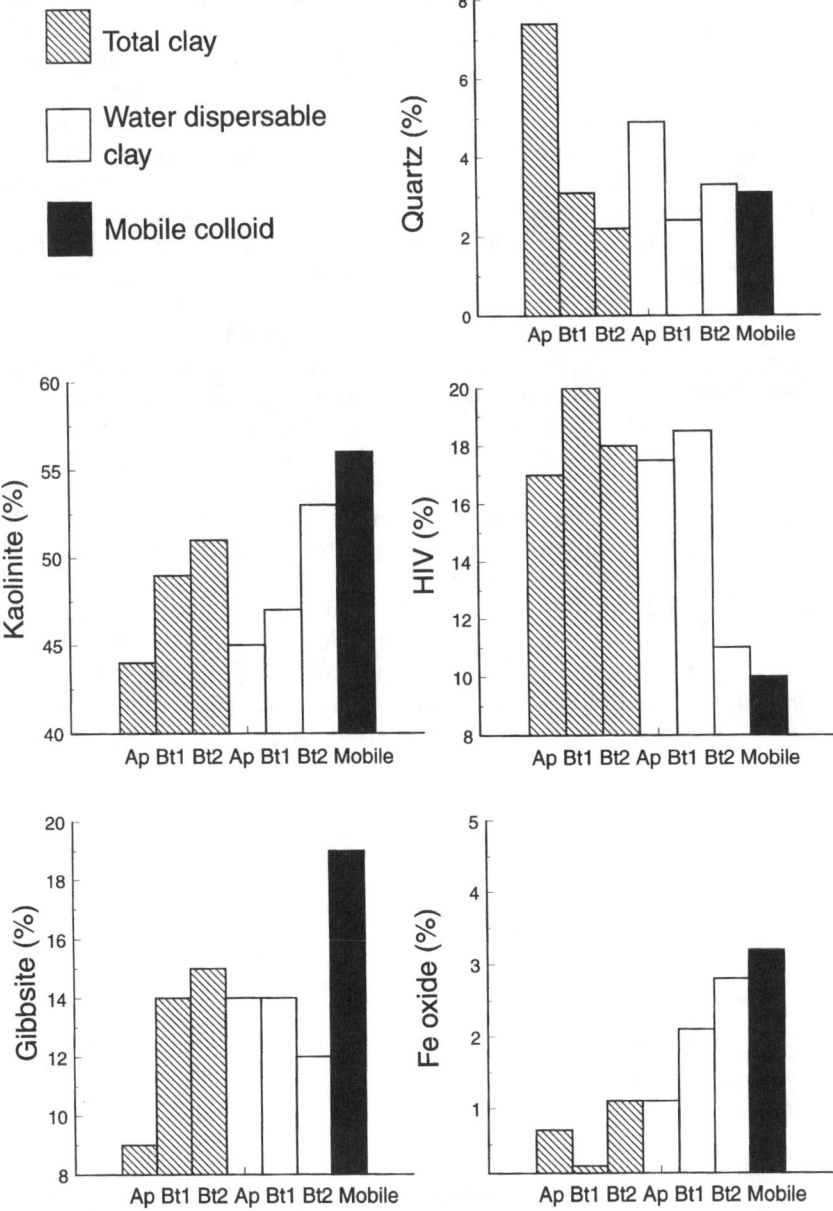

Fig. 1–9. Mineralogy of mobile colloids from soil columns relative to the mineralogical composition of the original Ultisol soil horizons (Kaplan et al., 1996).

sorption of P on soil surfaces particularly sesquioxides, which are the major sinks for this nutrient. The emphasis on P has been purely a reflection of the economic importance of this element in agriculture. The reason for this probably stemmed from the concern that a large part of the costly soluble P applied to soil as fertilizer was rapidly sorbed rendering it less available to roots. Early investigations (i) showed that P was sorbed and not precipitated and implicated divalent cations, acid conditions and organic matter in P sorption (Poszmann, 1927), (ii) developed the concept of anion exchange, (iii) showed that cations had an effect, and (iv) eliminated organic matter as a sink (Ravikovitch, 1934). Scarseth (1935) postulated correctly that phosphate could replace hydroxide and silicate groups on minerals that was subsequently verified on kaolinite (Low & Black, 1948) and aluminous surfaces (Muljadi et al., 1966); however today, we know that P can be sorbed on organic complexes but the reaction is not due to the organic surfaces themselves but rather to the complexed Fe and Al oxyhydroxides (Appelt et al., 1975; Kwong & Huang, 1978). The latter group of materials whether discrete or complexed with organic matter offers by far the most important category of P sorbing surfaces in soils. The effects of concentration, pH, time, temperature, and other variables on the sorption of P and also other elements have been extensively studied with a view to characterizing the kinetics and mechanisms of the reactions involved (Barrow, 1987; Harter, 1991). These effects are illustrated in Fig. 1–11 for P. There is an initial very rapid sorption reaction followed by a slow reaction that can continue for a substantial period of time. Various equations (Langmuir, Fruendlich, and others) have been fitted to the data from such studies with varying levels of success, but in any event, such equations are no more than means of organizing the data and give no *prima facie* indication of the mechanisms involved (Harter, 1991). Barrow (1987) summarized the effects that have been observed: (i) the amount of ion sorbed is characteristically related to the concentration in solution (Fig. 1–11A); (ii) the continuing reaction is a function of a fractional power of time; (iii) if the continuing reaction reflects slow conversion to a tightly-held form, the amount converted is proportional to the amount

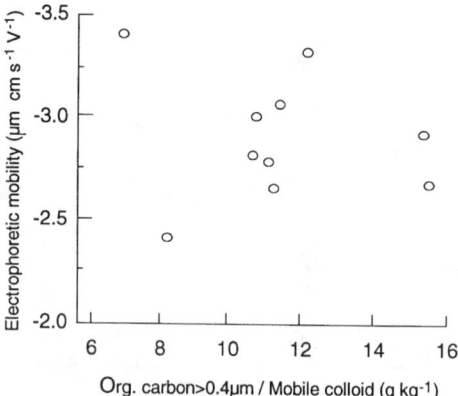

Fig. 1–10. Effect of the organic C content of mobile colloids in the leachate from a reconstructed profile of an Ultisol on their electrophoretic mobilities (Kaplan et al., 1993).

added; (iv) temperature and rates of sorption and desorption are positively related; (v) at slow rates, increasing temperature shifts the equilibrium to increase the solution concentration; (vi) pH affects the adsorption of each ion differently, and (vii) the nature and concentration of the background electrolyte affect pH. Barrow (1987) has developed a model that he claims reproduces the above reaction characteristics for variable charge oxide surfaces for P and some other ions; however, the applicability of this model to other ions and situations has not been fully established.

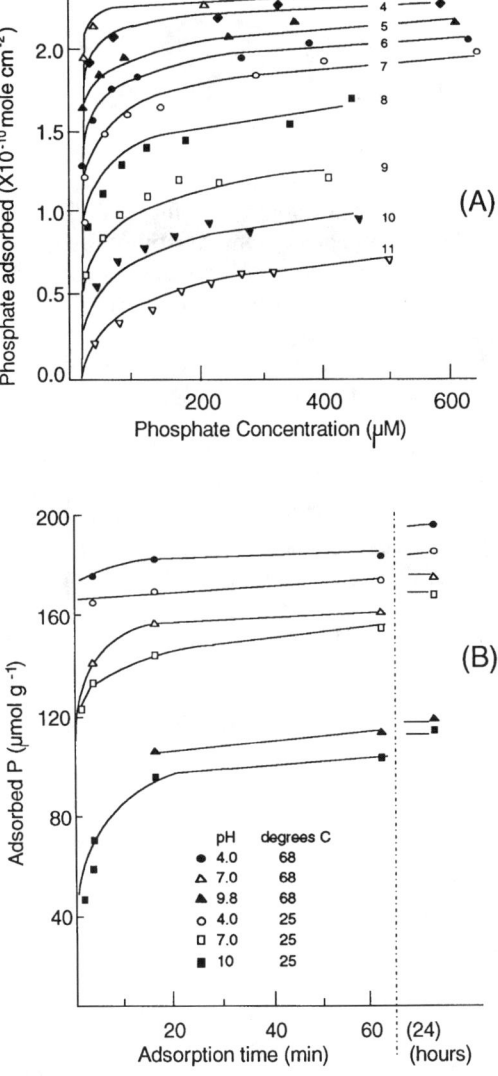

Fig. 1–11. Illustration of the effects of (A) concentration and pH, and (B) time on P sorption by iron oxyhydroxide surfaces (Bowden et al., 1980; Madrid & Posner, 1979).

The current view suggests that the most reactive sorption sites occur at valence-unsatisfied surface hydroxyls on the edges of clay minerals and on the surfaces of Fe, Al, and Mn hydroxyoxides and amorphous aluminosilicates with poorly ordered and amorphous materials contributing the most. Terminal OH^- groups on these surfaces readily protonate (Reaction [1]) entering into ligand exchange reactions with anions. Anions that bond strongly with a particular metal ion (M^{n+}) will also sorb strongly on M–OH groups of the same metal on surfaces (McBride, 1994). Sorption on these surfaces is thought to be by a binuclear bridging mechanism (Parfitt & Russell, 1977) illustrated below for Fe surfaces under acid conditions:

$$>[Fe]\text{-}OHH^+Cl^- \qquad\qquad >[Fe]\text{-}O\diagdown$$
$$+ X_nAO_y^{n-} \Rightarrow \qquad AO_{y-2}^{(n-2)-}X_{(n-2)}^+ + 2X^+Cl^- + 2\,H_2O \qquad [2]$$
$$>[Fe]\text{-}OHH^+Cl^- \qquad\qquad >[Fe]\text{-}O\diagup$$

This reaction has a high degree of nonreversibility stabilized by entropy considerations because of the high degree of improbability that two bonds will break simultaneously. The forward reaction may continue for extended periods with the reverse reaction taking place at a much slower rate. For example, in the case of the sorption of phosphate in Reaction [2], the charge on the surface would be reversed resulting in an increase in cation-exchange capacity (CEC; Wann & Uehara, 1978). This is consistent with the dispersive powers of phosphate discussed above.

As far as heavy metals are concerned, the large body of literature shows that soils have substantial capacities for sorption of heavy metals from solution. Sorption can take place on the surfaces of hydrous oxide (Kabata-Pendias, 1980; Johnson, 1986; Shuman, 1988; Manceau et al., 1993), organic matter (McLaren & Crawford, 1973; Petruzzelli et al., 1978; McLaren et al., 1981), clay mineral (Tiller & Hodgson, 1962; Schindler et al., 1976), and short-range ordered colloids (Wada, 1989; Xue & Huang, 1995) depending on the metal in question. Some examples of this behavior for Cu and Co are presented in Fig. 1–12. Sorption by the organic and oxide surfaces is much greater than by the clay minerals. Increasing pH increases metal sorption (Fig. 1–13). Most of the heavy metals undergo hydrolytic reactions as follows:

$$M(H_2O)_x^{m+} \rightleftharpoons [M(H_2O)_{x-y}(OH)_y]^{(m-y)+} + yH^+ \qquad [3]$$

Consequently, as the pH is raised, the metal becomes more hydroxylated, and therefore, becomes more strongly sorbed. The relative affinity of a metal for water vs. another complexing ligand determines whether complexation occurs. This can be deduced from the Misono softness parameter (Y; Misono et al., 1967), which characterizes hard and soft acid–base behavior:

$$Y = 10\,(I_n/I_{n+1})(r/\sqrt{n}) \qquad [4]$$

where r is ionic radius of the metal of valence n and ionization potential I_n, with subscripts n and $n+1$ referring to the degree of ionization. Metals with high values for Y (>0.32 nm) are soft Lewis acids and tend to form covalent bonds. Some examples of the mechanisms proposed to account for metal sorption on soil surfaces are summarized in Fig. 1–14. Sorption–desorption reactions at the surfaces of both organic and inorganic soil colloidal surfaces particularly the sesquioxides, control the metal concentration in solution.

The partition of the element of interest between the solid phase and solution is of importance in determining the supply to roots, the potential for leaching down the profile and transport in solution to surface waters. Because the preponderance of literature has dealt with sorption rather than desorption reactions, it is true to say that the former are better understood than the latter. In terms of

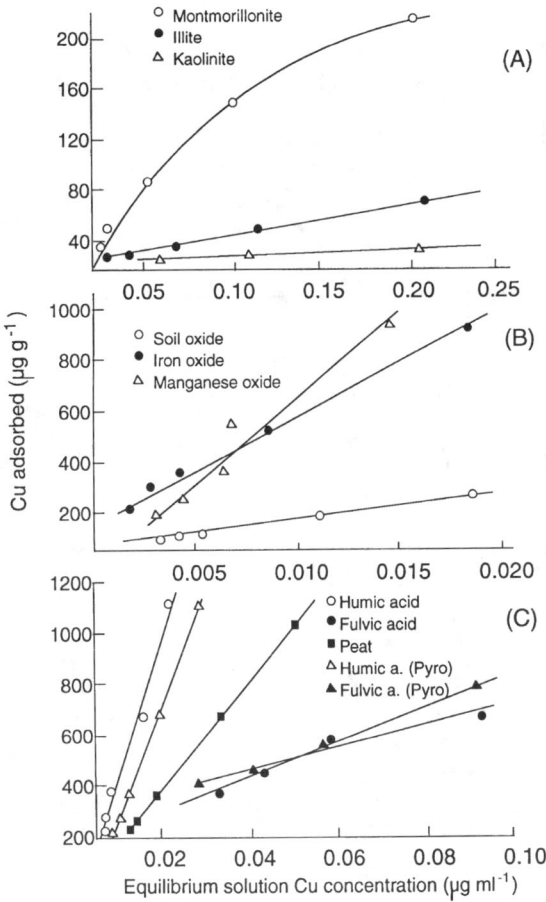

Fig. 1–12. Sorption of Cu by (A) clay minerals (B) soil oxide and (C) organic materials at low concentrations (0.01 M CaCl$_2$) and pH 6.0 [The humic and fulvic acids were extracted with either NaOH or sodium pyrophosphate (Pyro)] (McLaren et al., 1981, 1986).

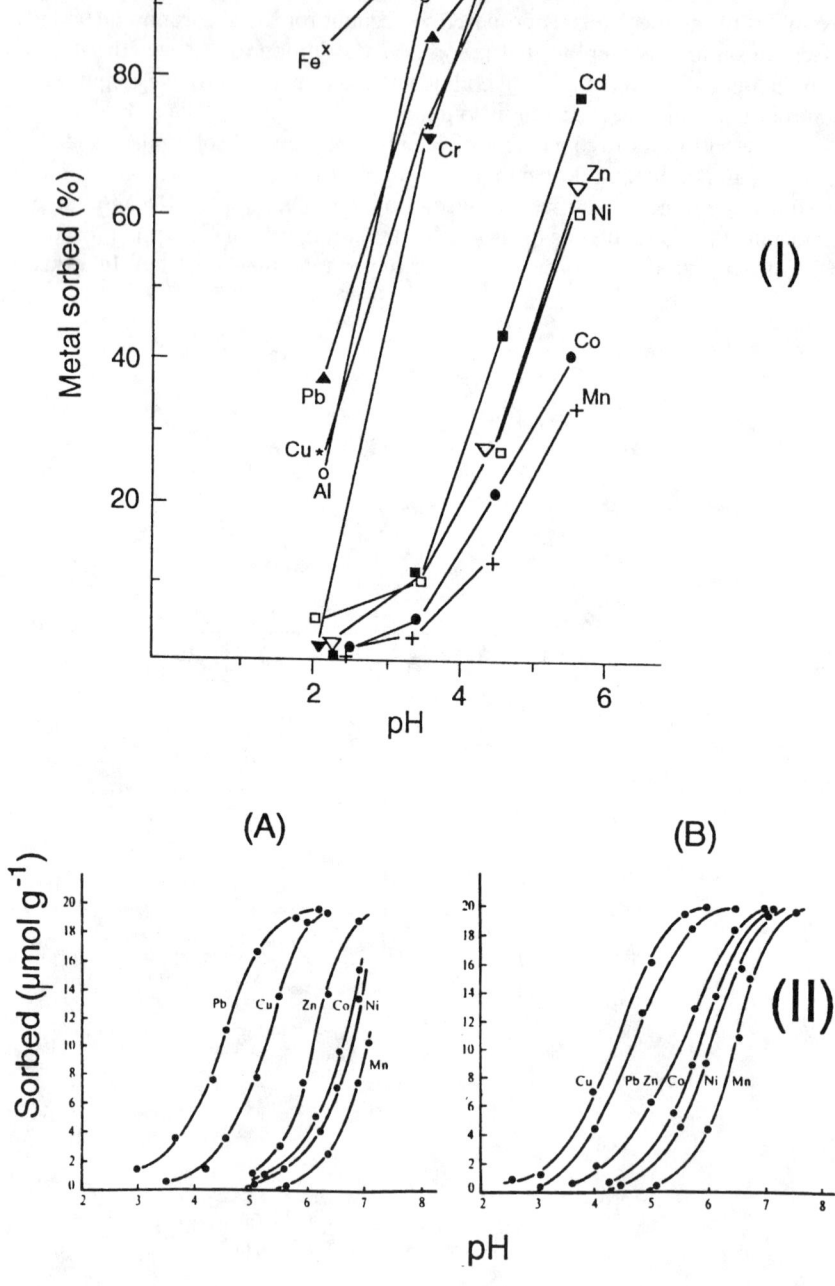

Fig. 1–13. Effect of pH on the adsorption of heavy metals on (i) humic acid and (ii) (A) hematite and (B) goethite (Kerndorff & Schnitzer, 1980; McKenzie, 1980).

the macro- and micronutrients and pollutant metals, this is a paradoxical situation because it is the desorption reaction that controls the rate and amount of an element released into solution for plant uptake and potential transport.

Some of the most recent studies of soils and soil components show that relatively little sorbed metal is desorbed back into solution (McLaren et al., 1986;

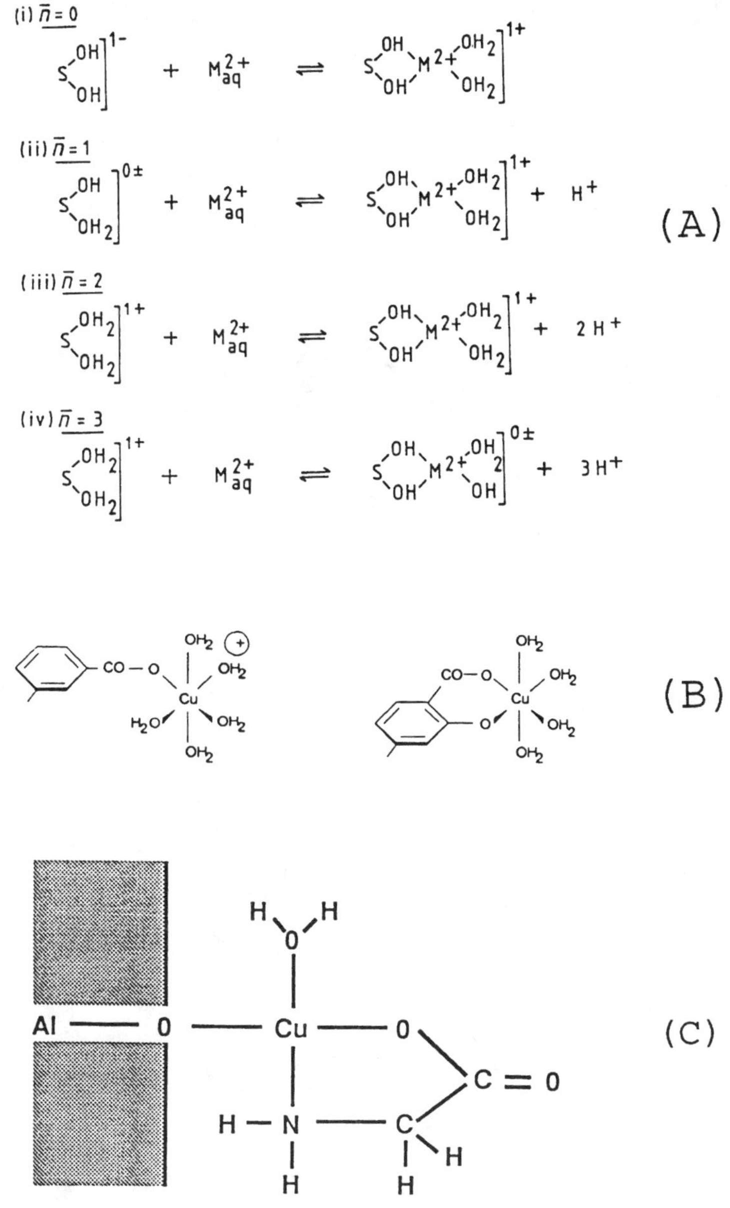

(i) $\bar{n} = 0$

$$\left[S\begin{matrix}OH\\OH\end{matrix} \right]^{1-} + M^{2+}_{aq} \rightleftharpoons \left[S\begin{matrix}OH\\OH\end{matrix}M^{2+}\begin{matrix}OH_2\\OH_2\end{matrix} \right]^{1+}$$

(ii) $\bar{n} = 1$

$$\left[S\begin{matrix}OH\\OH_2\end{matrix} \right]^{0\pm} + M^{2+}_{aq} \rightleftharpoons \left[S\begin{matrix}OH\\OH\end{matrix}M^{2+}\begin{matrix}OH_2\\OH_2\end{matrix} \right]^{1+} + H^+ \quad (A)$$

(iii) $\bar{n} = 2$

$$\left[S\begin{matrix}OH_2\\OH_2\end{matrix} \right]^{1+} + M^{2+}_{aq} \rightleftharpoons \left[S\begin{matrix}OH\\OH\end{matrix}M^{2+}\begin{matrix}OH_2\\OH_2\end{matrix} \right]^{1+} + 2H^+$$

(iv) $\bar{n} = 3$

$$\left[S\begin{matrix}OH_2\\OH_2\end{matrix} \right]^{1+} + M^{2+}_{aq} \rightleftharpoons \left[S\begin{matrix}OH\\OH\end{matrix}M^{2+}\begin{matrix}OH\\OH\end{matrix} \right]^{0\pm} + 3H^+$$

(B)

(C)

Fig. 1–14. Proposed mechanisms for the sorption of metals on (A) oxide, (B) organic surfaces in soils and (C) between oxide and organic surfaces (Kinniburgh, 1983; Stevenson, 1994; McBride, 1985).

Swift & McLaren, 1991; Hogg et al., 1993; Backes et al., 1995). Examples of this behavior are presented in Fig. 1–15. In some instances (Cu and Co on soil oxide), very little is recovered during desorption. It appears that increasing residence time in the soil reduces the ability of the soil to desorb the metal (Backes et al., 1995) and consequently, its availability to plants. This effect for Cd on goethite

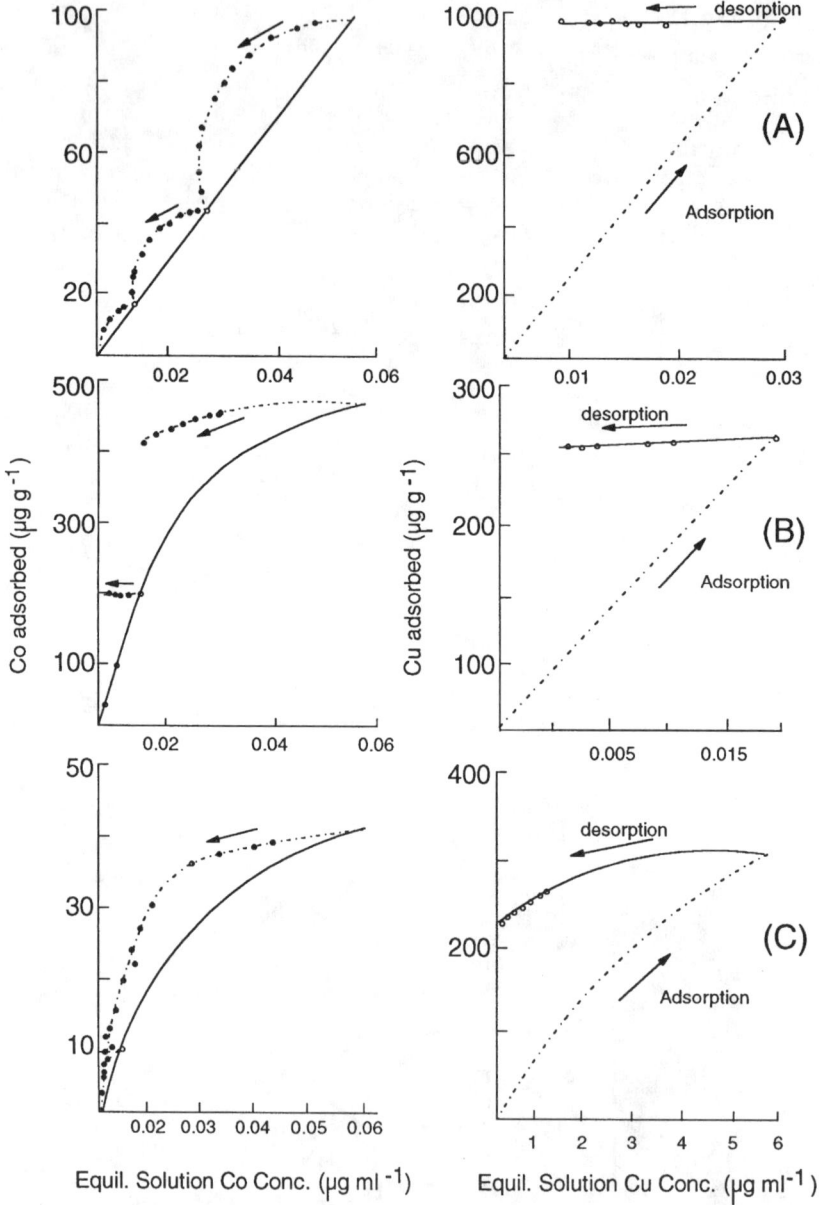

Fig. 1–15. Desorption of Co and Cu from (A) humic acid, (B) soil oxide material, and (C) montmorillonite (McLaren et al., 1983, 1986).

is illustrated in Fig. 1–16 where aging for 15 wk resulted in much lower recoveries. The mechanisms involved are not clearly understood, but may involve movement of the metals to sites with slower desorption reactions. Thus further work on desorption is required to study the processes involved with a view to managing the bioavailability of metals in the environment.

The movement of soluble organic materials in soils and their association with the movement of Fe and Al in the formation of the B horizons of Spodosols has been known for a long time (Bloomfield, 1955). Interest in this phenomenon has recently been highlighted (Harter & Naidu, 1996) in an excellent review in terms of the importance of dissolved organics on metal reactions with inorganic soil components and transport within the profile. Information on this topic is still relatively scarce. Although considerable work has been conducted on the complexation of metals by high molecular weight humic and fulvic acids, their role in metal retention by inorganic surfaces is not well understood. Even less information is available on low molecular weight organics because, in the past, these have been considered to be ephemeral; however, the improved ability to be able to measure organics in soil solutions has clearly indicated that they are indeed present at concentrations much higher than previously assumed. Low molecular weight organics are certainly sorbed on minerals surfaces such as sequioxides (McBride, 1987; McBride & Wesselink, 1988; Schwertmann & Taylor, 1989) causing changes in PZNCs and potentially providing sites for metal sorption. In addition, low molecular weight organics such as fulvic acids can promote the formation of short-range ordered Al and Fe oxides (Huang & Violante, 1986; Schwertmann et al., 1986) which, in turn, impacts surface reactivity and metal sorption (Xue & Huang, 1995). These soluble low molecular weight organics can therefore promote or retard metal reactions with inorganic surfaces depending on the surface and ligand properties and ambient conditions. For example, Chairidchai and Ritchie (1990, 1992) found that Zn retention by a variable charge Entisol was reduced or enhanced by certain ligands when the pH of the system was above or below the PZNC, respectively. When the soil was below the PZNC, the results were consistent with the sorption of a negatively charged Zn–citrate complex. Prasad and Sarangthem (1993) have reported that DTPA (diethylenetri-

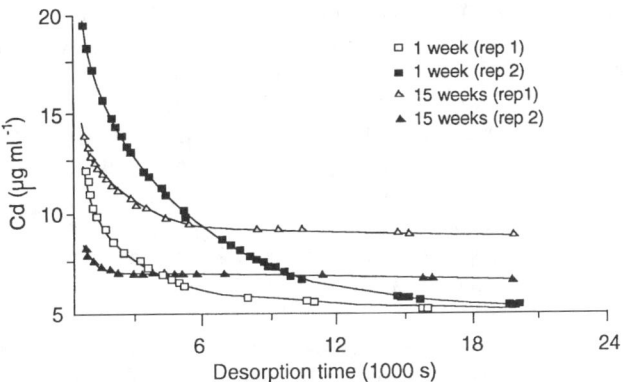

Fig. 1–16. Effect of initial period of sorption on desorption of Cd from goethite (Backes et al., 1995).

aminepentaacetic acid) and EDTA (ethylenediaminetetraacetic acid) prevented or reduced Zn sorption in a calcareous soil while Jardine et al. (1993) found that EDTA enhanced Co^{2+} transport through a column of material from the C horizon of a soil from the Oak Ridge Reservation. Dissolved organic C in a Spodosol complexed Cr and Cu resulting in substantial concentrations of these metals in B horizon leachate whereas Cd, which was not complexed, was sorbed in the inorganic form (Guggenberger et al., 1994). In addition, such low molecular weight compounds can increase the solubility of P in soils as a result of ligand exchange reactions that release P from mineral surfaces (Hue, 1991; Bolan et al., 1994). Furthermore because low molecular weight organic acids can perturb the crystallization of Al and Fe oxides (Huang & Violante, 1986; Schwertmann et al., 1986) and hence increase P sorption (Kwong & Huang, 1978, 1979), organics can either suppress or enhance P sorption depending on the particular mechanism involved. From the environmental point of view, complexation of metals by soluble organics can have a profound influence on phytoavailability (Hue et al., 1986), transport to surface and groundwater bodies (Kaplan et al., 1993), and toxicity to terrestrial, aquatic and marine organisms (Bernhard et al., 1986). The environmental effects of soluble organics are illustrated in Fig. 1–17. Much more work is needed to develop a body of literature that will permit the prediction of the behavior of heavy metals in the presence of both high and low molecular weight organic compounds.

In terms of the concerns relating to nonpoint pollution of surface waters with P, what is clearly important are the kinetics of P desorption from the surfaces of soil materials, particularly those that have been transported into water bodies. This is especially pertinent to the bioavailability of P to organisms responsible for algal blooms. Sharpley et al. (1991) have proposed a test utilizing 0.11 M NaOH for the measurement of potentially bioavailable particulate P (BPP) in runoff waters. In order to improve the prognostic power of such tests, P desorption studies are required that are likely to lead to the development of more precise tests in the future.

Work on the sorption of organic compounds in soils developed relatively much later than that on phosphate most probably because of the difficulties involved in studying organic sorption. Mattson (1932), who was one of the first to study organics, measured the sorption of proteins. Later, Hendricks (1941) who developed a mechanistic approach for studying adsorption of organic bases showed that this knowledge was useful in determining the size of the organic molecules and in studying the expansion of smectites. Bradley (1945) laid the definitive foundations for organic sorption and showed that organic liquids could be used to determine surface area and aid in clay mineral identification by x-ray diffraction; however, progress was slow until the 1960s. Major impetus for research in this area was provided by the development of organic pesticides during the war, their large scale use subsequently, and the development of suitable techniques such as infrared spectroscopy suitable for studying clay–organic interactions.

Much work has been conducted on the sorption of pesticides by soil materials that has included both **ad**sorption at solid-liquid interfaces and **ab**sorption into the interior of the sorbent matrix (organic matter). The former process is

SOIL CHEMISTRY: PAST, PRESENT, AND FUTURE

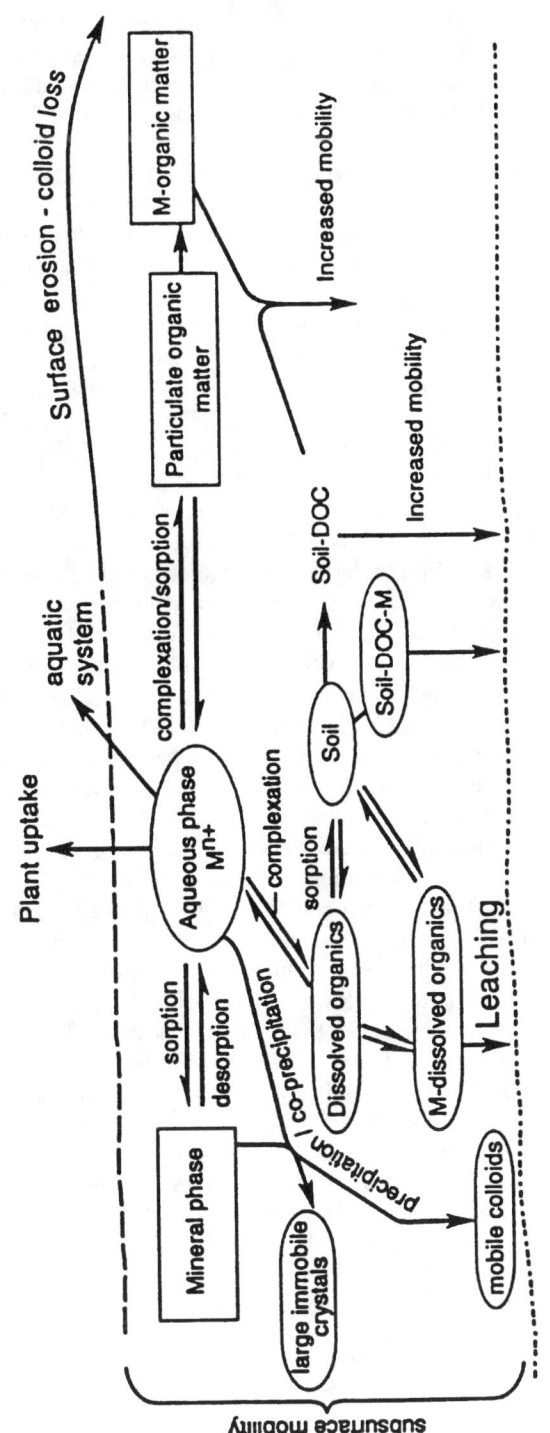

Fig. 1–17. Fate and transport of contaminants to ground water (Harter & Naidu, 1996).

important on mineral surfaces especially when the organic matter content is low (subsoils) and the pesticides are ionic (Mingelgrin & Gerstl, 1983), while the latter is important for nonpolar and nonionic chemicals on organic matter (Chiou et al., 1983). Overall, the latter process is probably the more important. Most of the work conducted has been of the batch type under laboratory conditions with nonionic pesticides receiving most attention. Partitioning of pesticides to the solid phase usually results in reduced bioavailability or bioactivity of the material, and therefore, these reactions are of importance in determining whether or not the pesticide will be effective, the changes in the type of interaction it undergoes on aging and the potential for its degradation in the soil. Generally, pesticide sorption has been modeled assuming that the sorption isotherm is linear and reversible and that instantaneous equilibrium has been reached (Rao & Alley, 1993). This is not always true due to nonlinear isotherms (Weber & Miller, 1989), high sorption enthalpies and competition for sorption sites (Mingelgrin & Gerstl, 1983), and nonattainment of instantaneous equilibrium (Kookana et al., 1992). As a result, the potential for transport and leaching is probably underestimated. Again most of the research work has concentrated on the sorption of pesticides whereas the kinetics of desorption of such materials will determine the levels that are active in the soil solution, the potential threat to the environment due to mobility and bioavailability for degradation. Desorption from soil also is biphasic with strong hysteresis in the sorption–desorption isotherms being observed for a number of pesticides (Pignatello & Huang, 1991). This means that resistance to leaching and persistence tend to increase with residence time in the soil illustrated in the data of Steinberg et al. (1987; Fig. 1–18). Recently applied EDB (ethylenedibromide) is more susceptible to degradation than that which had been present in the soil for many years. This effect has been attributed to the time-dependent entrapment of EDB in micropores. More work is required on desorption characteristics of pesticides at a basic level so that appropriate aspects of nonequilibrium sorption and hysteresis can be incorporated into models of pesticide transport.

Although microbial attack is the main process responsible for pesticide disappearance in soil, sorption by clay and organic matter can protect pesticides from biological attack by making the molecules less accessible to microbes and reducing solution concentrations below biodegradable levels (Scow, 1993). Ainsworth et al. (1993) demonstrated that surface–substrate interactions can increase, decrease or have no effect on rates of biodegradation; however, they highlighted the fact that there is a dearth of information relating sorption–desorption mechanisms to biodegradation rates. An example of the importance of desorption is illustrated in recent work on strychnine (heterocyclic) degradation in soils (Kookana et al., 1994). In an alkaline sandy soil where strychnine sorption was minimal because the molecule was largely uncharged, degradation was very rapid whereas in a heavier acid soil in which sorption was strong and desorption weak because of the positive charge carried by the strychnine molecule at low pH, virtually no degradation was recorded for 3-mo period. The microbial contribution to degradation simply adds another level of complexity to an already complex and ill-understood system. We need to elucidate how pesticides interact with the soil so that a rational basis for their effective use and for minimizing undesirable side effects can be developed (Schnitzer, 1991). We must solve the

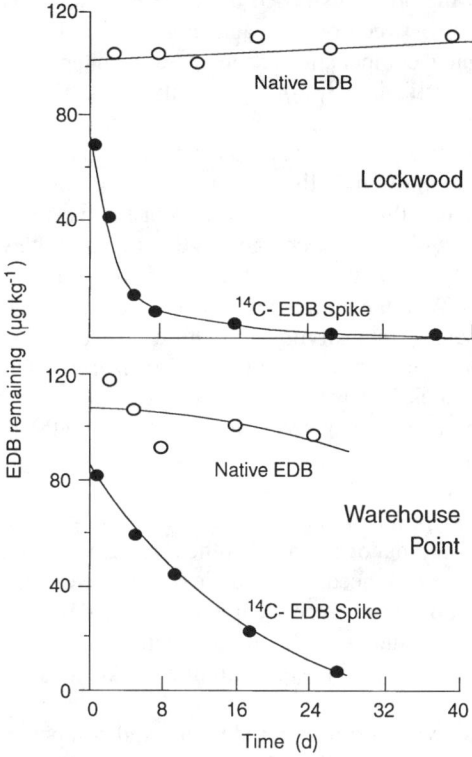

Fig. 1–18. Comparison of the degradation of native ethylene dibromide and a freshly added [^{14}C]ethylene dibromide spike in suspensions of two soils (Steinberg et al., 1987).

problems of the sorption–desorption process which determines the level of an organic pollutant available to the microbes before microbiologists will be able to solve the associated problems such as access to substrate, survival on low solution concentrations, and others.

1–3 THE FUTURE

In this presentation, I have only taken a look at the past and present in a very narrow segment of Soil Chemistry. Nevertheless based on the above observations and extension to areas that I have not dealt with, I would like to formulate my vision for the future of Soil Chemistry as follows:

1. The major thrust in the future will be in the area of Environmental Soil Science. Because of the ever-increasing production of anthropogenic wastes derived from agriculture (manures), industry (byproducts), and urban areas (sludges and composts) all of which ultimately find their way on to or through the soil, there will be a great need to redirect our current knowledge base as well as developing innovative techniques to

deal with the problems posed by these materials. Investigations of the interactions between organic and inorganic pollutants and soil organic and inorganic components that are already under way must continue. We need to investigate ways of beneficially using these materials by compounding them in ways that make them attractive for use as nutrient sources. In this way, we can reduce application rates to ensure minimal contamination of our soils, and surface and groundwaters.

2. Given the fact that a large proportion of the world's arable soils fall in soil orders with variable charge and that much of the world's population resides in these areas, we need to devote more attention to the study of such soils. With the high population density on such soils, threats to the environment by intensifying agriculture to meet the food and fiber needs of the population through the use of fertilizers, manures, and pesticides and the disposal of wastes will increase. The chemistry of such soils with respect to their interactions with these materials has hardly been investigated.

3. In all soil systems, the kinetics of the processes of sorption, and particularly, desorption of pollutants are exceedingly important. Recently, there has been major progress in the application of kinetics to soils, but much more work needs to be done. These needs basically involve the application of current knowledge to and the development of new techniques for the study of pollutant interactions with organic and inorganic ligands and surfaces in soils including the kinetics and factors affecting transport.

4. In the past, while studies on individual soil components have shed light on the nature of some chemical processes in soils, the use of single components, especially specimen minerals, has sometimes resulted in an inappropriate picture of the situation being obtained. While such studies should continue in the future, we should, wherever possible, focus on whole soil in a state as nearly representative of the field situation as we can obtain. The interactions between soil minerals and natural organics and microbes warrant further attention particularly with respect to impacts on environmental quality. Such efforts will ensure that our conclusions bear some relevance to the real world. Bartlett and James (1980) drew attention to the pitfalls of studying "laboratory dirt"—the dry, dead substitute for real soil that most of us insist on using in our investigations despite the fact that it often gives us the wrong answer!! We should always be cognizant of the fact that the soil in the field is alive and treat it accordingly.

5. It is quite clear that colloids play a substantial role in the transport of pollutants both within soil profiles and as overland flow. This will be an area of major importance in the future that will require improved techniques for the study of mobile colloids. In addition, the interactions between organic and inorganic colloids and the effects of ambient conditions (pH, electrolyte concentration, and others) on their charge and potential mobility need intensive investigation to give us a better perception of what takes place in nature. In addition, the potential for des-

orption of a pollutant from a transporting colloid at a new location needs investigation to ensure the safety of our water supplies.

6. Because the phytoavailability of a contaminant is determined by its concentration in the soil solution, the desorption of contaminants rather than their sorption reactions should receive the major emphasis. The aging effect of pesticides in soil on their desorption illustrated in Fig. 1–18 is an example of the type of investigation required in the future. There also is evidence that laboratory studies in which time scales are usually too short have been inappropriate for the study of pollutant behavior in the field where micropore diffusion that is exceedingly slow, dominates.

7. A vast array of new spectroscopic and microscopic techniques have recently become available for the study of metal and organic interactions with soil surfaces and will be more intensively used in the future. While these are valuable tools for investigation, they should not be considered the *be all and end all* of the investigation. Care must be exercised in translating the information obtained often under highly artificial conditions (vacuum) to the real world. Remember what happened when we placed too much weight on x-ray diffraction data! Some of the techniques, however, that use synchrotron light sources (EXAFS and XANES) and atomic force microscopy are particularly suitable for the study of pollutant interactions under near real world conditions.

8. Finally, we will need to work more closely with microbiologists, physicists, geologists, and ecologists all of whom have contributions to make.

REFERENCES

Adams, F., and Z. Rawajfih. 1977. Basaluminite and alunite: A possible cause of sulfate retention by acid soils. Soil Sci. Soc. Am. J. 41:686–692.

Ainsworth, C.C., J.K. Fredrickson, and S.C. Smith. 1993. Effect of sorption on the degradation of aromatic acids and bases. p. 125–144. *In* D.M. Linn. (ed.) Sorption and degradation of pesticides and organic chemicals in soil. SSSA Spec. Pub. 32. SSSA, Madison, WI.

Aldrich, D.G. 1987. Some personal reflections on Soil Science and Agriculture, 1936 to 1986. Soil Sci. Soc. Am. J. 51:1401–1405.

Alexander, M. 1991. Soil microbiology in the next 75 years: Fixed, flexible or mutable? Soil Sci. 151:35–40.

Alva, A.K., M.E. Sumner, and W.P. Miller. 1991a. Salt absorption in gypsum amended acid soils. p.93–97. *In* R.J. Wright et al. (ed.) Plant–soil interactions at low pH. Kluwer Academic Publ., Dordrecht, the Netherlands.

Alva, A.K., M.E. Sumner, and W.P. Miller. 1991b. Chemical effects of repeated equilibrations of variable-charge soils with phosphogypsum solution. Soil Sci. Soc. Am. J. 55:357–361.

Appelt, H., N.T. Coleman, and P.F. Pratt. 1975. Interactions between organic compounds, minerals and ions in volcanic-ash-derived soils: II. Effects of organic compounds on the adsorption of phosphate. Soil Sci. Soc. Am. Proc. 39:628–630.

Backes, C.A., R.G. McLaren, A.W. Rate, and R.S. Swift. 1995. Kinetics of cadmium and cobalt desorption from iron and manganese oxides. Soil Sci. Soc. Am. J. 59:778–785.

Barrow, N.J. 1987. Reactions with variable-charge soils. Martinus Nijhoff Publ., Dordrecht, the Netherlands.

Bartlett, R., and B. James. 1980. Studying air-dried, stored soil samples: Some pitfalls. Soil Sci. Soc. Am. J. 44:721–724.

Bear, F.E. 1964. Chemistry of the soil. Reinhold Publ. Corporation, New York.

Bellini, G., M.E. Sumner, and D.E. Radcliffe. 1996. Anion transport through columns of highly weathered acid soil: Adsorption and retardation. Soil Sci. Soc. Am. J. 60:132–137.

Bernhard, M., F.E. Brinckman, and P.J. Sadler. 1986. The importance of chemical speciation in environmental processes. Springer-Verlag, New York.
Bloomfield, C. 1955. A study of podzolization: VI. The immobilization of iron and aluminum. J. Soil Sci. 6:284–292.
Bolan, N.S., J.K. Syers, and M.E. Sumner. 1993. Calcium-induced sulfate adsorption by soils. Soil Sci. Soc. Am. J. 57:691–696.
Bolan, N.S., R. Naidu, S. Mahimairaja, and S. Baskaran. 1994. Influence of low molecular weight organic acids on the solubilization of phosphorus. Biol. Fert. Soils. 18:311–319.
Bolt, G.H., and M.G.M. Bruggenwert. 1976. Soil chemistry: 1. Basic elements. Elsevier Scientific Publ. Co., Amsterdam.
Bowden, J.W., S. Nagarajah, N.J. Barrow, A.M. Posner, and J.P. Quirk. 1980. Describing the adsorption of phosphate, citrate and selinite on a variable charge mineral surface. Aust. J. Soil Res. 18:49–60.
Bradfield, R. 1961. A quarter century in soil fertility research and a glimpse into the future. Soil Sci. Soc. Am. Proc. 25:439–442.
Bradley, W.F. 1945. Molecular associations between montmorillonite and some polyfunctional organic liquids. Am. Chem. Soc. J. 67:975–981.
Buddemeier, R.W., and J.R. Hunt. 1988. Transport of colloidal contaminants in groundwater: Radionuclide migration at the Nevada Test Site. Appl. Geochem. 3:535–548.
Chairidchai, P., and G.S.P. Ritchie. 1990. Zinc adsorption by a lateritic soil in the presence of organic ligands. Soil Sci. Soc. Am. J. 54:1242–1248.
Chairidchai, P., and G.S.P. Ritchie. 1992. The effect of pH on zinc adsorption by a lateritic soil in the presence of citrate and oxalate. J. Soil Sci. 43:723–728.
Chiou, C.T., R.L. Malcolm, T.I. Brinton, and E.E. Kile. 1986. Water solubility enhancement of some organic pollutants and pesticides by dissolved humic and fulvic acids. Environ. Sci. Technol. 20:502–508.
Chiou, C.T., P.E. Porter, and V.H. Freed. 1983. Partition equilibria of nonionic organic compounds between soil organic matter and water. Environ. Sci. Technol. 17:227–231.
Chorover, J., and G. Sposito. 1995a. Surface charge properties of kaolinitic tropical soils. Geochim. Cosmochim. Acta 59:875–884.
Chorover, J., and G. Sposito. 1995b. Colloid chemistry of kaolinitic tropical soils. Soil Sci. Soc. Am. J. 59:1558–1564.
Coleman, N.T., S.B. Weed, and R.J. McCracken. 1959. Cation exchange capacity and exchangeable cations in Piedmont soils of North Carolina. Soil Sci. Soc. Am. Proc. 23:146–149.
Cook, G.D., H.B. So, and R.C. Dalal. 1992. Structural degradation of two Vertisols under continuous cultivation. Soil Tillage Res. 24:47–64.
Cooke, G.W. 1979. Some priorities for British soil science. J. Soil Sci. 30:187–213.
Cresser, M., K. Killham, and T. Edwards. 1993. Soil chemistry and its applications. Cambridge Univ. Press, Cambridge, England.
Davidtz, J.C., and M.E. Sumner. 1965. Blocked charges on clay minerals in subtropical soils. J. Soil Sci. 16:270–274.
El-Swaify, S.A., and A.H. Sayegh. 1975. Charge characteristics of an Oxisol and an Inceptisol from Hawaii. Soil Sci. 120:49–56.
Emerson, W.W., and M.G. Dettman. 1960. The effect of pH on the wet strength of soil crumbs. J. Soil Sci. 11:149–158.
Espinoza, W., R.G. Gast, and R.S. Adams. 1975. Charge characteristics and nitrate retention by two andepts from south-central Chile. Soil Sci. Soc. Am. Proc. 39:842–846.
Farina, M.P.W., M.E. Sumner, and P. Channon. 1982. Lime induced yield depressions in maize (*Zea mays* L.) on highly weathered soils. p. 162–168. *In* Proc. 9th Int. Plant Nutr. Coll., Warwick, England. 22–27 August 1982. CAB, Slough, England.
Fey, M.V., and J. Le Roux. 1976. Electric charges on sesquioxidic soil clays. Soil Sci. Soc. Am. J. 40:359–363.
Fieldes, M., and R.K. Schofield. 1960. Mechanisms of ion adsorption by inorganic soil colloids. NZ J. Sci. 3:563–79.
Gallez, A., A.S.R. Juo, and A.J. Herbillon. 1976. Surface and charge characteristics of selected soils in the tropics. Soil Sci. Soc. Am. J. 40:601–607.
Gardner, W.R. 1991. Soil science as a basic science. Soil Sci. 151:2–6.
Gedroiz, K.K. 1925. Papers on soil reactions. USDA, Washington DC.
Gillman, G.P. 1974. The influence of net charge on water dispersible clay and sorbed sulphate. Aust. J. Soil Res. 12:173–176.

Gillman, G.P. 1979. A proposed method for the measurement of exchange properties of highly weathered soils. Aust. J. Soil Res. 17:129–139.

Gillman, G.P. 1981. Effects of pH and ionic strength on the cation exchange capacity of soils with variable charge. Aust. J. Soil Res. 19:93–96.

Gillman, G.P. 1984. Using variable charge characteristics to understand the exchangeable cation status of oxic soils. Aust. J. Soil Res. 22:71–80.

Gillman, G.P. 1991. The chemical properties of acid soils with emphasis on soils of the humid tropics. p. 3–14. *In* R.J. Wright et al. (ed.) Plant–soil interactions at low pH. Kluwer Academic Publ., Dordrecht, the Netherlands.

Gillman, G.P., and L.C. Bell. 1976. Surface charge characteristics of six weathered soils from tropical North Queensland. Aust. J. Soil Res. 14:351–360.

Gillman, G.P., and L.C. Bell. 1978. Soil solution studies on weathered soils from tropical North Queensland. Aust. J. Soil Res. 16:67–77.

Gillman, G.P., and G. Uehara. 1980. Charge characteristics of soils with variable and permanent charge minerals: II. Experimental. Soil Sci. Soc. Am. J. 44:252–255.

Gillman, G.P., and E.A. Sumpter. 1986a. Modification to the compulsive exchange method for measuring exchange characteristics of soils. Aust. J. Soil Res. 24:61–66.

Gillman, G.P., and E.A. Sumpter. 1986b. Surface charge characteristics and lime requirements of soils derived from basaltic, granitic and metamorphic rocks in high-rainfall tropical Queensland. Aust. J. Soil Res. 24:173–192.

Gillman, G.P., and D.F. Sinclair. 1987. The grouping of soils with similar charge properties as a basis for agrotechnology transfer. Aust. J. Soil Res. 25:275–285.

Gillman, G.P., and M.E. Sumner. 1987. Surface charge characterization and soil solution composition of four soils from the Southern Piedmont in Georgia. Soil Sci. Soc. Am. J. 51:589–594.

Gillman, G.P., and M.J. Hallman. 1988. Measurement of exchange properties of Andisols by the compulsive exchange method. Soil Sci. Soc. Am. J. 52:1196–1198.

Goldberg, S., and H.S. Forster. 1990. Flocculation of reference clays and arid-zone soil clays. Soil Sci. Soc. Am. J. 54:714–718.

Goldberg, S., B.S. Kapoor, and J.D. Rhoades. 1990. Effect of aluminum and iron oxides and organic matter on flocculation and dispersion of arid zone soils. Soil Sci. 150:588–593.

Green, R.E., and S.W. Karickhoff. 1990. Sorption estimates for modeling. p. 79–101. *In* H.H. Cheng. (ed.) Pesticides in the soil environment: Processes, impacts and modeling. SSSA Book Ser. 2. SSSA, Madison, WI.

Green, R.E., R.C. Schneider, R.T. Gavenda, and C.J. Miles. 1993. Utility of sorption and degradation parameters from the literature for site specific pesticide impact assessments. p. 209–225. *In* D.M. Linn et al. (ed.) Sorption and degradation of pesticides and organic chemicals in soil. SSSA Spec. Publ. 32. SSSA, Madison, WI.

Greenland, D.J. 1991. The contributions of soil science to society: Past, present and future. Soil Sci. 151:19–23.

Greenwood, D.J. 1993. The changing scene of British soil science. J. Soil Sci. 44:191–207.

Gregory, J. 1989. Fundamentals of flocculation. Crit. Rev. Environ. Contr. 19:185–230.

Grove, J.H., C.S. Fowler, and M.E. Sumner. 1982. Determination of the charge character of selected acid soils. Soil Sci. Soc. Am. J. 46:32–38.

Gu, B., J. Schmitt, Z. Chen, L. Liang, and J.F. McCarthy. 1994. Adsorption and desorption of natural organic matter on iron oxide: Mechanisms and models. Environ. Sci. Technol. 28:38–46.

Guggenberger, G., B. Glaser, and W. Zech. 1994. Heavy metal binding by hydrophobic and hydrophilic dissolved organic carbon fractions in a Spodosol A and B horizon. Water Air Soil Pollut. 72:111–127.

Harter, R.D. 1991. Kinetics of sorption–desorption processes in soil. p. 135–149. *In* D.L. Sparks and D.L. Suarez. (ed.) Rates of chemical processes. SSSA Spec. Publ. 27. SSSA, Madison, WI.

Harter, R.D., and R. Naidu. 1996. Role of metal-organic complexation in metal sorption by soils. Adv. Agron. 55:219–263.

Heil, D., and G. Sposito. 1993a. Organic matter role in illitic soil colloids flocculation: I. Counter ions and pH. Soil Sci. Soc. Am. J. 57:1241–1246.

Heil, D., and G. Sposito. 1993b. Organic matter role in illitic soil colloids flocculation: II. Surface charge. Soil Sci. Soc. Am. J. 57:1246–1253.

Heil, D., and G. Sposito. 1995. Organic matter role in illitic soil colloids flocculation: III. Scanning force microscopy. Soil Sci. Soc. Am. J. 59:266–269.

Hendershot, W.H., and L.M. Lavkulich. 1978. The use of zero point of charge (ZPC) to assess pedogenic development. Soil Sci. Soc. Am. J. 42:468–472.

Hendricks, S.B. 1941. Base exchange of the clay mineral montmorillonite for organic cations and its dependence upon adsorption due to van der Waals forces. J. Chem. Phys. 45:65–81.
Hendricks, S.B., and W.H. Fry. 1930. The results of x-ray and mineralogical examination of soil colloids. Soil Sci. 29:457–476.
Hillel, D. 1991. Research in soil physics: A review. Soil Sci. 151:30–34.
Hissink, D.J. 1924. Base exchange in soils: General views. Trans. Faraday Soc. 20:551–566.
Hogg, D.S., R.G. McLaren, and R.S. Swift. 1993. Desorption of copper from some New Zealand soils. Soil Sci. Soc. Am. J. 57:361–366.
Holdridge, D.A. 1959. Isomorphous replacements in kaolinite. The A.T. Green Book. British Ceramic Research Association, Stoke on Trent, England.
Huang, P.M., and A. Violante. 1986. Influence of organic acids on crystallization and surface properties of the precipitation products of aluminum. p. 159–221. In P.M. Huang and M. Schnitzer. (ed.) Interactions of soil minerals with natural organics and microbes. SSSA Spec. Publ. 17. SSSA, Madison, WI.
Hue, N.V., G.R. Craddock, and F. Adams. 1986. Effects of organic acids on aluminum toxicity in subsoils. Soil Sci. Soc. Am. J. 50:28–34.
Hue, N.V. 1991. Effect of organic acid–anion on P sorption and phytoavailability in soils with different mineralogies. Soil Sci. 152:463–471.
Jackson, M.L. 1963. Aluminum bonding in soils: A unifying principle in soil science. Soil Sci. Soc. Am. Proc. 27:1–10.
James, B.R. 1993. The future of soil chemistry in the Northeast: Lessons from colonial New England. In J.T. Sims. (ed.) Agricultural research in the Northeastern United States: Critical review and future perspectives. ASA, Madison, WI.
Jardine, P.M., G.K. Jacobs, and J.D. O'Dell. 1993. Unsaturated transport processes in undisturbed heterogeneous porous media: II. Co-contaminants. Soil Sci. Soc. Am. J. 57:954–962.
Jenny, H. 1941. Factors of soil formation. McGraw-Hill, New York.
Johnson, C.A. 1986. The regulation of trace element concentrations in river and estuarine waters contaminated with acid mine drainage: The adsorption of Cu and Zn on amorphous Fe oxyhydroxides. Geochim. Cosmochim. Acta. 50:2433–2438.
Kabata-Pendias, A. 1980. Heavy metal sorption by clay minerals and oxides of iron and manganese. Mineral. Polonica 11:3–13.
Kamprath, E.J., W.L. Nelson, and J.W. Fitts. 1956. The effect of pH, sulfate and phosphate concentrations on the adsorption of sulfate by soils. Soil Sci. Soc. Am. Proc. 20:463–466.
Kaplan, D.I., P.M. Bertsch, D.C. Adriano, and W.P. Miller. 1993. Soil-borne mobile colloids as influenced by water flow and organic carbon. Environ. Sci. Technol. 27:1193–1200.
Kaplan, D.I., M.E. Sumner, P. Bertsch, and D. Adriano. 1996. Chemical conditions conducive to the release of mobile colloids from Ultisol profiles. Soil Sci. Soc. Am. J. 60:269–274.
Katsura, T., I. Kushiro, S. Akimoto, J.L. Walker, and G.D. Sherman. 1962 Titanomagnetite and titanomaghemite in a Hawaiian soil. J. Sediment. Petrol. 32:299–308.
Kelley, W.P. 1948. Cation exchange in soils. Reinhold Publ. Corp., New York.
Kellog, C.E. 1961. A challenge to American soil scientists: On the occasion of the 25th Anniversary of the Soil Science Society of America. Soil Sci. Soc. Am. Proc. 25:419–423.
Kerndorff, H., and M. Schnitzer. 1980. Sorption of metals on humic acid. Geochim. Cosmochim. Acta 44:1701–1708.
Kinniburgh, D.G. 1983. The H^+/M^{2+} exchange stoichiometry of calcium and zinc adsorption by ferrihydrate. J. Soil Sci. 34:759–768.
Kirkham, D. 1961. Soil physics 1936–1961 and a look ahead. Soil Sci. Soc. Am. Proc. 25:423–427.
Kookana, R.S., R.G. Gerritse, and L.A.G. Aylmore. 1992. A method for studying nonequilibrium sorption during transport of pesticides in soil. Soil Sci. 154:344–349.
Kookana, R.S., S. Rogers, and D. Oliver. 1994. Behaviour of strychnine in soils: 1. Retention, release and degradation studies. Coop. Res. Ctr. for Soil and Land Management Rep., Glen Osmond, Australia.
Kretzschmar, R., W.P. Robarge, and S.B. Weed. 1993. Flocculation of kaolinitic soil clays: Effects of humic substances and iron oxides. Soil Sci. Soc. Am. J. 57:1277–1283.
Kretzschmar, R., W.P. Robarge, and A. Amoozegar. 1994. Filter efficiency of three saprolites for natural clay and iron oxides. Environ. Sci. Technol. 28:1907–1915.
Kretzschmar, R., W.P. Robarge, and A. Amoozegar. 1995. Influence of natural organic matter on colloid transport through saprolite. Water Resour. Res. 31:435–445.
Kruyt, H.R. 1949. Colloid science: II. Reversible systems. Elsevier, New York.
Kruyt, H.R. 1952. Colloid science: I. Irreversible Systems. Elsevier, Amsterdam, the Netherlands.

Kwong, F.K., and P.M. Huang. 1978. Sorption of phosphate by hydrolytic reaction products of aluminum. Nature (London) 271:336–337.

Kwong, F.K., and P.M. Huang. 1979. Surface reactivity of aluminum hydroxides precipitated in the presence of low molecular weight organic acids. Soil Sci. Soc. Am. J. 43:1107–1113.

Läuchli, A. 1987. Soil Science in the next twenty-five years: Does biotechnology play a role? Soil Sci. Soc. Am. J. 51:1405–1409.

Looney, B.B., M.W. Grant, and C.M. King. 1987. Estimation of geochemical parameters for assessing subsurface transport at the Savannah River Site. Westinghouse Savannah River Co., Aiken, SC.

Low, P.F., and C.A. Black. 1948. Phosphate-induced decomposition of kaolinite. Soil Sci. Soc. Am. Proc. 12:180–184.

MacCracken, R.J. 1987. Soils, soil scientists and civilization. Soil Sci. Soc. Am. J. 51:1395–1400.

Madrid, L., and A.M. Posner. 1979. Desorption of phosphate from goethite. J. Soil Sci. 30:697–707.

Manceau, A., L. Charlet, M.C. Boisset, B. Didier, and L. Spandini. 1993. Sorption and speciation of heavy metals on hydrous Fe and Mn oxides: From microscopic to macroscopic. Appl. Clay Sci. 7:201–223.

Mattson, S. 1927. Anionic and cationic adsorption by soil colloidal materials of varying $SiO_2/(Al_2O_3 + Fe_2O_3)$ ratio. p. 199–211. *In* Trans. 1st. Int. Cong. Soil Sci., Washington, DC. 13–22 June 1927. USDA, Washington, DC.

Mattson, S. 1931. The laws of soil colloidal behavior: VI. Amphoteric behavior. Soil Sci. 32:343–365.

Mattson, S. 1932. The laws of soil colloidal behavior: VII. Proteins and proteinated complexes. Soil Sci. 33:41–72.

Mattson, S., and K. Hou. 1937. The laws of soil colloidal behavior: XX. The neutral salt effect and the amphoteric points of soils. Soil Sci. 44:151–166.

Mattson, S., and A.J. Pugh. 1934. The laws of soil colloidal behavior: XIV. The electrokinetics of hydrous oxides and their ionic exchange. Soil Sci. 38:299–313.

McBride, M.B. 1985. Influence of glycine on Cu^{2+} adsorption by microcrystalline gibbsite and boehmite. Clays Clay Min. 33:397–402.

McBride, M.B. 1987. Adsorption and oxidation of phenolic compounds by iron and manganese oxides. Soil Sci. Soc. Am. J. 51:1466–1472.

McBride, M.B. 1994. Environmental chemistry of soils. Oxford Univ. Press, New York.

McBride, M.B., and L.G. Wesselink. 1988. Chemisorption of catechol on gibbsite, boehmite, and noncrystalline alumina surfaces. Environ. Sci. Technol. 22:703–708.

McCarthy, J.F., and J.M. Zachara. 1989. Subsurface transport of contaminants. Environ. Sci. Technol. 23:496–502.

McIntyre, D.S. 1958. Permeability measurements on soil crusts formed by raindrop impact. Soil Sci. 85:185–189.

McKenzie, R.M. 1980. The adsorption of lead and other heavy metals on oxides of manganese and iron. Aust. J. Soil Res. 18:61–73.

McLaren, R.G., and D.V. Crawford. 1973. Studies on soil copper: II. The specific sorption of copper by soils. J. Soil Sci. 24:443–452.

McLaren, R.G., R.S. Swift, and J.G. Williams. 1981. The adsorption of copper by soil materials at low equilibrium solution concentrations. J. Soil Sci. 32:247–256.

McLaren, R.G., J.D. Williams, and R.S. Swift. 1983. Some observations on the desorption and distribution behavior of copper with soil components. J. Soil Sci. 34:325–331.

McLaren, R.G., J.G. Williams, and R.S. Swift. 1986. Sorption and desorption of cobalt by soils and soil components. J. Soil Sci. 37:413–426.

Melamed, R., J.J. Jurinak, and L.M. Dudley. 1994. Anion exclusion-pore water velocity interaction affecting transport of bromine through an Oxisol. Soil Sci. Soc. Am. J. 58:1405–1410.

Menzel, R.G. 1991. Soil science: The environmental challenge. Soil Sci. 151:24–29.

Miller, W.P., and M.K. Baharuddin. 1986. Relationship of soil dispersibility to infiltration and erosion of southeastern soils. Soil Sci. 142:235–240.

Miller, W.P., D.E. Radcliffe, and M.E. Sumner. 1986. The effect of soil amendment with phosphogypsum on clay dispersion and implications for soil conservation and environmental quality. *In* Int. Symp. on Phosphogypsum, Miami, FL. December 1986. Florida Inst. of Phosphate Res., Bartow.

Miller, W.P., H. Frenkel, and K.D. Newman. 1990. Flocculation concentration and sodium–calcium exchange of kaolinitic soil clays. Soil Sci. Soc. Am. J. 54:346–351.

Mingelgrin, U., and Z. Gerstl. 1983. Reevaluation of partitioning as a mechanism of nonionic chemical adsorption in soils. J. Environ. Qual. 12:1–11.

Misono, M., E. Ochiai, Y. Saito, and Y. Yoneda. 1967. A new dual parameter scale for the strength of Lewis acids and bases with the evaluation of their softness. J. Inorg. Nucl. Chem. 29:2685–2691.

Morais, F.I., A.L. Page, and L.J. Lund. 1976. The effect of pH, salt concentration, and nature of electrolytes on the charge characteristics of Brazilian tropical soils. Soil Sci. Soc. Am. J. 40:521–527.

Muljadi, D., A.M. Posner, and J.P. Quirk. 1966. The mechanism of phosphate absorption by kaolinite, gibbsite and pseudoboehmite: II. The location of the absorption sites. J. Soil Sci. 17:230–237.

Naidu, R., R.S. Kookana, M.E. Sumner, R.D. Harter, and K.G. Tiller. 1997. Cadmium sorption and transport in variable charge soils: A review. J. Environ. Qual. 26:602-617.

Newman, M.E., A.W. Elzerman, and B.B. Looney. 1993. Facilitated transport of selected metals in aquifer material packed columns. J. Contamin. Hydrol. 14:233–246.

Nielsen, D.R. 1987. Emerging frontiers in soil science. Geoderma 40:267–273.

Nightingale, H.I., and W.L. Bianchi. 1977. Ground-water turbidity resulting from artificial recharge. Ground Water 15:146–152.

Oster, J.D. 1982. Gypsum use in irrigated agriculture: A review. Fert. Res. 3:73–89.

Oster, J.D. 1993. Sodic soil reclamation. p. 485–490. In H. Leith and A. Al Massom (ed.) Toward the rational use of high salinity tolerant plants. Kluwer Academic Publ., Dordrecht, the Netherlands.

Oster, J.D., I. Shainberg, and J.D. Wood. 1980. Flocculation value and gel structure of sodium–calcium montmorillonite and illite suspensions. Soil Sci. Soc. Am. J. 44:955–959.

Overstreet, R., J.C. Martin, and H.M. King. 1951. Gypsum, sulfur and sulfuric acid for reclaiming an alkali soil of the Fresno series. Hilgardia 21:113–128.

Parfitt, R.L., and J.D. Russell. 1977. Adsorption on hydrous oxides: IV. Mechanisms of adsorption of various ions on goethite. J. Soil Sci. 28:297–305.

Pearce, R., and M.E. Sumner. 1997. Apparent salt sorption reactions in an unfertilized acid subsoil. Soil Sci. Soc. Am. J. 61:765–772.

Penrose, W.R., W.L. Polzer, E.H. Essington, D.W. Nelson, and K.A. Orlandini. 1990. Mobility of plutonium and americium through a shallow aquifer in a semiarid region. Environ. Sci. Technol. 24:228–234.

Petruzzelli, G., G. Guidi, and L. Lubrano. 1978. Organic matter as an influencing factor on copper and cadmium adsorption by soils. Water Air Soil Pollut. 9:263–269.

Philip, J.R. 1991. Soils, natural science and models. Soil Sci. 151:91–98.

Pignatello, J.J., and L.Q. Huang. 1991. Sorptive reversibility of atrazine and metolachlor residues in field soil samples. J. Environ. Qual. 22:222–230.

Poszmann, C.A. 1927. Retention of phosphorus by soil colloids. Soil Sci. 38:219–239.

Prasad, B., and I. Sarangthem. 1993. Adsorption of zinc as affected by its sources in calcareous soils. J. Ind. Soc. Soil Sci. 41:261–265.

Rajan, S.S.S. 1978. Sulfate adsorbed on hydrous alumina, ligands displaced and changes in surface charge. Soil Sci. Soc. Am. J. 42:39–44.

Rao, P.S.C., and W.M. Alley. 1993. Pesticides. p. 345–382. In W.M. Alley. (ed.) Regional groundwater quality. Van Nostrand Reinhold, New York.

Ravikovitch, S. 1934. Anion exchange: I. Adsorption of the phosphoric acid ions by soils. Soil Sci. 38:219–239.

Rengasamy, P. 1983. Clay dispersion in relation to changes in the electrolyte composition of dialysed red-brown earths. J. Soil Sci. 34:723–732.

Rich, C.I. 1968. Hydroxy interlayers in expansible phyllosilicates. Clays Clay Miner. 16:15–30.

Sanchez, P.A., and J.G. Salinas. 1981. Low-impact technology for managing oxisols and ultisols in tropical America. Adv. Agron. 34:279–406.

Scarseth, G.D. 1935. The mechanism of phosphate retention by natural alumino-silicate colloids. J. Am. Soc. Agron. 27:595–616.

Schindler, P.W., B. Furst, R. Dick, and P.U. Wolf. 1976. Ligand properties of surface silanol groups: I. Surface complex formation with Fe, Cu, Cd, and Pb. J. Coll. Interf. Sci. 55:469–475.

Schnitzer, M. 1991. Soil organic matter: The next 75 years. Soil Sci. 151:41–58.

Schofield, R.K. 1939. The electrical charges on clay particles. Soils Fert. 2:1–5.

Schofield, R.K. 1947. The electric charge on soil particles. Roth. Exp. Stn. Ann. Rep. 95-100. Rothamsted Agric. Exp. Sn., Harpenden, England.

Schofield, R.K. 1949. Effect of pH on electric charges carried by clay particles. J. Soil Sci. 1:1–8.

Schuffelen, A.C. 1974. A few aspects of 50 years of soil chemistry. Geoderma 12:281–297.

Schwertmann, U., H. Kodama, and W.R. Fisher. 1986. Mutual interactions between organics and iron oxides. p. 223–250. *In* P.M. Huang and M. Schnitzer. (ed.) Interactions of soil minerals with natural organics and microbes. SSSA Spec. Publ. 17. SSSA, Madison, WI.

Schwertmann, U., and R.M. Taylor. 1989. Iron oxides in the laboratory. p. 379–438. *In* J.B. Dixon and S.B. Weed. (ed.) Minerals in soil environments. SSSA, Madison, WI.

Scow, K.M. 1993. Effect of sorption–desorption and diffusion processes on the kinetics of biodegradation of organic chemicals in soils. p. 73–114. *In* D.M. Linn. (ed.) Sorption and degradation of pesticides and organic chemicals in soil. SSSA Spec. Publ. 32. SSSA, Madison, WI.

Seaman, J.C. 1994. Physicochemical and mineralogical controls on colloid generation and transport within the highly weathered alluvial sediments of the Upper Coastal Plain. Ph.D. diss. Univ. of Georgia, Athens.

Shainberg, I., M.E. Sumner, W.P. Miller, M.P.W. Farina, M.A. Pavan, and M.V. Fey. 1989. Use of gypsum on soils: A review. Adv. Soil Sci. 9:1–111.

Sharpley, A.N., W.W. Troeger, and S.J. Smith. 1991. The measurement of bioavailable phosphorus in agricultural runoff. J. Environ. Qual. 20:235–238.

Sharpley, A.N., S.C. Chapra, R. Wedephol, J.T. Sims, T.C. Daniel, and K.R. Reddy. 1994. Managing agricultural phosphorus for protection of surface waters: Issues and options. J. Environ. Qual. 23:437–451.

Shuman, L.M. 1988. Effect of removal of organic matter and iron- and manganese-oxides on zinc adsorption by soil. Soil Sci. 146:248–254.

Simonson, R.W. 1991. Soil science: Goals for the next 75 years. Soil Sci. 151:7–18.

Sparks, D.L. 1986. Soil physical chemistry. CRC Press, Boca Raton, FL.

Sparks, D.L. 1989. Kinetics of soil chemical processes. Academic Press, New York.

Sparks, D.L. 1993. Soil and environmental chemistry research in the northeastern USA: Challenges and opportunities for the 1990s. p. *In* J.T. Sims. (ed.) Agricultural research in the Northeastern United States: Critical review and future perspectives. ASA, Madison, WI.

Sparks, D.L. 1995. Environmental soil chemistry. Academic Press, San Diego, CA.

Sposito, G. 1981. Cation exchange in soils: An historical and theoretical perspective. p. 13–30. *In* R.H. Dowdy. (ed.) Chemistry in the soil environment. ASA, Madison, WI.

Sposito, G. 1989. The chemistry of soils. Oxford Univ. Press, New York.

Steinberg, S.M., J.J. Pignatello, and B.L. Sawhney. 1987. Persistence of 1,2 dibromoethane in soils: Entrapment in interparticle micropores. Environ. Sci. Technol. 21:1201–1208.

Stevenson, F.J. 1994. Humus chemistry, genesis, composition, reactions. 2nd ed. John Wiley & Sons, New York.

Sumner, M.E. 1963a. Effect of alcohol washing and pH value of leaching solution on positive and negative charges in ferruginous soils. Nature (London) 198:1018–1019.

Sumner, M.E. 1963b. Effect of iron oxides on positive and negative charge in clays and soils. Clay Min. Bull. 5:218–226.

Sumner, M.E. 1993. Sodic soils: New perspectives. Aust. J. Soil Res. 31:683–750.

Sumner, M.E., and J.C. Davidtz. 1965. Positive and negative charges in some Natal soils. S. Afr. J. Agric. Sci. 8:1045–1050.

Sumner, M.E., L. West, and J. Leal. 1993. Suelos de la agroindustria cafetalera del Guatemala: Region Sur. Univ. of Georgia, Athens.

Swift, R.S., and R.G. McLaren. 1991. Micronutrient adsorption by soils and soil colloids. p. 257–292. *In* G.H. Bolt et al. (ed.) Interactions at the soil colloid-soil solution interface. Kluwer Academic Publ., Dordrecht, the Netherlands.

Swindale, L.S., and M.L. Jackson. 1956. Genetic processes in some residual podzolized soils of New Zealand. p. 233–239. *In* Trans. 6th. Int. Congr. Soil Sci., Paris. 28 Mar. 1956. Laboureur, Paris.

Theng, B.K.G. 1980. Soils with variable charge. New Zealand Soc. of Soil Sci., Lower Hutt, New Zealand.

Theng, B.K.G. 1991. Soil science in the tropics: The next 75 years. Soil Sci. 151:76–90.

Thomas, G.W. 1977. Historical developments in soil chemistry: Ion exchange. Soil Sci. Soc. Am. J. 41:230–238.

Thompson, H.S. 1850. On the absorbent power of soils. J. Royal Agric. Soc. England. 11:68–74.

Tiller, K.G., and J.F. Hodgson. 1962. The specific sorption of cobalt and zinc by layer silicates. Clays Clay Min. 9:393–403.

Tinker, P.B. 1985. Soil science in a changing world. J. Soil Sci. 36:1–8.

Uehara, G., and G. Gillman. 1981. The mineralogy, chemistry, and physics of tropical soils with variable charge clays. Westview Press, Boulder, CO.

van Olphen, H. 1977. An introduction to clay colloid chemistry. 2nd ed. John Wiley & Sons, New York.

van Raij, B., and M. Peech. 1972. Electrochemical properties of some Oxisols and Alfisols of the tropics. Soil Sci. Soc. Am. Proc. 36:587–593.

van Raij, B., H. Cantarella, and P.R. Furlani. 1988. Efeito na reação do solo da absorção de amônio e nitrato pelo sorgo na presença e na ausência de gesso. R. Bras. Ci. Solo. 12:131–136.

Wada, K. 1977. Allophane and imogolite. p. 603–638. *In* J.B. Dixon nad S.B. Weed. (ed.) Minerals in soil environments. SSSA Book Ser. 1. SSSA, Madison, WI.

Wada, K. 1989. Allophane and imogolite. p. 1051–1088. *In* J. Dixon and S.B. Weed. (ed.) Minerals in soil environments. SSSA Book Ser. 1. SSSA, Madison, WI.

Wada, K., and Y. Okamura. 1983. Net charge characteristics of Dystrandept B and theoretical prediction. Soil Sci. Soc. Am. J. 47:902–905.

Wann, S.S., and G. Uehara. 1978. Surface charge manipulation of constant surface potential soil colloids: I. Relation to sorbed phosphorus. Soil Sci. Soc. Am. J. 42:565–570.

Wauchope, R.D., T.M. Buttler, A.G. Hornsby, P.W.M. Augstijn Beckers, and J.P. Burt. 1992. The SCS/ARS/CES pesticide properties database for environmental decision-making. Rev. Environ. Contamin. Toxicol. 123:1–157.

Way, J.T. 1850. On the power of soils to absorb manure. J. Royal Agric. Soc. England 11:313–379.

Way, J.T. 1852. On the power of soils to absorb manure. J. Royal Agric. Soc. England 13:123–143.

Weber, J.B., and C.T. Miller. 1989. Organic chemical movement over and through soil. p. 305–334. *In* B.L. Sawhney and K. Brown. (ed.) Reactions and movement of organic chemicals in soils. SSSA Spec. Publ. 22. SSSA, Madison, WI.

Wild, A. 1989. Soil scientists as members of the scientific community. J. Soil Sci. 40:209–221.

Wong, M.T.F., A. Wild, and A.S.R. Juo. 1987. Retarded leaching of nitrate measured in monolith lysimeters in south-east Nigeria. J. Soil Sci. 38:511–518.

Wong, M.T.F., R. Hughes, and D.L. Rowell. 1990a. Retarded leaching of nitrate in acid soils from the tropics: Measurement of the effective anion exchange capacity. J. Soil Sci. 41:655–663.

Wong, M.T.F., R. Hughes, and D.L. Rowell. 1990b. The retention of nitrate in acid soils from the tropics. Soil Use Manage. 6:72–74.

Xue, J., and P.M. Huang. 1995. Zinc adsorption-desorption on short-range ordered iron oxide as influenced by citric acid during its formation. Geoderma 64:343–356.

Yu, T.R. 1997. Chemistry of variable charge soils. Oxford Univ. Press, New York.

Zhang, G.Y., X.N. Zhang, and T.R. Yu. 1991. Adsorption of sulfate and fluoride in relation to some surface chemical properties of Oxisols. Pedosphere 1:17–28.

Zhang, X.N., G.Y. Zhang, A.Z. Zhao, and T.R. Yu. 1989. Surface electrochemical properties of the B horizon of a Rhodic Ferralsol, China. Geoderma 44:275–286.

2 Computational Chemistry in the Future of Soil Chemistry

Brian J. Teppen

Advanced Analytical Center for Environmental Sciences
Savannah River Ecology Laboratory
University of Georgia
Aiken, South Carolina

2–1 INTRODUCTION

2–1.1 The Quest for Mechanism

The purest drive toward science is our sense of wonder about how nature works. The increasingly common utilitarian viewpoint is that science should focus on improving the technologies that make our lives easier. These two impulses toward science are often at odds, since the applicability of basic, wonder-driven research is often difficult to predict or visualize. As soil chemists, we are fortunate that our curiosity about the fundamental interactions between soil colloids and aqueous solutes also provides useful knowledge that can be quickly applied to real-world problems. That is, an understanding of the detailed reactions of a fertilizer, pesticide, or contaminant in a soil system is the key to our ability to control (manage) efficient application or effective remediation of these chemicals. Molecular understanding of an element's speciation also is key to understanding availability of the element to biological systems (i.e., us) and its toxicology within those systems. The computational chemistry techniques that I wish to discuss in this chapter are primarily those for performing molecular simulations, in which the structures and/or dynamics of atoms in a system are theoretically modeled. As such, these computational techniques can help us describe and visualize the mechanisms of interaction among model colloids, water, and solutes.

Surface reaction mechanisms or micro-scale solute distributions may not be validly inferred from any macroscopic measurements of bulk systems, as G. Sposito has so tirelessly reminded us (i.e., Sposito, 1965, 1984; Johnston & Sposito, 1987), although violations of this precept continue to occur. Rather, direct observation of surface and interfacial structures through diffraction, nanoscale imaging, and spectroscopic techniques is the only way to make any

---

Copyright © 1998. Soil Science Society of America, 677 S. Segoe Rd., Madison, WI 53711, USA.
Future Prospects for Soil Chemistry. SSSA Special Publication no. 55.

claims about adsorption mechanisms or surface speciation (Johnston & Sposito, 1987).

2–1.2 Interpreting Spectroscopic Results

As soil chemists adapt and advance experimental techniques for probing environmentally relevant systems at the atomic scale, it becomes evident that the data are often difficult to interpret unambiguously. Therefore, our understanding of the underlying phenomena still depends on frequently uncertain assumptions. Atomic-scale, theoretical models of colloid–solution interfaces will facilitate the interpretation of diffraction, nanoscale imaging, and spectroscopic data through probing the energetic consequences of the atomic arrangements observed hypothesized by experimentalists.

2–1.3 Constraining Thermodynamic Modeling

A variety of models calculate the equilibrium speciation of a given solution based upon an experimental chemical analysis and a thermodynamic database. Some of these models can also compute the distributions of aqueous species sorbed to solid phases. When adsorbates are speciated in this manner according to diffuse double layer or surface complexation theory, the model is always based upon a postulated atomic-scale mechanism for the equilibrium adsorbent–adsorbate configuration. When such a model is calibrated, equilibrium adsorption constants and activity coefficients are fit to the data and can only be regarded as empirical parameters with little physical meaning (Sposito, 1981). Indeed, the adjustable parameters result in multiple models that each adequately describe the same macroscopic data even though they are based on mutually contradictory atomic-scale postulates (Westall & Hohl, 1980). Such models are nonetheless useful in that they fit the calibration data and can even have predictive power for the same system (Sposito, 1990). Unfortunately, the model must be recalibrated through extensive experimentation if it is to be applied to a new system, since a new set of empirical parameters must be fit.

Another future promise of molecular modeling is that simulations will add new physicochemical constraints on the use of thermodynamic models for adsorption to soil surfaces. In the first place, the model chosen for a given adsorbate–adsorbent system will be constrained by the simulated distribution of the adsorbate with respect to the adsorbent surface. Only thermodynamic speciation models that postulate similar adsorbate distributions could then be justifiably used. Secondly, simulations will be able to provide physically based estimates for many of the parameters needed to apply the model. This will result in fewer parameters that need to be fit empirically for a given system, and the parameters that must be fit will have more physical meaning.

While I believe that molecular modeling tools offer almost limitless opportunity to investigate soil systems, the toolbox is not yet complete, nor is it well tested. Figure 2–1 may serve as a reminder that it is best to begin with a sober eye on the longer-term benefits of modeling. An overzealous evangelist who promises too many miracles may damage the credibility of all modeling efforts.

2–2 THE POTENTIAL ENERGY SURFACE

Chemical thermodynamics is built upon the recognition that natural systems proceed to increase the entropy of the universe and thereby tend toward states of lower overall energy (when no work is being done). In this spirit, all molecular modeling techniques are based upon calculating the energy of the system under investigation. For a collection of many (n) atoms, the potential energy as a function of atomic positions is a ($3n - 6$)-dimensional surface. We will now consider the various approximations that may be used to calculate the energy of a collection of atoms, and thus trace the potential energy surface.

2–2.1 Ab Initio Quantum Mechanics

Since molecular modeling treats atomic-scale systems, the only theoretical method that may be rigorously used to compute energies is quantum mechanics. Quantum mechanical methods are often termed ab initio because they are based on no empirical data other than the masses of elementary particles, the elementary charge, the speed of light, and Planck's constant. Quantum theory has never been falsified, but practical methods of quantum mechanics are always approximate when more than one electron is included in the calculation. Three different sorts of approximations are employed. First, relativistic effects are almost always neglected and will be discussed no further here. Second, interactions among the electrons of the system may be treated with various degrees of rigor. Third, it would take an infinitely long time to compute an exact result for a given model of electronic interaction, so the series of functions comprising the molecular wavefunction are truncated, often brutally.

In the usual approach to computing a molecular wavefunction (Szabo & Ostlund, 1982; Hehre et al., 1986; Levine, 1991), nuclear and electronic motions are considered to be decoupled, and all energies are calculated for stationary nuclei. The electrons interact with the nuclei, but also with each other. In the

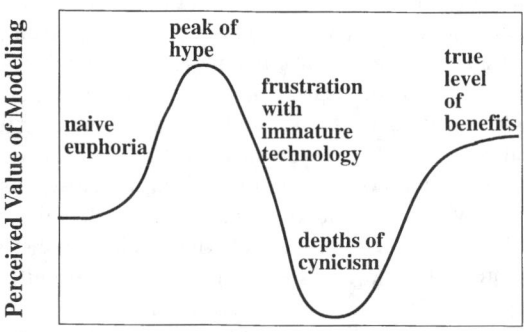

Fig. 2–1. Schematic for the fate of a molecular modeling program that makes too many initial promises. It is true that one can visualize interactions, compute geometries in mechanistic detail, and calculate energies, but simulation of realistic systems for reasonable lengths of time is difficult.

Hartree-Fock approximation (Levine, 1991), each electron is considered somewhat independently and interacts with other electrons only in an average way. The electrostatic potential felt by the electron under consideration is computed from the stationary spatial charge densities due to all the other electrons in their orbitals. In reality, we know that electrons repel each other, so electrons will avoid each other more effectively than can be modeled by the Hartree-Fock approximation. Less repulsion between electrons will result, and the system energy will decrease. Thus, electron motions must be allowed to correlate with each other in order for ab initio calculations to become more accurate (Löwdin, 1959; Borden & Davidson, 1996; Mok et al., 1996; Raghavachari & Anderson, 1996).

Several methods for introducing electron correlation are used (Pople et al., 1976; Levine, 1991) and all are considerably more expensive than Hartree-Fock methods. The most common is to treat the electron correlation as a perturbation on the Hartree-Fock solution; the theory was developed by Møller and Plesset (1934) so the technique is called MPx where x is the order of the perturbation treatment. Second order (MP2) is simplest and has been used very often because it accounts for >80% of correlation effects (Levine, 1991) at modest computational cost (about 10 times that of a Hartree-Fock calculation on the same system). Another major, and more rigorous, means of treating electron correlation effects is through configuration interaction, in which the ground-state Hartree-Fock solution is linearly combined with as many excited states of the proper symmetry as is computationally feasible. If infinitely many configurations are combined, then the exact quantum mechanical solution is obtained. This is impossible, of course, but still allows one the excitement of knowing that one could get arbitrarily close to an exact solution if one could afford a large enough calculation (Cremer & He, 1996).

2–2.2 Basis Function Nomenclature

It is difficult to read quantum mechanical papers without knowing the definition of a basis set and the conventions for naming one, so a short description is given here. The molecular electron density is the sum of the electron densities in each molecular orbital. The electron density in each molecular orbital is, in turn, the square of the wavefunction for that molecular orbital. The wavefunction for each molecular orbital is a linear combination of atom-centered orbitals, and the set of coefficients for these atomic orbitals is what the quantum mechanical computer program solves as it builds up the molecular orbitals. The optimal coefficients are those for which no change in the electron density results in a lower molecular energy. Now, each atomic orbital is approximated as a linear combination of one or more basis functions, each of which is in turn the linear combination of one or more primitive gaussian functions. Each primitive, in cartesian coordinates, is a polynomial in x, y, and z (the angular dependence) multiplied by a gaussian exponential (the radial dependence) centered on a nucleus. Thus, a basis set is the set of functions used to model the wavefunction near a given atom.

To give a concrete example of how the jargon is normally used in the quantum literature, suppose some property of a hydroxyl-containing molecule is being computed and the basis set is specified as 6-311G. Each molecular orbital can contain contributions from the 1s, 2s, and three 2p orbitals on oxygen and the 1s orbital on H. The basis set designation tells us that the nonvalence (left of the hyphen in 6-311G) 1s orbital on oxygen is fashioned from one basis function that is the sum of six primitive gaussian functions. The valence (right of the hyphen on 6-311G) 2s and 2p orbitals on oxygen and the 1s orbital on H will each be a linear combination of three basis functions. One basis function contributing to each of these orbitals will be a combination of three primitive gaussians, while the other two basis functions will simply consist of one primitive gaussian function each.

Both the angular and the radial portions of these gaussian functions impose restrictions on the electron distribution (the bonds) within a system, so larger basis sets allow for more electronic (and hence, molecular) flexibility at a given level of approximation. A common method for increasing the flexibility of the electron distribution is to augment the basis set with polarization functions of higher angular momentum. Thus, the 6-311G(d,p) basis set, also called 6-311G**, is the 6-311G basis set with the addition of d-functions on nonhydrogen atoms and p-functions on H.

For multielectron systems of interest, the quantum mechanical solution would only be exact in the limit of an infinite number of basis functions used to compute the electron density. Again, such a calculation is impossible, but one is comforted by the realization that quantum calculations are systematically improvable.

2–2.3 Density Functional Theory

Density functional theory (Hohenberg & Kohn, 1964; Kohn & Sham, 1965; Kohn et al., 1996) provides a means of treating electron correlation at much less computational cost than is incurred through MP2 or configuration interaction methods. The savings are gained by replacing the ab initio electron-electron exchange terms (Szabo & Ostlund, 1982; Levine, 1991) with an exchange-correlation term computed for an electron gas of the proper electron density. The difficulty arises because the exchange-correlation energy is precisely known only for uniform electron densities, while real electron densities are continually changing on the atomic scale. Systematic comparisons of density functional and MP2 geometry optimizations (Johnson et al., 1993; Hobza et al., 1995; Saint-Amant et al., 1995; Ma et al., 1996; Pudzianowski, 1996) indicate that the two methods predict observable properties at similar levels of accuracy, with the density functional methods doing so in roughly a tenth of the time (Pudzianowski, 1996). Unfortunately, it is not understood quite why density functional theory works so well. Thus, it is not clear how to systematically improve upon a result, and the methods have been mistrusted when applied to new systems (Schaefer, 1996). It does seem that density functional calculations, like MP2 and configuration interaction, improve with the size of the basis set used to model the atomic orbitals.

Despite their simplicity and speed, density functional methods are in principle exact provided that the exact exchange-correlation terms are used (Kohn et al., 1996). Thus, functional representation of these terms is the subject of vigorous research. The simplest estimate for the exchange-correlation energy treats the local electron density as if it were a uniform electron gas, and is called the local-density approximation. While this approximation produces generally good molecular geometries, frequencies, and charge densities, the addition of exchange-correlation corrections for the gradient of the electron density improve the thermodynamic predictions of density functional theory (Kohn et al., 1996). When nonbonded interactions or transition states are to be modeled, one must move beyond these local approximations to employ nonlocal exchange-correlation terms (Kohn et al., 1996; Pudzianowski, 1996).

Density functional methods would seem to offer the only tractable means of including electron correlation effects in systems approaching 100 atoms. Another strength relevant to soil chemical research is that transition metals are much more amenable to treatment by density functional methods than by the Hartree-Fock approximation.

2–2.4 Semi-Empirical Quantum Mechanics

An alphabet soup of methods have been cooked up to decrease the time needed to run quantum-like calculations. These semi-empirical methods employ a quantum-mechanical formalism but either ignore or approximate many of the inter-electron terms that would be computed in a rigorously ab initio calculation. Examples of semi-empirical methods that have been employed in recent years are the extended Huckel, MINDO/3, MNDO, AM1, and PM3. The essential approximations inherent in each of these methods, as well as some strengths and weaknesses of each, are reviewed by Levine (1991). The latter four methods can all be used to calculate reasonable geometries and heats of formation for gas-phase molecules. Geometric predictions from these semiempirical methods tend to be qualitative rather than quantitative, but heats of formation are often more accurate than those derived from ab initio calculations, since heats of formation are included in the database used to parameterize the electron–electron interactions. The utility of semiempirical methods is that they achieve their results in 1/1000th the time of a moderate-quality ab initio calculation. The two major niches for semiempirical methods are (i) to qualitatively predict the geometries, energies, electronic structures, or vibrational spectra for systems of molecules too large to submit to rigorously ab initio calculations, and (ii) to speed up ab initio calculations by providing reasonable initial guesses for the geometries and force constants of new systems. An important recent development is the extension of semi-empirical techniques to several heavier elements, including the transition metals (Ignatov et al., 1996; Malagoli & Thiel, 1996; Thiel & Voityuk, 1996).

2–2.5 Empirical Molecular Mechanics

Even semiempirical methods are too slow to simulate large systems because the number of electrons grows so rapidly. The only feasible way to compute the static properties of thousand-atom systems or the dynamic properties of much smaller systems is to ignore the electrons. The primary method has been to attempt to subsume all electronic effects near a given nucleus into an effective atom type that depends on the element and its chemical environment. The system under consideration is then reduced to a set of atom types whose interactions are calculated using the formalism of classical Newtonian physics. Such a method is termed molecular mechanics. Since electrons are not explicitly present in the molecular mechanics model, no quantum processes may be simulated. Thus, bond formation or bond breakage cannot be easily modeled (although it can be done (Taylor & Garrison, 1995)), and the bonding pattern for a system usually must be determined before a simulation begins.

This crude approximation of physical reality (i.e., that the atomic scale may be modeled in the absence of electrons) allows molecular mechanics methods to simulate many thousands of atoms. The price paid to gain this power is that the atomic potential energy surface now must be approximated by functions of the nuclear coordinates alone. Analytical potential energy functions are created that satisfy our notions of how the electrons would behave if they were present. Then, the parameters of these analytical functions are adjusted in an attempt to recreate the true molecular potential energy surface. Hence, molecular mechanics is often called empirical molecular modeling. Ideally, molecular mechanics can reproduce the atomic-scale geometries, vibrational motions, and interaction energies of a collection of atoms.

Since the 1930s, vibrational spectroscopists used force fields to simulate the spectra they measured for simple molecules, with harmonic, polynomial, or Morse descriptions of bond stretching. In the 1940s, several researchers (see reviews by Rasmussen, 1985; Allinger, 1992) proposed that, if nonbonded potentials (i.e., van der Waals and/or Coulombic terms) were incorporated, then these types of calculations could additionally be used to compute molecular structures and even relative energies of different conformations of a given molecule. These prescient scientists had to wait for computers that could make their ideas practical.

The first molecular geometry optimizations by computer were done by Hendrickson (1961) for several cycloalkane molecules. Soon thereafter, Wiberg (1965) published the essentials of a molecular mechanics program, and Verlet (1967) created the molecular dynamics algorithms that are still used widely (Allen & Tildesley, 1987). The program developed at the Weizmann Institute in Rehovot, Israel (Lifson & Warshel, 1968; Warshel & Lifson, 1970) deserves special mention, because its progeny dominate the $100 million (Gelin, 1996) molecular mechanics software market today. Martin Karplus, Arnold Hagler, and Kjeld Rasmussen all worked with Schneior Lifson in the early 1970s, and each took a current copy of the Lifson group's program home with them. Karplus used it as a basis for CHARMM (Brooks et al., 1983) the rights to which were bought by Polygen, a precursor to the present Molecular Simulations, Hagler used his copy to develop Discover (Molecular Simulations, 1995) which became the computational core of the commercial Biosym software suite. Rasmussen, too, has continued to improve the code but still gives his software away. His CFF program (Niketic & Rasmussen, 1977; Rasmussen et al., 1993; Engelsen et al., 1994) is a special tool for optimizing new molecular mechanics parameters based on a variety of experimental and quantum mechanical data.

The classical mechanics functions that have been used to model the molecular potential energy surface are a diverse group. A given set of potential energy functions and their associated parameters is still called a force field because of the vibrational spectroscopic heritage preceeding molecular mechanics. There are two major types of force fields in use today. The valence force fields are used in situations where the directional nature of covalent bonding is thought to dominate. On the other hand, many inorganic crystal structures are modeled respectably well by so-called ionic force fields that ignore most covalent effects.

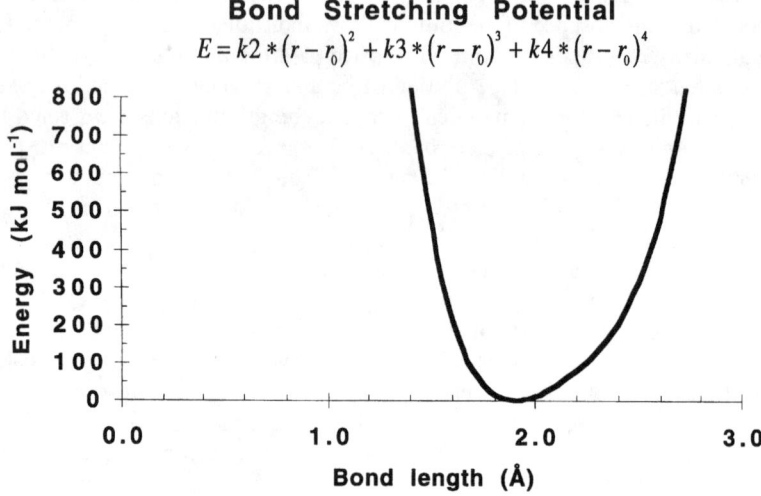

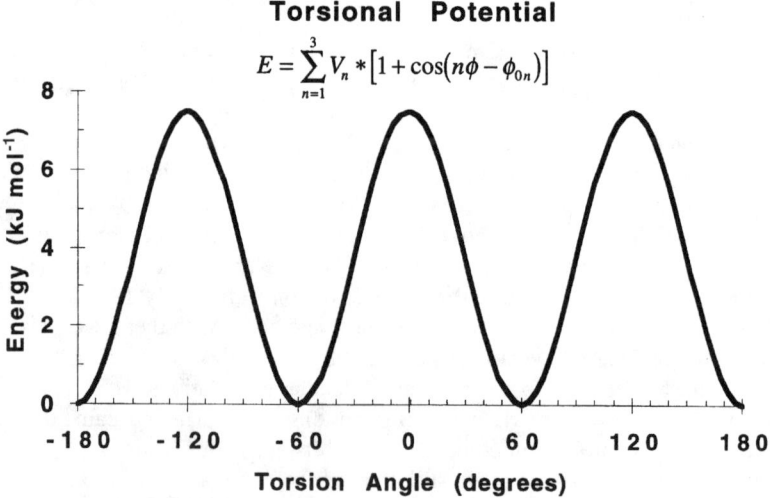

Fig. 2–2. A representative set of analytic functions used to perform classical mechanics simulations of atomic-scale systems. The functions and their associated parameters are collectively called a force field because the sum of their gradients yields the forces on each atom. At each step of a simulation, the atomic positions are used to calculate r, θ, ϕ, and r_{ij} and the energies and/or forces are reevaluated. The parameters that must be determined include all force constants (k, V), equilibrium internal coordinates (r0, θ0, ϕ0), van der Waals repulsion and dispersion terms, and atomic partial charges (q). Then, energies and forces evolve during a simulation as r, θ, ϕ, and r_{ij} change.

As an illustrative example of a valence force field, consider the functional forms used by cff91 (Maple et al., 1988, 1994). Figure 2–2 displays mathematical and graphical representations of the general molecular mechanics functions used by cff91 for computing the potential energy as a function of atomic positions. There are bond-stretching and angle-bending functions that are familiar from vibrational analysis; since these include cubic and quartic terms, a desirable anharmonicity is built into the force field (i.e., it is usually easier to lengthen a bond than to compress one). Torsions are relative rotations of substituent bonds around a central bond that connects the two substituents. A classic example of a torsional potential is the energy of internal rotation for ethane, which is three-fold-periodic due to the nonbonded interactions between the H on opposing

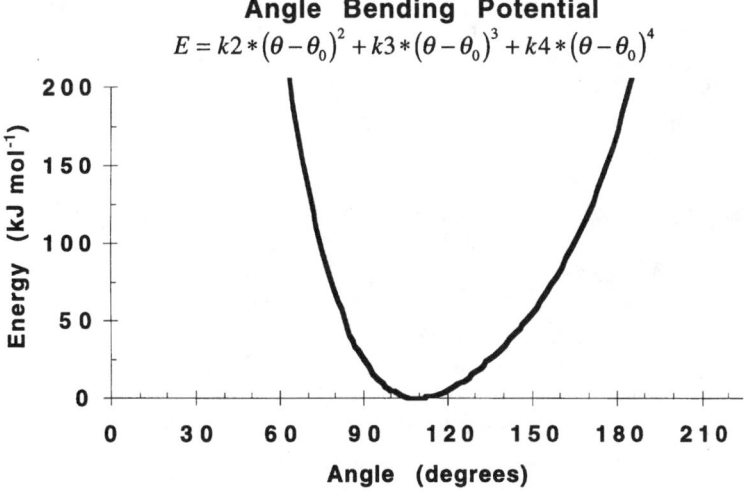

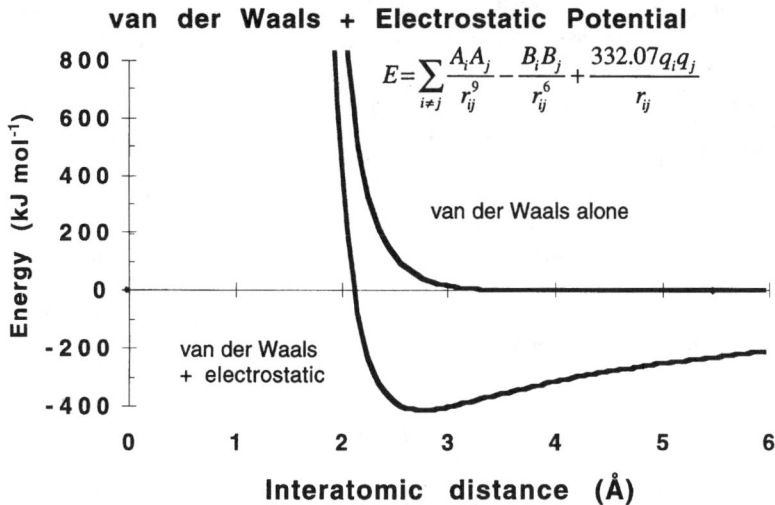

Fig. 2–2. Continued.

methyl groups. Twofold- and fourfold-periodicity may be required in other bonding situations. These three valence bonding terms may be augmented by cross-terms (not illustrated) that reflect interactions between them. For example, crystal structures of silicates show that, when an Si–O–Si angle is compressed, the Si–O bonds that form that angle lengthen (Hill & Gibbs, 1979). To faithfully represent this reality, a stretch-bend cross-term may be needed in a valence force field for silicates.

In a valence force field, all interactions between atoms in a common bond or bond angle are generally considered to be included in the valence terms. This convention appears ripe for revision, as a mounting body of evidence indicates that 1,3-nonbonded interactions help reproduce physical reality (Cundari et al., 1995; Derreumaux & Vergoten, 1995; MacKerell et al., 1995a). All other pairs of atoms in the system exchange nonbonded interactions comprised of Coulombic and van der Waals potentials based on the distance r between atomic centers. The van der Waals potential shown in Fig. 2–2 represents the short-range repulsion of electron clouds by an r^{-9} function, while dispersion forces (attraction of electron clouds due to mutually induced dipoles) are modeled by the r^{-6} function. Note that the Coulombic potential not only has a much greater magnitude than the van der Waals potential, but also acts over much greater distance owing to its dependence on r^{-1}.

An ionic force field would begin with only nonbonded potentials, calculated between all atom pairs. Catlow's group has been the major developer of ionic force field methodology. They (Sanders et al., 1984) found it necessary to add some angle-bending terms back into the force field in order to better reproduce a variety of crystal properties. Also, they consistently employ the shell model (Dick & Overhauser, 1958) of electronic polarizability. In this model, oxygen centers are represented by a negatively charged, massless spherical shell attached by a harmonic spring to a massive, positively charged core. Thus, Coulombic forces can cause displacement of the shell center relative to the core, simulating polarization of oxygen's valence electrons. Not only is the center of oxygen's charge displaced, but a dipole is created within the atom itself. A recent shell- model ionic potential for silicates and zeolites (Schröder & Sauer, 1996) appears to be the most accurate inorganic force field to date. Its self-consistent reproduction of both structural and vibrational data must be attributed to the combination of oxygen polarizability and 1,3-Coulombic interactions.

Once a force field is chosen for a system, it may be applied in several different ways (Allen & Tildesley, 1987). The three most common techniques are known as energy minimization, molecular dynamics, and Monte Carlo. Each of these techniques is widely used, depending on the properties one wants to simulate. They are large topics in themselves, but their differences are not crucial to the discussion in this chapter.

2–3 USES OF QUANTUM MECHANICS IN SOIL CHEMISTRY

The application of quantum mechanical methods to study mineral structures and geochemical solutions has been reviewed several times recently (Sauer,

1989; Lasaga, 1990; Tossell & Vaughan, 1992; Bleam, 1993; Bukowinski, 1994). Rather than recapitulate these reviews, my goals are to assess current trends in the field and to offer an impression of future directions where ab intio calculations will benefit our understanding of soil chemical systems.

2–3.1 Quantum Mechanical Investigations of Bonding and Reactions

The model mineral used as the preeminent testing ground for applying quantum methodology to crystals is quartz. Silicon and oxygen are main-group elements, for which quantum techniques are best understood, and the tetrahedral unit makes a convenient, small model for the crystal (Sauer, 1989). Even though the silicon tetrahedron has been undergoing quantum investigation for 25 yr (Collins et al., 1972), there is not yet a consensus on how the Si–O bond forms (Gibbs et al., 1994). Indeed, there is not even agreement on whether the bond is predominantly ionic or covalent (Tsirelson et al., 1990; Gibbs et al., 1994)! One problem is that, as the size of the basis set used increases, or as electron correlation effects are treated more completely, the quantum results do not yet converge to a constant solution (Nicholas et al., 1992a; Bär & Sauer, 1994; Teppen et al., 1994b). This is bound to be even more of a problem for models of other oxides, since heavy elements and transition metals in particular are notoriously difficult to treat (Salahub & Zerner, 1989).

There is, then, no generally agreed-upon method for computing the electronic properties of models for oxide minerals. One simply uses the best method and the largest model that one can afford. If one wants to examine interactions between a large number of crystal structural units (i.e., tetrahedra and octahedra), then one must use a low to moderate level of theory (Hill & Sauer, 1994, 1995). Parenthetically, since computer speed is doubling every year, the number of tetrahedra considered to be a large silicate model has changed rapidly from two (Lasaga & Gibbs, 1988, 1991) to 24 (Hill & Sauer, 1994). If one wants to examine the electronic properties in great detail, then one is limited to a few structural units (Gibbs et al., 1994; Teppen et al., 1994b).

The ultimate goal of the quantum study of crystals would be to compute a wavefunction for the entire crystal, as Felix Bloch did in crude fashion (Bloch, 1928). Today, crystal wavefunctions may be approximated through semiempirical means (Hoffmann, 1988), and extended Huckel techniques have been applied to clay minerals (Bleam & Hoffmann, 1988; Bleam, 1990a,b). Also, a program (CRYSTAL) for computing even more accurate crystal wavefunctions using Hartree-Fock ab initio methods has been developed (Dovesi et al., 1988; Pisani et al., 1988). The program has been applied to many minerals (Dovesi et al., 1992) of increasing complexity, including kaolinite (Hess & Saunders, 1992) and several large-unit-cell zeolites (White & Hess, 1993a,b; Nicholas & Hess, 1994; Stave & Nicholas, 1995). Unfortunately, the computational demands are staggering. For example, holding the kaolinite crystal rigid and optimizing only the orientation of the inner hydroxyl with a minimal basis set required a supercomputer (Hess & Saunders, 1992).

Removal of computational bottlenecks for periodic systems requires more than faster computers; a reevaluation of quantum mechanical approximations and computational algorithms is needed. Such a reevaluation has been undertaken by Payne and co-workers (Payne et al., 1992), resulting in the program package CASTEP (CAmbridge Serial Total Energy Package). As Bloch realized long ago, the natural choice of basis set for a periodic crystal is a Fourier series of plane waves. A single parameter (the highest frequency) controls the quality of the basis (Wimmer, 1993). A problem arises because the electronic potential decreases very steeply near each nucleus in the crystal, requiring high frequencies (large plane-wave basis sets) for an accurate description of the core electrons. Since bonding, polarizability, and other electronic characteristics of crystals result primarily from the distribution of the valence electrons, CASTEP makes the approximation that the core electrons for each nucleus may be regarded as fixed (Payne et al., 1992). Replacing the steep potential near each nucleus and its core electrons by a shallower pseudopotential allows much smaller plane-wave basis sets to describe the valence electronic structure of a crystal. Furthermore, pseudopotential methods can be readily extended to the heavy elements.

Soil minerals, aqueous solutions, and their interfaces are generally very polar environments, so polarization, hydrogen bonding, and other nonlocal phenomena due to electron correlation are expected to contribute strongly to molecular-scale structures and energetics. To include a degree of electron correlation in its calculations, CASTEP uses density functional theory (Hohenberg & Kohn, 1964; Kohn & Sham, 1965). To make these density functional calculations tractable for complex periodic systems, CASTEP makes several algorithmic innovations (Payne et al., 1992). As a result, the program has already been applied to first-principles studies of oxide mineral surfaces (Manassidis & Gillan, 1994; Goniakowski et al., 1996) and adsorption of Na^+ (Ursenbach et al., 1995), H_2O (Winkler et al., 1994; Refson et al., 1995; Goniakowski & Gillan, 1996), NH_3 (Pugh & Gillan, 1994), and CH_3OH (Shah et al., 1996a,b) to oxide surfaces. Despite the many advances made with CASTEP, the computational demands of these simulations remain daunting. For example, computing the ground state geometry of a 36-atom unit cell, beginning with an experimentally refined crystal structure, required about one CPU-month on a Cray T3D (Shah et al., 1996a).

While CASTEP offers tremendous promise for the future, many computational geochemists will be forced to limit their present quantum investigations of structures and processes to small systems. To take the structure of water for an example, very high-quality calculations can only be done on a few water molecules (Liu et al., 1996). As is typical in geochemical systems, hydrogen bonding determines the water structure. Since hydrogen bonding is a nonlocal phenomenon, electron correlation must be included in calculations that are to be useful (Frey et al., 1992; Teppen et al., 1994a). A comparison of results from MP2–MP4 quantum theory (Suhai, 1994) and density-functional techniques (Suhai, 1995) indicate that water can be modeled reasonably well by the latter, in much less time than required by the former. Advances in laser spectroscopy (Liu et al., 1996) provide experimental data against which to test quantum results on small water clusters, and the comparisons are favorable.

Quantum calculations also are being used to investigate the properties of solutions, often by undertaking calculations of a solute and its first hydration shell. Systematic Hartree-Fock and MP2–MP4 investigations of cation solvation complexes reported recently include Li$^+$ (Feller et al., 1994), Mg^{2+} (Bock et al., 1994), and the metals of the first transition period (Magnusson & Moriarty, 1993). Again, density functional calculations of solvated cations, anions (Combariza & Kestner, 1995), and polar organics (Lee et al., 1995) indicate that results are comparable to an MP2 treatment of electron correlation at much less computational cost.

Since many water molecules would be needed to compute a true picture of solvation, and since high-quality methods must be restricted to small systems, there has been a move (reviewed in Varnek et al., 1995) to qualitatively account for aqueous solvation effects on the solute's electronic structure through a high-quality calculation of the solute wavefunction, but in the context of a polarizable continuum with dielectric constant 78.3. The polarizable continuum polarizes, in turn, the solute wavefunction until a balance is struck between favorable electrostatic interactions with the continuum and unfavorable distortion of the gas-phase molecular wavefunction. This self-consistent reaction field (SCRF) methodology can be used to compute free energies of solvation and solvation effects on reactions in solution (Tunon et al., 1996). So far, these models have been applied mostly to organic solutes, but a similar methodology has been recently used to study the solvation of metal cations (Furuki et al., 1995). The shortcoming of SCRF methods is that energy contributions due to reorientation of waters in the solvation shell and the polarization of individual water molecules is ignored.

For capturing these contributions to the free energy, either explicit water molecules must be incorporated or a model for their incorporation must be developed. The AM1-SM2 and PM3-SM3 models (Cramer & Truhlar, 1992, 1995) have been successful in doing the latter. These models treat the solute with semiempirical (AM1 or PM3) quantum mechanical methods, using an SCRF methodology. In addition, free energy contributions due to solvent cavitation (i.e., the energy required to make a water surface around the solute), solute-solvent dispersion forces, and first-shell solvent rearrangements have been parameterized. Since these methods are semiempirical, they can treat large solutes. While no water molecules are explicitly present in the model, the parameterization of the solvation model is good enough that it computes absolute free energies of aqueous solvation to within a mean deviation of 0.75 kcal mol^{-1} from experimental values obtained for 150 solutes (Cramer & Truhlar, 1995). While SM2 and SM3 were developed expressly for aqueous solvation, the SM4 model was recently introduced for computing free energies of solvation in n-hexadecane (Giesen et al., 1995b) and in general nonpolar solvents (Giesen et al., 1995a).

Computational techniques in the spirit of the PM3–SM3 model should find applicability to studying adsorption processes in the near future. Many theoretical studies have used ab initio methods to examine the gas-phase sorption of protons (Brand et al., 1992, 1993; Nicholas et al., 1992b; Blaszkowski et al., 1994), water (Lasaga, 1990; Ugliengo et al., 1990; Himei et al., 1995), ammonia (Ugliengo et al., 1989; Kassab et al., 1993; Kyrlidis et al., 1995), and organics (Blaszkowski & van Santen, 1995; Haase & Sauer, 1995; O'Malley & Brathwaite, 1995; Farnsworth & O'Malley, 1996) onto molecular models for aluminosilicate surfaces. These studies have generally suffered in two regards. First, for computational convenience, the molecule representing the crystal surface has often been chosen too small. The use of silanol (H_3SiOH) and disiloxane ($H_3SiOSiH_3$) analogues as models for silicates has been common, but generally results in computed free energies of adsorption that are far too low (Ugliengo et al., 1989, 1990). The use of models in which Si, Al, and other cations are fully coordinated by oxygen would be more appropriate (Teppen et al., 1994b), despite the additional computational burden. A second shortcoming of quantum adsorption studies heretofore, especially for systems of interest to soil chemists, is that solvation effects have not been included. Here is a situation where the PM3–SM3 model should be useful, because (i) it is fast enough that a substantial crystal fragment could be used to represent the surface, (ii) transition states and full reaction pathways could be modeled, and (iii) both long- and short-range solvation effects are included. Semiempirical methods that include d-orbitals and that are parameterized for transition metals have recently become available (Ignatov et al., 1996; Thiel & Voityuk, 1996). Unfortunately, the solvation models (i.e., SM3) have not yet been parameterized except for H, C, N, O, F, P, S, Cl, Br, and I (Cramer & Truhlar, 1995), but the previous success of the model ensures that it will be expanded to more of the periodic table.

In some cases, such as outer-sphere complex formation, parameterized solvation models will be inadequate and explicit water will have to be included. As an illustrative example, Glendening and co-workers (Glendening et al., 1994) recently completed substantial Hartree-Fock and MP2 calculations on the gas-phase selectivity of the ether 18-crown-6 for the alkali cations. Unfortunately, the gas-phase selectivity series was completely different from the experimentally observed series in aqueous solution. Variations in the relative strength of hydration from Li+ to Cs+ cause this difference, and it is unlikely that even a well-parameterized continuum solvation model could account for the differences in hydrated cation complexation with the crown ether.

2–3.2 Quantum Mechanical Force Field Development

A major use for quantum mechanics in soil chemical investigations will be for developing molecular mechanics parameters that can then be used for simulations of larger, more complex systems. Again, the purpose of molecular mechanics is to have a fast technique for sampling the potential energy surface of a system, so the force field must approximate that surface. Some aspects of the surface may be gleaned from experimental data. For example, diffraction experiments yield interatomic distances, which indicate where the minima are on the surface. Vibrational spectroscopy measures frequencies that result from the curvature of the potential energy surface near each minimum. Often however, the available set of atomic-scale experimental data is too sparse to determine all the parameters needed to define a force field. In this situation, quantum mechanical calculations on well-chosen small systems can provide estimates for both the positions and curvatures of minima on the atomic-scale potential energy surfaces of larger systems (Clementi, 1980; Maple et al., 1988; Momany et al., 1992).

Inorganic systems, for which gas-phase structural data are often unavailable, seem to be especially fertile ground for ab initio parameter development. It has been demonstrated that a successful force field for quartz and the zeolites could be derived entirely from quantum calculations, with no input from empirical data (Boisen et al., 1994; Hill & Sauer, 1994; Schröder & Sauer, 1996). Most other recent force fields for crystalline systems also have been developed with at least some input from quantum mechanics (i.e., Gale et al., 1992; de Boer et al., 1995; Hill & Sauer, 1995; Ermoshin et al., 1996).

In the geochemical study of adsorption or solvation by molecular mechanics methods, interactions are determined by nonbonded forces, which in turn are dominated by the electrostatic terms. Thus, for the polar systems of interest to soil chemists, the charges assigned to the atoms are arguably the most important step of the molecular mechanics parameterization. These charges are not physical observables that can be determined by any experiment (Tossell & Vaughan, 1992; Gibbs et al., 1994), but qualitative estimates may be gathered from so-called deformation electron densities (Tsirelson et al., 1990) based on x-ray diffraction data. Quantum mechanical methods can be used to estimate these charges, too. Using the optimized nuclear positions and the computed electron densities, the electrostatic potential at a grid of points outside the molecular van der Waals surface can be computed (Naray-Szabo & Ferenczy, 1995). Automated methods such as CHELPG (Chirlian & Francl, 1987; Breneman & Wiberg, 1990) can then be used to assign charges to each atom in order to best reproduce this electrostatic potential grid, although there are some questions about the uniqueness of this solution (Francl et al., 1996). When this procedure is followed using high-quality wavefunctions for small silicate models, the computed charges (Teppen et al., 1994b) on silicon and oxygen fall near the center of the range postulated on the basis of experimental deformation densities (Tsirelson et al., 1990).

Walker and Mezey (Walker & Mezey, 1994, 1995) have created a means for approximating the electron density of a large molecule based on summing the densities of its constituent functional groups. As a prerequisite, one must build a database of quantum mechanical wavefunctions for a group of small molecules that, together, represent all of the fragments needed to construct the larger molecule. Then, an additive fuzzy density fragmentation (AFDF) scheme chooses appropriate fragments from the database and uses their wavefunctions to construct an overall electron density for the large molecule. These electron densities could be useful in a number of ways.

For example, a database of quantum mechanical structures representative of structural units in minerals, such as various tetrahedra and octahedra, could be constructed. The small molecular structures are typically cut out of some mineral and then the dangling oxygen are protonated to satisfy their valences before solving for the optimized geometry and the molecular wavefunction. It seems that one could then use the AFDF approach to build up the electron density of a much larger mineral fragment or even a unit cell of 20 to 100 atoms, that would be replicated through periodic boundary conditions to form an infinite crystal. Also, ab initio databases could be constructed for simulating the electron densities of models (i.e., Schulten & Schnitzer, 1993; Leenheer et al., 1995) for humic materials. The technique has been used for proteins containing up to 1600 atoms (Walker & Mezey, 1995).

These electron densities could then be used to compute the magnitude of the electrostatic potential at a given location, such as near a specific humic acid functional group or at a given site near a mineral surface. Further, molecular mechanics charge parameters could probably be derived by using standard techniques (i.e., CHELPG) to find the atomic partial charges that best reproduce the all-electron electrostatic potential surface. The promise of the AFDF method is that the AFDF electron densities are qualitatively indistinguishable from truly ab initio electron densities for test molecules of up to 24 atoms (Walker & Mezey, 1994).

2–4 USES OF MOLECULAR MECHANICS IN SOIL CHEMISTRY

Molecular mechanics promises to allow soil chemists to perform atomistic simulations on very large systems in the near future. To give a state-of-the-art example of the system sizes that are achievable now, it takes a substantial crystal fragment to be able to trace the evolution of a fracture in the structure. Simulations that studied dynamic crack propagation in systems comprising 100 000 (Nakano et al., 1995) and 300 000 (Holian & Ravelo, 1995) atoms have been recently reported. The development of massively parallel computers and integrated molecular modeling software systems to run on them (Foster et al., 1996; Harrison et al., 1996; Plimpton & Hendrickson, 1996) is an exciting prospect for the future of soil chemistry.

While the primary usage of molecular mechanics methods has been to investigate the conformations and interactions of organic molecules, their use in studying systems that include mineral surfaces is accelerating rapidly. The first step in simulating a crystal is to create an atomistic model of the bulk crystal. Once an acceptable bulk structure is created, one must make a surface on that mineral, which is more difficult. As discussed by Bleam (Bleam, 1993), proper computation of bulk or surface structures demands that long-range contributions to the nonbonded potential be correctly accounted for through Ewald summation. Next, one must find or create a force field for the system under consideration. The choice must be made between ionic and valence representations of the system, keeping in mind that the parameters must work in conjunction with the force fields chosen for water and/or adsorbates in the system to be simulated. Since the force field is the crux of all molecular mechanics methods, this step is crucial to the success of any subsequent simulations. Testing and development of force fields is thus a very active focus of research. Not only are new inorganic force fields appearing regularly (de Boer et al., 1995; Hill & Sauer, 1995; Ermoshin et al., 1996; Schröder & Sauer, 1996), but organic force fields that have been under development for 30 yr are being revised (Cornell et al., 1995; MacKerell et al., 1995a).

Finally, computational methods that use molecular mechanics simulations to calculate useful quantities must be developed and tested. In the testing phase, these quantities must be observables, but once a model has been validated, it will undoubtedly be valuable to know things that cannot be measured but can be simulated. The discussion of current trends and future prospects for applying molecular mechanics methods to soil chemical systems will be subdivided into structural studies and adsorption studies.

2–4.1 Molecular Mechanics in Structural Studies

The bulk structures of most minerals important to soil chemists are well known from x-ray and neutron diffraction studies. Exceptions to this generality are the various amorphous oxides and mixed oxide precipitates. Methods pioneered by Catlow's group should find increasing applicability to the structural study of these amorphous soil minerals and mixed phases. Their approach has been to develop ionic-type force fields for performing static energy minimizations of general oxide and halide minerals. Since ionic force fields require far fewer parameters than valence force fields, they have generally been able to derive their force fields from experimental data for simple oxides (Sanders et al., 1984; Collins & Catlow, 1990), although they have had occasion to employ quantum calculations too (Gale et al., 1992; Purton et al., 1993). Recently, they have tested their overall force field for its applicability to mixed oxides (Bush et al., 1992, 1994; Battle et al., 1995; Lewis et al., 1995), mica (Collins & Catlow, 1990, 1992; Collins et al., 1993), and clay minerals (Breu & Catlow, 1995). Their methods have proven a valuable aid to diffraction and nuclear magnetic resonance spectroscopy (NMR) determinations of the structures of novel materials (Morris et al., 1994; Sankar et al., 1996).

In ionic force fields, cations and anions interact strictly through nonbonded terms, except for O-cation-O angle bending terms and sometimes an O–H bond in hydroxyl groups. These force fields have been criticized (Kramer et al., 1991; de Vos Burchart et al., 1992) for poorly predicting lattice vibrations, but the use of quantum calculations in the parameterization seems to help in that regard (Purton et al., 1993; Schröder & Sauer, 1996). An advantage of the ionic approach is that the coordination numbers of each cation do not need to be known at the outset, and they could change during either minimization or dynamics. Such a force field is very useful in the study of amorphous solids, because an initial structure with the proper composition can be heated during a molecular dynamics run, then rapidly quenched (Smith et al., 1995). An adequate number of these simulations can support inferences about the most likely local structures to be found in the glassy solid. A much slower quenching results in a simulated annealing that can be a useful technique for solving new crystal structures (Deem & Newsam, 1992).

Other techniques may be used to provide educated guesses for structures of amorphous materials, and Faulon has a notably creative effort. His SIGNATURE program (Faulon, 1994) begins with a variety of data gathered for the amorphous system. The program generates a wide variety of connectivity networks between likely functional groups, and those networks that best satisfy the analytical data, NMR results, density, surface area, and porosity, are used as candidates for further study by molecular mechanics. This methodology has been applied to humic acids (Hatcher et al., 1994) and to amorphous silicates (Faulon et al., 1995).

For studying interfaces of minerals whose bulk structure is already known, it may be problematic to create a model for the mineral surface. A variety of surface-sensitive technologies can be used to constrain the model, as can knowledge-based models for the speciation of surface groups (i.e., Dzombak & Morel, 1990). Also, Bleam has applied the bond-valence sum methodology (Brown,

1981; Brown & Altermatt, 1985) to compute self-consistent surface structures for clay mineral edges (Bleam et al., 1993). Once a rational but crude model for the surface has been created, either valence or ionic forcefields may be used to relax the structure of the surface and its transition to the bulk mineral (i.e., Alvarez et al., 1995). Another kind of surface that can be simulated is be the interface between two different bulk minerals (Blonski & Garofalini, 1996).

Since experimental structures are always time averages of lattice- or molecular-dynamics, it would seem reasonable that molecular dynamics methods should more accurately mimic experimental data than do energy minimization methods. This has been recently tested and verified for the structure of crystalline cholesterol (Liang et al., 1995).

Vibrational data tend to be easier to gather than they are to interpret, yet can provide valuable insight into mineral surface structures. Simulated crystal vibrational spectra can be gathered by a number of methods (Iyer & Singer, 1994; Ishioka et al., 1995) and interpreted easily because individual normal modes can be visualized. While these theoretical spectra tend not to be quantitatively accurate, trends are more often realistic. Thus, a calculated vibrational shift due to adsorption (Jousse et al., 1996) may suggest mechanistic explanations for frequency shifts that are observed experimentally.

2–4.2 Molecular Mechanics Studies of Adsorption

As stated above, a key limitation of most valence force fields is their inability to treat bond breakage or bond formation. Thus, simple oxide surface reactions such as adsorption of a proton cannot be simulated with these forcefields. Ionic-type force fields, on the other hand, do allow cation–anion dissociations, and an ionic force field has been recently adapted to study the hydroxylation of various CaO and MgO surfaces (de Leeuw et al., 1995). Another surface complexation reaction that has been studied with ionic force fields is the interaction of phosphonates with barite surfaces (Rohl et al., 1996). There is no reason why soil chemists could not use these techniques to examine proton and oxyanion sorption to Fe and Al oxides as well.

For this purpose, the ideal ionic force field would be consistent with a molecular mechanics model for dissociable water. Such a force field is indeed being developed and has been applied to the study of silicate (Rustad & Hay, 1995) and Fe(III) (Rustad et al., 1995) hydrolysis in solution, to bulk Fe oxyhydroxide structures (Rustad et al., 1996a), and to the protonation of goethite surfaces (Rustad et al., 1996b).

In many instances, we imagine that the forces causing adsorption are almost strictly of a nonbonded nature, as in the case of clays interacting with hydrated cations or nonpolar organics. For these cases, valence force fields are ideally suited for simulating adsorption, and they have the advantage that good parameters already exist for water, organic solutes, and some inorganic solutes. Water dynamics (Ohmine, 1995; Luzar & Chandler, 1996), interactions of water with proteins (Gu & Schoenborn, 1995; Kuhn et al., 1995; Muegge & Knapp, 1995), and binding between substrates and proteins (Ajay & Murcko, 1995;

Honig & Nicholls, 1995; MacKerell et al., 1995b) continue to be active research areas that provide us with examples of molecular mechanics methodology.

In addition, molecular mechanics methods have been broadly applied to the study of zeolites because of their economic importance as cracking catalysts and molecular sieves. The literature is extensive, but recent simulations have modeled interactions between zeolites and rare gases (Pellenq & Nicholson, 1994), O_2 (Jousse et al., 1996), hydrocarbons (Nicholas et al., 1993; Klein et al., 1994; Snurr et al., 1994; Bandyopadhyay & Yashonath, 1995; Smit, 1995; Smit & Maesen, 1995), quaternary amines (Koelmel et al., 1994), and hydrated alkali metals (Edwards et al., 1996). The structure of the interior surfaces of zeolites is similar in many ways to that of clay mineral basal surfaces, so many of these methods should be directly applicable to studying soil systems.

Among soil mineral surfaces, these basal surfaces of clays have received most of the study by both experimental and theoretical methods. Several substrates have now been simulated in the clay interlayer space, including aqueous alkali cations (Delville, 1991, 1992, 1993; Skipper et al., 1991, 1995a,b; Delville & Sokolowski, 1993; Boek et al., 1995; Chang et al., 1995; Karaborni et al., 1996), several metal chelates (Sato et al., 1992a,b, 1996; Breu & Catlow, 1995), and alkanes (Keldsen et al., 1994). These simulations have normally been done by holding the clay lattice rigid and allowing the interlayer space to relax through various Monte Carlo or molecular dynamics methods. On the other hand, Breu and Catlow (1995) observed chiral recognition of an organometallic intercalate by a smectite when they allowed the entire system, including the clay lattice, to relax.

If dynamic properties of adsorbates are to be studied, it may be desirable for the mineral substrate to be dynamic, too. Modeling of organic diffusion rates within zeolite channels (Kawano et al., 1992; Bandyopadhyay & Yashonath, 1995) indicates that diffusion can be much too rapid if the zeolite cage is held rigid, especially when the size of the adsorbate approaches the size of the pore; however, diffusivities that were too slow have also been blamed on the rigidity of the lattice (Snurr et al., 1994).

One interesting idea for future dynamics simulations would be to estimate the dielectric constant of water near various soil mineral surfaces. This value can be computed from the dipole moment autocorrelation function (Karasawa & Goddard, 1995), and would be useful to know for applications of Poisson-Boltzmann theory, such as surface complexation modeling (Sposito, 1984).

Finally, besides the structural, mechanistic, and dynamic properties of soil systems that could be computed using molecular mechanics, one can simulate thermodynamic aspects of a given process. For example, in biomolecular modeling, a hot topic is the rigorous calculation of binding free energies between proteins and substrates (Brady & Sharp, 1995; Pearlman & Connelly, 1995; Wang et al., 1995; Shen & Wendoloski, 1996). It is only a matter of time before these ideas are applied to inorganic systems to study relative free energies of adsorption. Indeed, the adsorption enthalpies of a series of alkanes on a model clay mineral have already been computed through molecular dynamics (Keldsen et al., 1994).

2–5 COMBINED QUANTUM MECHANICAL: MOLECULAR MECHANICAL METHODS

We have seen above that quantum mechanical and molecular mechanical methods offer some complementary aspects to the computational chemist. On the one hand, quantum mechanics can study electron distributions and reaction pathways for gas-phase molecules, but becomes too cumbersome to study condensed phases. On the other hand, molecular mechanics ignores electronic detail for the sake of having all atoms explicitly present. An exciting current development in molecular modeling is the attempt to combine these techniques. The basic model is that of a manageably small quantum compartment within the context of a broader molecular mechanics simulation (Field et al., 1990). The quantum mechanical machinery allows processes such as chemical reactions to be studied, and the explicit molecular mechanics context ensures that solvation and other environmental influences are felt by the quantum processes. The current literature shows examples where the hybrid techniques are being applied to study solvation (Liu et al., 1995; Tunon et al., 1996) and adsorption (Teunissen et al., 1995). The quantum mechanical model could be at the semiempirical (Stanton et al., 1995c), Hartree-Fock (Stanton et al., 1995b), or density-functional (Stanton et al., 1995a) level of theory, depending on the problem at hand. These methods should prove very flexible for the investigation of soil chemical systems. For example, a small fragment of a mineral structure, containing the desired functional group(s) could be constructed and embedded in the larger mineral surface. The fragment and a specifically adsorbing ligand could be examined quantum mechanically as they undergo ligand exchange. Assuming that the molecular mechanics parameters are good, especially the nonbonded terms, this ligand exchange pathway will occur in a context of proper interfacial forces. Then, the free energies, enthalpic and entropic components, and activation energies calculated for the quantum compartment may be meaningful results, giving rise to predictions for both equilibrium and rate constants.

2–6 SUMMARY AND CONCLUSIONS

The confluence of readily available computing power and the development of software for condensed-phase simulations has created opportunities for theoretical studies of soil solutions, soil colloids, and the interfaces between them. As an example, consider the modeling of adsorption. Studies for which all tools are presently available range from rigorously ab initio explorations of specific adsorption, catalysis, or electron transfer at mineral surfaces to strictly empirical molecular dynamics predictions of adsorbate conformations, diffusion rates, vibrational frequencies, or adsorption enthalpies. The boundaries between these two extremes are beginning to blur, as techniques for embedding quantum mechanical regions within empirical simulations are developed.

Molecular simulations can be fascinating on their own, as reflections of human ingenuity and as glimpses into atomic-scale behavior. Insofar as they accurately represent nature, simulations also have practical value. It is evident that efficient management of many chemicals in soils depends on a mechanistic understanding of chemical speciation, and molecular modeling seems poised to contribute to that understanding. Simulations have already assisted in the interpretation of spectroscopic data, suggested new experiments, and generated some surprising results that have been neither verified nor disproven.

ACKNOWLEDGMENT

This research was supported by Financial Assistance Award Number DE-FC09- 96SR18546 between the U.S. Department of Energy and the University of Georgia. Kjeld Rasmussen, Paul Mezey, P.M. Huang, and an anonymous reviewer improved the chapter through constructive criticism.

REFERENCES

Ajay, and M.A. Murcko. 1995. Computational methods to predict binding free energy in ligand-receptor complexes. J. Med. Chem. 38:4953–4967.

Allen, M.P., and D.J. Tildesley. 1987. Computer simulation of liquids. Oxford Univ. Press, Oxford, England.

Allinger, N.L. 1992. Molecular mechanics. p. 336–354. *In* A. Domenicano and I. Hargittai (ed.) Accurate molecular structures: Their determination and importance. International Union of Crystallography Monographs on Crystallography. Oxford Univ. Press, New York.

Alvarez, L.J., L.E. Leon, J.F. Sanz, M.J. Capitan, and J.A. Odriozola. 1995. Computer simulation of γ-Al_2O_3 crystal. J. Phys. Chem. 99:17872–17876.

Bär, M.R., and J. Sauer. 1994. Ab initio calculations of the structure and properties of disiloxane. The effect of electron correlation and basis set extension. Chem. Phys. Lett. 226:405–412.

Bandyopadhyay, S., and S. Yashonath. 1995. Diffusion anomaly in silicalite and VPI-5 from molecular dynamics simulations. J. Phys. Chem. 99:4286–4292.

Battle, P.D., T.S. Bush, and C.R.A. Catlow. 1995. Structures of quaternary Ru and Sb oxides by computer simulation. J. Am. Chem. Soc. 117:6292–6296.

Blaszkowski, S.R., and R.A. van Santen. 1995. Density functional theory calculations of the activation of methanol by a Bronsted zeolitic proton. J. Phys. Chem. 99:11728–11738.

Blaszkowski, S.R., A.P.J. Jansen, M.A.C. Nascimento, and R.A. van Santen. 1994. Density functional theory calculations of the transition states for hydrogen exchange and dehydrogenation of methane by a Bronsted zeolitic proton. J. Phys. Chem. 98:12938–12944.

Bleam, W.F. 1990a. Electrostatic potential at the basal (001) surface of talc and pyrophyllite as related to tetrahedral sheet distortions. Clays Clay Miner. 38:522–526.

Bleam, W.F. 1990b. The nature of cation-substitution sites in phyllosilicates. Clays Clay Miner. 38:527–536.

Bleam, W.F. 1993. Atomic theories of phyllosilicates: Quantum chemistry, statistical mechanics, electrostatic theory, and crystal chemistry. Rev. Geophys. 31:51–73.

Bleam, W.F., and R. Hoffmann. 1988. Orbital interactions in phyllosilicates: Perturbations of an idealized, two-dimensional, infinite silicate frame. Phys. Chem. Miner. 15:398–408.

Bleam, W.F., G.J. Welhouse, and M.A. Janowiak. 1993. The surface coulomb energy and proton coulomb potentials of pyrophyllite {010}, {110}, {100}, and {130} edges. Clays Clay Miner. 41:305–316.

Bloch, F. 1928. Über den Quantenmechanik der Electronen in Kristallgittern. Z. Phys. 52:555–600.

Blonski, S., and S.H. Garofalini. 1996. Molecular dynamics study of silica-alumina interfaces. J. Phys. Chem. 100:2201–2205.

Bock, C.W., A. Kaufman, and J.P. Glusker. 1994. Coordination of water to magnesium cations. Inorg. Chem. 33:419–427.

Boek, E.S., P.V. Coveney, and N.T. Skipper. 1995. Molecular modeling of clay hydration: A study of hysteresis loops in the swelling curves of sodium montmorillonites. Langmuir 11:4629–4631.

Boisen, M.B., Jr., G.V. Gibbs, and M.S.T. Bukowinski. 1994. Framework silica structures generated using simulated annealing with a potential energy function based on an $H_6Si_2O_7$ molecule. Phys. Chem. Miner. 21:269–284.

Borden, W.T., and E.R. Davidson. 1996. The importance of including dynamic electron correlation in ab initio calculations. Acc. Chem. Res. 29:67–75.

Brady, G.P., and K.A. Sharp. 1995. Decomposition of interaction free energies in proteins and other complex systems. J. Mol. Biol. 254:77–85.

Brand, H.V., L.A. Curtiss, and L.A. Iton. 1992. Computational studies of acid sites in ZSM-5: Dependence on cluster size. J. Phys. Chem. 96:7725–7732.

Brand, H.V., L.A. Curtiss, and L.E. Iton. 1993. *Ab initio* molecular orbital cluster studies of the zeolite ZSM-5: 1. Proton affinities. J. Phys. Chem. 97:12773–12782.

Breneman, C.M., and K.B. Wiberg. 1990. Determining atom-centered monopoles from molecular electrostatic potentials. The need for high sampling density in formamide conformational analysis. J. Comp. Chem. 11:361–373.

Breu, J., and C.R.A. Catlow. 1995. Chiral recognition among tris(diimine)-metal complexes: 4. Atomistic computer modeling of a monolayer of $[Ru(bpy)_3]^{2+}$ intercalated into a smectite clay. Inorg. Chem. 34:4504–4510.

Brooks, B.R., R.E. Bruccoleri, B.D. Olafson, D.J. States, S. Swaminathan, and M. Karplus. 1983. CHARMM: A program for macromolecular energy, minimization, and dynamics calculations. J. Comp. Chem. 4:187–217.

Brown, I.D. 1981. The bond-valence method: An empirical approach to chemical structure and bonding. p. 1–30. *In* M. O'Keeffe and A. Navrotsky (ed.) Structure and bonding in crystals. Vol. II. Academic Press, New York.

Brown, I.D., and D. Altermatt. 1985. Bond-valence parameters obtained from a systematic analysis of the inorganic crystal structure database. Acta Cryst. B41:244–247.

Bukowinski, M.S.T. 1994. Quantum geophysics. Annu. Rev. Earth Planet. Sci. 22:167–205.

Bush, T.S., C.R.A. Catlow, A.V. Chadwick, M. Cole, R.M. Geatches, G.N. Greaves, and S.M. Tomlinson. 1992. Studies of cation dopant sites in metal oxides by EXAFS and computer-simulation techniques. J. Mater. Chem. 2:309–316.

Bush, T.S., J.D. Gale, C.R.A. Catlow, and P.D. Battle. 1994. Self-consistent interatomic potentials for the simulation of binary and ternary oxides. J. Mater. Chem. 4:831–837.

Chang, F.-R.C., N.T. Skipper, and G. Sposito. 1995. Computer simulation of interlayer molecular structure in sodium montmorillonite hydrates. Langmuir 11:2734–2741.

Chirlian, L.E., and M.M. Francl. 1987. Atomic charges derived from electrostatic potentials: A detailed study. J. Comp. Chem. 8:894–905.

Clementi, E. 1980. Computational aspects for large chemical systems. Lecture Notes in Chemistry. 19. Springer-Verlag, Berlin.

Collins, D.R., and C.R.A. Catlow. 1990. Interatomic potentials for micas. Mol. Simul. 4:341–346.

Collins, D.R., and C.R.A. Catlow. 1992. Computer simulation of structures and cohesive properties of micas. Am. Mineral. 77:1172–1181.

Collins, D.R., W.G. Stirling, C.R.A. Catlow, and G. Rowbotham. 1993. Determination of acoustic phonon dispersion curves in layer silicates by inelastic neutron scattering and computer simulation techniques. Phys. Chem. Miner. 19:520–527.

Collins, G.A.D., D.W.J. Cruickshank, and A. Breeze. 1972. *Ab initio* calculations on the silicate ion, orthosilicic acid, and the $L_{2,3}$ X-ray spectra. J. Chem. Soc., Faraday Trans. 686:1189–1195.

Combariza, J.E., and N.R. Kestner. 1995. Density functional study of short-range interaction forces between ions and water molecules. J. Phys. Chem. 99:2717–2723.

Cornell, W.D., P. Cieplak, C.I. Bayly, I.R. Gould, K.M. Merz Jr., D.M. Ferguson, D.C. Spellmeyer, T. Fox, J.W. Caldwell, and P.A. Kollman. 1995. A second generation force field for the simulation of proteins, nucleic acids, and organic molecules. J. Am. Chem. Soc. 117:5179–5197.

Cramer, C.J., and D.G. Truhlar. 1992. An SCF solvation model for the hydrophobic effect and absolute free energies of aqueous solvation. Science (Washington, DC) 256:213–217.

Cramer, C.J., and D.G. Truhlar. 1995. Continuum solvation models: Classical and quantum mechanical implementations. p. 1–72. *In* K.B. Lipkowitz and D.B. Boyd (ed.) Reviews in computational chemistry. Vol. 6. VCH Publ., New York.

Cremer, D., and Z. He. 1996. Sixth-order Møller-Plesset perturbation theory: On the convergence of the MP*n* series. J. Phys. Chem. 100:6173–6188.

Cundari, T.R., E.W. Moody, and S.O. Sommerer. 1995. Computer-aided design of metalloporphyrins: A molecular mechanics force field for gadolinium complexes. Inorg. Chem. 34:5989–5999.

de Boer, K., A.P.J. Jansen, and R.A. van Santen. 1995. Structure-stability relationships for all-silica structures. Phys. Rev. B52:12579–12590.

Deem, M.W., and J.M. Newsam. 1992. Framework crystal structure solution by simulated annealing: Test application to known zeolite structures. J. Am. Chem. Soc. 114:7189–7198.

de Leeuw, N.H., G.W. Watson, and S.C. Parker. 1995. Atomistic simulation of the effect of dissociative adsorption of water on the surface structure and stability of calcium and magnesium oxide. J. Phys. Chem. 99:17219–17225.

Delville, A. 1991. Modeling the clay–water interface. Langmuir 7:547–555.

Delville, A. 1992. Structure of liquids at a solid interface: An application to the swelling of clay by water. Langmuir 8:1796–1805.

Delville, A. 1993. Structure and properties of confined liquids: A molecular model of the clay–water interface. J. Phys. Chem. 97:9703–9712.

Delville, A., and S. Sokolowski. 1993. Adsorption of vapor at a solid interface: A molecular model of clay wetting. J. Phys. Chem. 97:6261–6271.

Derreumaux, P., and G. Vergoten. 1995. A new spectroscopic molecular mechanics force field. Parameters for proteins. J. Chem. Phys. 102:8586–8605.

de Vos Burchart, E., V.A. Verheij, H. van Bekkum, and B. van de Graaf. 1992. A consistent molecular mechanics force field for all-silica zeolites. Zeolites 12:183–189.

Dick, B.G., and A.W. Overhauser. 1958. Theory of the dielectric constants of alkali halide crystals. Phys. Rev. 112:90–113.

Dovesi, R., C. Pisani, C. Roetti, M. Causa, and V.R. Saunders. 1988. CRYSTAL88. An *ab initio* all-electron LCAO-Hartree-Fock program for periodic solids. Program no. 577. QCPE, Bloomington, IN.

Dovesi, R., C. Roetti, C. Freyria-Fava, E. Aprà, V.R. Saunders, and N.M. Harrison. 1992. *Ab initio* Hartree-Fock treatment of ionic and semi-ionic compounds: State of the art. Phil. Trans. Royal Soc. Lond. Ser. A 341:203–210.

Dzombak, D.A., and F.M.M. Morel. 1990. Surface complexation modeling: Hydrous ferric oxide. John Wiley & Sons, New York.

Edwards, P.P., P.A. Anderson, and J.M. Thomas. 1996. Dissolved alkali metals in zeolites. Acc. Chem. Res. 29:23–29.

Engelsen, S.B., J. Fabricius, and K. Rasmussen. 1994. The consistent force field: 1. Methods and strategies for optimization of empirical potential energy functions. Acta Chem. Scand. 48:548–552.

Ermoshin, V.A., K.S. Smirnov, and D. Bougeard. 1996. Ab initio generalized valence force field for zeolite modeling: 1. Siliceous zeolites. Chem. Phys. 202:53–61.

Farnsworth, K.J., and P.J. O'Malley. 1996. A density functional study of acidic hydroxyl groups in zeolites and their interaction with carbon monoxide. J. Phys. Chem. 100:1814–1819.

Faulon, J.-L. 1994. Stochastic generator of chemical structure: 1. Application to the structure elucidation of large molecules. J. Chem. Inform. Comp. Sci. 34:1204–1218.

Faulon, J.-L., D.A. Loy, G.A. Carlson, and K.J. Shea. 1995. Computer-aided structure elucidation for arylene-bridged polysilsesquioxanes. Comput. Mater. Sci. 3:334–346.

Feller, D., E.D. Glendening, R.A. Kendall, and K.A. Peterson. 1994. An extended basis set *ab initio* study of Li$^+$(H$_2$O)$_n$, n=1-6. J. Chem. Phys. 100:4981–4997.

Field, M.J., P.A. Bash, and M. Karplus. 1990. A combined quantum mechanical and molecular mechanical potential for molecular dynamics simulations. J. Comp. Chem. 11:700–733.

Foster, I.T., J.F. Tilson, A.F. Wagner, R.L. Shepard, R.J. Harrison, R.A. Kendall, and R.L. Littlefield. 1996. Toward high-performance computational chemistry: I. Scalable Fock matrix contruction algorithms. J. Comp. Chem. 17:109–123.

Francl, M.M., C. Carey, L.E. Chirlian, and D.M. Gange. 1996. Charges fit to electrostatic potentials: II. Can atomic charges be unambiguously fit to electrostatic potentials? J. Comp. Chem. 17:367–383.

Frey, R.F., J. Coffin, S.Q. Newton, M. Ramek, V.K.W. Cheng, F.A. Momany, and L. Schäfer. 1992. Importance of correlation-gradient geometry optimization for molecular conformational analyses. J. Am. Chem. Soc. 114:5369–5377.

Furuki, T., M. Sakurai, and Y. Inoue. 1995. An application of the reaction field theory to hydrated metal cations in the framework of the MNDO, AM1, and PM3 methods. J. Comp. Chem. 16:378–384.

Gale, J.D., C.R.A. Catlow, and W.C. Mackrodt. 1992. Periodic *ab initio* determination of interatomic potentials for alumina. Model. Simul. Mater. Sci. Eng. 1:73–81.

Gelin, B.E. 1996. Fifth annual industry report: Computational chemistry and molecular modeling. MMCC Publ., Cambridge, MA.

Gibbs, G.V., J.W. Downs, and M.B. Boisen Jr. 1994. The elusive SiO bond. Rev. Mineralogy 29:331–368.

Giesen, D.J., C.J. Cramer, and D.J. Truhlar. 1995a. A semiempirical quantum mechanical solvation model for solvation free energies in all alkane solvents. J. Phys. Chem. 99:7137–7146.

Giesen, D.J., J.W. Storer, C.J. Cramer, and D.G. Truhlar. 1995b. General semiempirical quantum mechanical solvation model for nonpolar solvation free energies. N-hexadecane. J. Am. Chem. Soc. 117:1057–1068.

Glendening, E.D., D. Feller, and M.A. Thompson. 1994. An ab initio investigation of the structure and alkali metal cation selectivity of 18-crown-6. J. Am. Chem. Soc. 116:10657–10669.

Goniakowski, J., and M.J. Gillan. 1996. The adsorption of H_2O on TiO_2 and SnO_2 studied by first principles calculations. Surface Sci. 350:145–158.

Goniakowski, J., J.M. Holender, L.N. Kantorovich, M.J. Gillan, and J.A. White. 1996. Influence of gradient corrections on the bulk and surface: Properties of TiO_2 and SnO_2. Phys. Rev. B 53:957–960.

Gu, W., and B.P. Schoenborn. 1995. Molecular dynamics simulation of hydration in myoglobin. Proteins 22:20–26.

Haase, F., and J. Sauer. 1995. Interaction of methanol with Bronsted acid sites of zeolite catalysts: An ab initio study. J. Am. Chem. Soc. 117:3780–3789.

Harrison, R.J., M.F. Guest, R.A. Kendall, D.E. Bernholt, A.T. Wong, M. Stave, J.L. Anchell, A.C. Hess, R.J. Littlefield, G.L. Gann, J. Nieplocha, G.S. Thomas, D. Elwood, J.L. Tilson, R.L. Shepard, A.F. Wagner, I.T. Foster, E. Lusk, and R. Stevens. 1996. Toward high-performance computational chemistry: II. A scalable self-consistent field program. J. Comp. Chem. 17:124–132.

Hatcher, P.G., J.-L. Faulon, D.A. Clifford, and J.P. Mathews. 1994. A three-dimensional structure model for humic acids from oxidized soil. p. 133–138. *In* N. Senesi and T.M. Miano (ed.) Humic substances in the global environment and implications on human health. Elsevier, Amsterdam.

Hehre, W.J., L. Radom, P.v.R. Schleyer, and J.A. Pople. 1986. *Ab initio* molecular orbital theory. John Wiley & Sons, New York.

Hendrickson, J.B. 1961. Molecular geometry: I. Machine computation of the common rings. J. Am. Chem. Soc. 83:4537–4547.

Hess, A.C., and V.R. Saunders. 1992. Periodic ab initio Hartree-Fock calculations of the low-symmetry mineral kaolinite. J. Phys. Chem. 96:4367–4374.

Hill, J.-R., and J. Sauer. 1994. Molecular mechanics potential for silica and zeolite catalysts based on ab initio calculations: 1. Dense and microporous silica. J. Phys. Chem. 98:1238–1244.

Hill, J.-R., and J. Sauer. 1995. Molecular mechanics potential for silica and zeolite catalysts based on ab initio calculations: 2. Aluminosilicates. J. Phys. Chem. 99:9536–9550.

Hill, R.J., and G.V. Gibbs. 1979. Variations in d(T-O), d(T⋯T) and <TOT in silica and silicate minerals, phosphates and aluminates. Acta Cryst. B35:25–30.

Himei, H., M. Yamadaya, M. Kubo, R. Vertrivel, E. Broclawik, and A. Miyamoto. 1995. Study of the activity of Ga-ZSM-5 in the de-NOx process by a combination of quantum chemistry, molecular dynamics, and computer graphics methods. J. Phys. Chem. 99:12461–12465.

Hobza, P., J. Sponer, and T. Reschel. 1995. Density functional theory and molecular clusters. J. Comp. Chem. 16:1315–1325.

Hoffmann, R. 1988. Solids and surfaces: A chemist's view of bonding in extended structures. VCH Publ., New York.

Hohenberg, P., and W. Kohn. 1964. Inhomogeneous electron gas. Phys. Rev. B 136:864–871.

Holian, B.L., and R. Ravelo. 1995. Fracture simulations using large-scale molecular dynamics. Phys. Rev. B51:11275–11288.

Honig, B., and A. Nicholls. 1995. Classical electrostatics in biology and chemistry. Science (Washington DC) 268:1144–1149.

Ignatov, S.K., A.G. Razuvaev, V.N. Kokorev, and Y.A. Alexandrov. 1996. Extension of the PM3 method on s,p,d basis. Test calculations on organochromium compounds. J. Phys. Chem. 100:6354–6358.

Ishioka, T., S. Murotani, I. Kanesaka, and S. Hayashi. 1995. Molecular dynamics simulation of infrared spectra for potassium palmitate B-form crystal. J. Chem. Phys. 103:1999–2005.

Iyer, K.A., and S.J. Singer. 1994. Local-mode analysis of complex zeolite vibrations: Sodalite. J. Phys. Chem. 98:12670–12678.

Johnson, B.G., P.M.W. Gill, and J.A. Pople. 1993. The performance of a family of density functional methods. J. Chem. Phys. 98:5612–5626.

Johnston, C.T., and G. Sposito. 1987. Disorder and early sorrow: Progress in the chemical speciation of soil surfaces. p. 89–99. *In* L.L. Boersma (ed.) Future developments in soil science research. SSSA, Madison, WI.

Jousse, F., A.V. Larin, and E.C. De Lara. 1996. Induced infrared absorption of molecular oxygen sorbed in exchanged A zeolites: 2. Frequency shift calculation. J. Phys. Chem. 100:238–244.

Karaborni, S., B. Smit, W. Heidug, J. Urai, and E. van Oort. 1996. The swelling of clays: Molecular simulations of the hydration of montmorillonite. Science (Washington DC) 271:1102–1104.

Karasawa, N., and W.A. Goddard III. 1995. Dielectric properties of poly(vinylidene fluoride) from molecular dynamics simulations. Macromolecules 28:6765–6772.

Kassab, E., J. Fouquet, M. Allavena, and E.M. Evleth. 1993. Ab initio study of proton-transfer surfaces in zeolite models. J. Phys. Chem. 97:9034–9039.

Kawano, M., B. Vessal, and C.R.A. Catlow. 1992. A molecular dynamics simulation of the temperature dependence of the diffusion of methane in silicalite. J. Chem. Soc., Chem. Commun. 1992:879–880.

Keldsen, G.L., J.B. Nicholas, K.A. Carrado, and R.E. Winans. 1994. Molecular modelling of the enthalpies of adsorption of hydrocarbons on smectite clay. J. Phys. Chem. 98:279–284.

Klein, H., C. Kirschhock, and H. Fuess. 1994. Adsorption and diffusion of aromatic hydrocarbons in Zeolite Y by molecular mechanics calculation and x-ray powder diffraction. J. Phys. Chem. 98:12345–12360.

Koelmel, C.M., Y.S. Li, C.M. Freeman, S.M. Levine, M.-J. Hwang, J.R. Maple, and J.M. Newsam. 1994. Quantum and molecular mechanics study of the tris(quaternary ammonium) cation used as the zeolite ZSM-18 synthesis template. J. Phys. Chem. 98:12911–12918.

Kohn, W., and L.J. Sham. 1965. Self-consistent equations including exchange and correlation effects. Phys. Rev. A140:1133–1138.

Kohn, W., A.D. Becke, and R.G. Parr. 1996. Density functional theory of electronic structure. J. Phys. Chem. 100:12974–12980.

Kramer, G.J., N.P. Farragher, B.H.W. van Beest, and R.A. van Santen. 1991. Interatomic force fields for silicas, aluminophosphates, and zeolites: Derivation based on *ab initio* calculations. Phys. Rev. B 43:5068–5080.

Kuhn, L.A., C.A. Swanson, M.E. Pique, J.A. Tainer, and E.D. Getzoff. 1995. Atomic and residue hydrophobicity in the context of folded protein structures. Proteins 23:536–547.

Kyrlidis, A., S.J. Cook, A.K. Chakraborty, A.T. Bell, and D.N. Theodorou. 1995. Electronic structure calculations of ammonia adsorption in H-ZSM-5 zeolites. J. Phys. Chem. 99:1505–1515.

Lasaga, A.C. 1990. Atomic treatment of mineral-water surface reactions. Rev. Mineralogy 23:17–85.

Lasaga, A.C., and G.V. Gibbs. 1988. Quantum mechanical potential surfaces and calculations on minerals and molecular clusters: I. STO-3G and 6-31G* results. Phys. Chem. Miner. 16:29–41.

Lasaga, A.C., and G.V. Gibbs. 1991. Quantum mechanical Hartree-Fock potential surfaces and calculations on minerals: II. 6-31G* results. Phys. Chem. Miner. 17:485–491.

Lee, C., E. Stahlberg, and G. Fitzgerald. 1995. Chemical structure of urea in water. J. Phys. Chem. 99:17737–17741.

Leenheer, J.A., R.L. Wershaw, and M.M. Reddy. 1995. Strong-acid, carboxyl-group structures in fulvic acid from the Suwannee River, Georgia: 2. Major structures. Environ. Sci. Technol. 29:399–405.

Levine, I.N. 1991. Quantum chemistry. 4th ed. Prentice Hall, Englewood Cliffs, NJ.

Lewis, D.W., C.R.A. Catlow, G. Sankar, and S.W. Carr. 1995. Structure of iron-substituted ZSM-5. J. Phys. Chem. 99:2377–2383.

Liang, C., L. Yan, J.-R. Hill, C.S. Ewig, T.R. Stouch, and A.T. Hagler. 1995. Force field studies of cholesterol and cholesteryl acetate crystals and cholesterol-cholesterol intermolecular interactions. J. Comp. Chem. 16:883–897.

Lifson, S., and A. Warshel. 1968. Consistent force field for calculations of conformations, Vibrational spectra, and enthalpies of cyclohexane and *n*-alkane molecules. J. Chem. Phys. 49:5116–5129.

Liu, H., F. Mueller-Plathe, and W.F. van Gunsteren. 1995. A molecular dynamics simulation study with a combined quantum mechanical and molecular mechanical energy function: Solvation effects on the conformational equilibrium of dimethoxyethane. J. Chem. Phys. 102:1722–1730.

Liu, K., J.D. Cruzan, and R.J. Saykally. 1996. Water clusters. Science (Washington DC) 271:929–933.

Löwdin, P.-O. 1959. Correlation problem in many-electron quantum mechanics: I. Review of different approaches and discussion of some current ideas. Adv. Chem. Phys. 2:207–322.

Luzar, A., and D. Chandler. 1996. Effect of environment on hydrogen bond dynamics in liquid water. Phys. Rev. Lett. 76:928–931.

Ma, B., J.-H. Lii, H.F. Schaefer, and N.L. Allinger. 1996. Systematic comparison of experimental, quantum mechanical, and molecular mechanical bond lengths for organic molecules. J. Phys. Chem. 100:8763–8769.

MacKerell, A.D., J. Wiorkiewicz-Kuczera, and M. Karplus. 1995a. An all-atom empirical energy function for the simulation of nucleic acids. J. Am. Chem. Soc. 117:11946–11975.

MacKerell, A.D., M.S. Sommer, and M. Karplus. 1995b. pH dependence of binding reactions from free energy simulations and macroscopic continuum electrostatic calculations: Application to 2'GMP/3'GMP binding to ribonuclease T1 and implications for catalysis. J. Mol. Biol. 247:774–807.

Magnusson, E., and N.W. Moriarty. 1993. Computational convergence of electronic structure calculations of transition metal ligand complexes. J. Comp. Chem. 14:961–969.

Malagoli, M., and W. Thiel. 1996. A semiempirical approach to nonlinear optical properties of large molecules at the MNDO and MNDO/d level. Chem. Physics 206:73–85.

Manassidis, I., and M.J. Gillan. 1994. Structure and energetics of alumina surfaces calculated from first principles. J. Am. Ceram. Soc. 77:335–338.

Maple, J.R., U. Dinur, and A.T. Hagler. 1988. Derivation of force fields for molecular mechanics and dynamics from ab initio energy surfaces. Proc. Natl. Acad. Sci. USA 85:5350–5354.

Maple, J.R., M.-J. Hwang, T.P. Stockfish, U. Dinur, M. Waldman, C.S. Ewig, and A.T. Hagler. 1994. Derivation of Class II force fields: I. Methodology and quantum force field for the alkyl functional group and alkane molecules. J. Comp. Chem. 15:162–182.

Mok, D.K.W., R. Neumann, and N.C. Handy. 1996. Dynamical and nondynamical correlation. J. Phys. Chem. 100:6225–6230.

Møller, C., and M.S. Plesset. 1934. Note on an approximate treatment for many-electron systems. Phys. Rev. 46:618–622.

Momany, F.A., R. Rone, R.F. Frey, and L. Schäfer. 1992. On the use of correlation ab initio studies in the development of the CHARMm empirical molecular force field. Chem. Design Automation News 7(1):38–41.

Morris, R.E., S.J. Weigel, N.J. Henson, L.M. Bull, M.T. Janicke, B.F. Chmelka, and A.K. Cheetham. 1994. A synchrotron x-ray diffraction, neutron diffraction, ^{29}Si MAS-NMR, and computational study of the siliceous form of a zeolite ferrierite. J. Am. Chem. Soc. 116:11849–11855.

Molecular Simulations. 1995. Discover user guide. Versions 2.9.5, 94.0, and 3.0.0. Molecular Simulations, San Diego, CA.

Muegge, I., and E.W. Knapp. 1995. Residence times and lateral diffusion of water at protein surfaces: Application to BPTI. J. Phys. Chem. 99:1371–1374.

Nakano, A., R.K. Kalia, and P. Vashishta. 1995. Dynamics and morphology of brittle cracks: A molecular-dynamics study of silicon nitride. Phys. Rev. Lett. 75:3138–3141.

Naray-Szabo, G., and G.G. Ferenczy. 1995. Molecular electrostatics. Chem. Rev. 95:829–847.

Nicholas, J.B., and A.C. Hess. 1994. Ab initio periodic hartree-fock investigation of a zeolite acid site. J. Am. Chem. Soc. 116:5428–5436.

Nicholas, J.B., R.E. Winans, R.J. Harrison, L.E. Iton, L.A. Curtiss, and A.J. Hopfinger. 1992a. An ab initio investigation of disiloxane using extended basis sets and electron correlation. J. Phys. Chem. 96:7958–7965.

Nicholas, J.B., R.E. Winans, R.J. Harrison, L.E. Iton, L.A. Curtiss, and A.J. Hopfinger. 1992b. Ab initio molecular orbital study of the effects of basis set size on the calculated structure and acidity of hydroxyl groups in framework molecular sieves. J. Phys. Chem. 96:10247–10257.

Nicholas, J.B., F.R. Trouw, J.E. Mertz, L.E. Iton, and A.J. Hopfinger. 1993. Molecular dynamics simulation of propane and methane in silicalite. J. Phys. Chem. 97:4149–4163.

Niketic, S.R., and K. Rasmussen. 1977. The consistent force field: A documentation. Lecture Notes in Chemistry. 3. Springer-Verlag, Heidelberg.

Ohmine, I. 1995. Liquid water dynamics: Collective motions, fluctuations, and relaxation. J. Phys. Chem. 99:6767–6776.

O'Malley, P.J., and C.J. Brathwaite. 1995. Ab initio molecular orbital and molecular graphics studies of benzene adsorption in NaY zeolite. Zeolites 15:198–201.

Payne, M.C., M.P. Teter, D.C. Allan, T.A. Arias, and J.D. Joannopoulos. 1992. Iterative minimization techniques for ab initio total energy calculations: Molecular dynamics and conjugate gradients. Rev. Mod. Phys. 64:1045–1097.

Pearlman, D.A., and P.R. Connelly. 1995. Determination of the differential effects of hydrogen bonding and water release on the binding of FK506 to native and Tyr 82 —> Phe 82 FKBP-12 proteins using free energy simulations. J. Mol. Biol. 248:696–717.

Pellenq, R.J.-M., and D. Nicholson. 1994. Intermolecular potential function for the physical adsorption of rare gases in silicalite. J. Phys. Chem. 98:13339–13349.

Pisani, C., R. Dovesi, and C. Roetti. 1988. Hartree-Fock ab initio treatment of crystalline systems. Lecture Notes in Chemistry. 48. Springer-Verlag, Berlin.

Plimpton, S., and B. Hendrickson. 1996. A new parallel method for molecular dynamics simulation of macromolecular systems. J. Comp. Chem. 17:326–337.
Pople, J.A., J.S. Binkley, and R. Seeger. 1976. Theoretical models incorporating electron correlation. Int. J. Quantum Chem. Symp. 10:1–19.
Pudzianowski, A.T. 1996. Systematic appraisal of density functional methodologies for hydrogen bonding in binary ionic complexes. J. Phys. Chem. 100:4781–4789.
Pugh, S., and M.J. Gillan. 1994. The energetics of NH_3 adsorption at the MgO (001) surface. Surface Sci. 320:331–343.
Purton, J., R. Jones, C.R.A. Catlow, and M. Leslie. 1993. Ab initio potentials for the calculation of the dynamical and elastic properties of α-quartz. Phys. Chem. Miner. 19:392–400.
Raghavachari, K., and J.B. Anderson. 1996. Electron correlation effects in molecules. J. Phys. Chem. 100:12960–12973.
Rasmussen, K. 1985. Potential energy functions in conformational analysis. Lecture Notes in Chemistry. 37. Springer-Verlag, Heidelberg.
Rasmussen, K., S.B. Engelsen, J. Fabricius, and B. Rasmussen. 1993. The consistent force field: Development of potential energy functions for conformational analysis. p. 381–419. *In* R. Fausto (ed.) Recent experimental and computational advances in molecular spectroscopy. NATO ASI Series C: Mathematical and Physical Sciences. Vol. 406. Kluwer Academic Publ., Dordrecht, the Netherlands.
Refson, K., R.A. Wogelius, D.G. Fraser, M.C. Payne, M.H. Lee, and V. Milman. 1995. Water chemisorption and reconstruction of the MgO surface. Phys. Rev. B 52:10823–10826.
Rohl, A.L., D.H. Gay, R.J. Davey, and C.R.A. Catlow. 1996. Interactions at the organic–inorganic interface: Molecular modeling of the interaction between diphosphonates and the surfaces of barite crystals. J. Am. Chem. Soc. 118:642–648.
Rustad, J.R., and B.P. Hay. 1995. A molecular dynamics study of solvated orthosilicic acid and orthosilicate anion using parameterized potentials. Geochim. Cosmochim. Acta 59:1251–1257.
Rustad, J.R., B.P. Hay, and J.W. Halley. 1995. Molecular dynamics simulation of iron(III) and its hydrolysis products in aqueous solution. J. Chem. Phys. 102:427–431.
Rustad, J.R., A.R. Felmy, and B.P. Hay. 1996a. Molecular statics calculations for iron oxide and hydroxide minerals: Toward a flexible model of the reactive mineral-water interface. Geochim. Cosmochim. Acta 60:1553–1562.
Rustad, J.R., A.R. Felmy, and B.P. Hay. 1996b. Molecular statics calculations of proton binding to goethite surfaces: A new approach to estimation of stability constants for multisite surface complexation models. Geochim. Cosmochim. Acta 60:1563–1576.
Saint-Amant, A., W.D. Cornell, P.A. Kollman, and T.A. Halgren. 1995. Calculation of molecular geometries, relative conformational energies, dipole moments, and molecular electrostatic potential fitted charges of small organic molecules of biochemical interest by density functional theory. J. Comp. Chem. 16:1483–1506.
Salahub, D.R., and M.C. Zerner. 1989. The challenge of d and f electrons. ACS Symp. Ser. 354. Am. Chem. Soc., Washington, DC.
Sanders, M.J., M. Leslie, and C.R.A. Catlow. 1984. Interatomic potentials for SiO_2. J. Chem. Soc., Chem. Commun. 1984:1271–1273.
Sankar, G., R.G. Bell, J.M. Thomas, M.W. Anderson, P.A. Wright, and J. Rocha. 1996. Determination of the structure of distorted TiO_2 units in the titanosilicate ETS-10 by a combination of x-ray absorption spectroscopy and computer modeling. J. Phys. Chem. 100:449–452.
Sato, H., A. Yamagishi, and S. Kato. 1992a. Theoretical study on racemic adsorption of tris(1,10-phenanthroline)metal(II) by a clay: Monte Carlo simulations. J. Phys. Chem. 96:9382–9387.
Sato, H., A. Yamagishi, and S. Kato. 1992b. Theoretical study on the interactions between a metal chelate and a clay: Monte Carlo simulations. J. Phys. Chem. 96:9377–9382.
Sato, H., A. Yamagishi, K. Nata, and S. Kato. 1996. Monte Carlo simulations on intercalation of tris(1,10- phenanthroline)metal(II) by saponite clay. J. Phys. Chem. 100:1711–1717.
Sauer, J. 1989. Molecular models in ab initio studies of solids and surfaces: From ionic crystals and semiconductors to catalysts. Chem. Rev. 89:199–255.
Schaefer, H.F., III. 1996. Molecular quantum mechanics: Methods and applications. Concluding conference remarks. J. Phys. Chem. 100:6004–6006.
Schröder, K.-P., and J. Sauer. 1996. Potential functions for silica and zeolite catalysts based on ab initio calculations: 3. A shell model ion pair potential for silica and aluminosilicates. J. Phys. Chem. 100:11043–11049.
Schulten, H.R., and M. Schnitzer. 1993. A state of the art structural concept for humic substances. Naturwissenschaften 80:29–30.

Shah, R., J.D. Gale, and M.C. Payne. 1996a. Methanol adsorption in zeolites: A first principles study. J. Phys. Chem. 100:11688–11697.

Shah, R., M.C. Payne, M.-H. Lee, and J.D. Gale. 1996b. Understanding the catalytic behavior of zeolites: A first- principles study of the adsorption of methanol. Science (Washington, DC) 271:1395–1397.

Shen, J., and J. Wendoloski. 1996. Electrostatic binding energy calculation using the finite difference solution to the linearized Poisson-Boltzmann equation: Assessment of its accuracy. J. Comp. Chem. 17:350–357.

Skipper, N.T., K. Refson, and J.D.C. McConnell. 1991. Computer simulation of interlayer water in 2:1 clays. J. Chem. Phys. 94:7434–7445.

Skipper, N.T., F.-R.C. Chang, and G. Sposito. 1995a. Monte Carlo simulation of interlayer molecular structure in swelling clay minerals: 1. Methodology. Clays Clay Miner. 43:285–293.

Skipper, N.T., G. Sposito, and F.-R.C. Chang. 1995b. Monte Carlo simulation of interlayer molecular structure in swelling clay minerals: 2. Monolayer hydrates. Clays Clay Miner. 43:294–303.

Smit, B. 1995. Simulating the adsorption isotherms of methane, ethane, and propane in the zeolite silicalite. J. Phys. Chem. 99:5597–5603.

Smit, B., and T.L.M. Maesen. 1995. Commensurate 'freezing' of alkanes in the channels of a zeolite. Science (Washington DC) 267:42–44.

Smith, W., G.N. Greaves, and M.J. Gillan. 1995. Computer simulation of sodium disilicate glass. J. Chem. Phys. 103:3091–3097.

Snurr, R.Q., A.T. Bell, and D.N. Theodorou. 1994. Investigation of the dynamics of benzene in silicalite using transition-state theory. J. Phys. Chem. 98:11948–11961.

Sposito, G. 1965. On the general theory of the clay–water interaction. Ph.D. diss. Univ. of California, Berkeley (Diss. Abstr. 66-3702).

Sposito, G. 1981. Cation exchange in soils: An historical and theoretical perspective. p. 13–30. In R.H. Dowdy et al. (ed.) Chemistry in the soil environment. ASA, Madison, WI.

Sposito, G. 1984. The surface chemistry of soils. Oxford Univ. Press, New York.

Sposito, G. 1990. Molecular models of ion adsorption on mineral surfaces. Rev. Mineralogy 23:261–279.

Stanton, R.V., D.S. Hartsough, and K.M. Merz Jr. 1995a. An examination of a density functional/molecular mechanical coupled potential. J. Comp. Chem. 16:113–128.

Stanton, R.V., L.R. Little, and K.M. Merz Jr. 1995b. An examination of a Hartree-Fock/molecular mechanical coupled potential. J. Phys. Chem. 99:17344–17348.

Stanton, R.V., L.R. Little, and K.M. Merz, Jr. 1995c. Quantum free energy perturbation study within a PM3/MM coupled potential. J. Phys. Chem. 99:483–486.

Stave, M.S., and J.B. Nicholas. 1995. Density functional studies of zeolites: 2. Structure and acidity of [T]-ZSM-5 models (T = B, Al, Ga, and Fe). J. Phys. Chem. 99:15046–15061.

Suhai, S. 1994. Cooperative effects in hydrogen bonding: Fourth-order many body perturbation theory studies of water oligomers and of an infinite water chain as a model for ice. J. Chem. Phys. 101:9766–9782.

Suhai, S. 1995. Density functional studies of the hydrogen-bonded network in an infinite water polymer. J. Phys. Chem. 99:1172–1181.

Szabo, A., and N.S. Ostlund. 1982. Modern quantum chemistry: Introduction to advanced electronic structure theory. Macmillan Publ., New York.

Taylor, R.S., and B.J. Garrison. 1995. Molecular dynamics simulations of reactions between molecules: High-energy particle bombardment of organic films. Langmuir 11:1220–1228.

Teppen, B.J., M. Cao, R.F. Frey, C. van Alsenoy, D.M. Miller, and L. Schäfer. 1994a. An investigation into intramolecular hydrogen bonding: Impact of basis set and electron correlation on the *ab initio* conformational analysis of 1,2-ethanediol and 1,2,3-propanetriol. J. Mol. Struct. (Theochem) 314:169–190.

Teppen, B.J., D.M. Miller, S.Q. Newton, and L. Schäfer. 1994b. Choice of computational techniques and molecular models for *ab initio* calculations pertaining to solid silicates. J. Phys. Chem. 98:12545–12557.

Teunissen, E.H., A.P.J. Jansen, and R.A. van Santen. 1995. Ab initio embedded cluster study of the adsorption of NH_3 and NH_4^+ in chabazite. J. Phys. Chem. 99:1873–1879.

Thiel, W., and A.A. Voityuk. 1996. Extension of MNDO to d orbitals: Parameters and results for the second-row elements and for the zinc group. J. Phys. Chem. 100:616–626.

Tossell, J.A., and D.J. Vaughan. 1992. Theoretical geochemistry: Application of quantum mechanics in the earth and mineral sciences. Oxford Univ. Press, New York.

Tsirelson, V.G., O.A. Evdokimova, E.L. Belokoneva, and V.S. Urusov. 1990. Electron density distribution and bonding in silicates: A review of recent data. Phys. Chem. Miner. 17:275–292.

Tunon, I., M.T.C. Martins-Costa, C. Millto, M.F. Ruiz-Lopez, and J.L. Rivail. 1996. A coupled density functional: Molecular mechanics Monte Carlo simulation method: The water molecule in liquid water. J. Comp. Chem. 17:19–29.

Ugliengo, P., V.R. Saunders, and E. Garrone. 1989. Silanol as a model for the free hydroxyl of amorphous silica: Ab initio calculations of the interaction with ammonia. Surface Sci. 224:498–514.

Ugliengo, P., V. Saunders, and E. Garrone. 1990. Silanol as a model for the free hydroxyl of amorphous silica: Ab initio calculations of the interaction with water. J. Phys. Chem. 94:2260–2267.

Ursenbach, C.P., P.A. Madden, I. Stich, and M.C. Payne. 1995. Cluster formation in sodium-doped zeolite Y: Ab initio simulation study. J. Phys. Chem. 99:6697–6714.

Varnek, A.A., G. Wipff, A.S. Glebov, and D. Feil. 1995. An application of the Miertus-Scrocco-Tomasi solvation model in molecular mechanics and molecular dynamics simulations. J. Comp. Chem. 16:1–19.

Verlet, L. 1967. Computer "experiments" on classical fluids: I. Thermodynamical properties of Lennard-Jones molecules. Phys. Rev. 159:98–103.

Walker, P.D., and P.G. Mezey. 1994. Ab initio quality electron densities for proteins: A MEDLA approach. J. Am. Chem. Soc. 116:12022–12032.

Walker, P.D., and P.G. Mezey. 1995. A new computational microscope for molecules: High resolution MEDLA images of taxol and HIV-1 protease, using additive electron density fragmentation principles and fuzzy set methods. J. Math. Chem. 17:203–234.

Wang, J., Z. Szewczuk, S.-Y. Yue, Y. Tsuda, Y. Konishi, and E.O. Purisima. 1995. Calculation of relative binding free energies and configurational entropies: A structural and thermodynamic analysis of the nature of non-polar binding of thrombin inhibitors based on Hirudin 55-65. J. Mol. Biol. 253:473–492.

Warshel, A., and S. Lifson. 1970. Consistent force field calculations: II. Crystal structures, sublimation energies, molecular and lattice vibrations, molecular conformations, and enthalpies of alkanes. J. Chem. Phys. 53:582–594.

Westall, J., and H. Hohl. 1980. A comparison of electrostatic models for the oxide/solution interface. Adv. Colloid Interface Sci. 12:265–294.

White, J.C., and A.C. Hess. 1993a. An examination of the electrostatic potential of silicalite using periodic Hartree-Fock theory. J. Phys. Chem. 97:8703–8706.

White, J.C., and A.C. Hess. 1993b. Periodic Hartree-Fock study of siliceous mordenite. J. Phys. Chem. 97:6398–6404.

Wiberg, K.B. 1965. A scheme for strain energy minimization. Application to the cycloalkanes. J. Am. Chem. Soc. 87:1070–1078.

Wimmer, E. 1993. Simulations and the design of electronic, optic, and magnetic materials. Chem. Design Automation News 8(1):33–39.

Winkler, B., V. Milman, and M.C. Payne. 1994. Orintation, location, and total energy of hydration of channel H_2O in cordierite investigated by *ab initio* total energy calculations. Am. Mineral. 79:200–204.

3 Thermodynamics of Soil Systems: A Personal Perspective

P. F. Low (deceased)
Department of Agronomy
Purdue University
West Lafayette, Indiana

3–1 INTRODUCTION

Predicting the future is always fraught with uncertainty; however, the uncertainty can be reduced if the past is known because, to some extent, predictions usually involve extrapolations of the past. In other words, before we know where we are going, we must know where we have been. For this reason it will be worthwhile to consider the state of thermodynamics in soil chemistry in the past. In doing so, personal experiences will be cited for the purpose of illustration.

3–2 HISTORY OF THERMODYNAMICS IN SOIL SCIENCE

Although occasional references have been made to thermodynamics in the soils literature over the years, only a few soil scientists have given sufficient attention to the subject to become identified with it. Prominent among them are N.E. Edlefsen, A.B.C. Anderson, L.E. Davis, R. Overstreet, K.L. Babcock, G. Sposito, W. Gardner, S.A. Taylor, G. Bolt, W.L. Lindsay, and P.F. Low. Edlefsen and Anderson (1943) wrote the classic monograph on Thermodynamics of Soil Moisture. Of the 11 scientists mentioned, the first six were associated with the University of California. Thus, once established at an institution, a scientific tradition tends to persist. Unfortunately, however, a thermodynamic tradition in soil science has not been established elsewhere.

World War II ushered in the age of science and technology. When the war was over, mature young men and women flooded college campuses on the G.I. bill. Many of them had received technical training while they were in military service and were well prepared for further training. I was one of these. I attended California Institute of Technology and received an M.S. degree in Meteorology under the auspices of the U.S. Army Air Corps. Immediately after victory was achieved in Europe, I was released from military service. My release occurred before the graduate schools of the nation had resumed their normal activities and

---

Copyright © 1998. Soil Science Society of America, 677 S. Segoe Rd., Madison, WI 53711, USA. *Future Prospects for Soil Chemistry.* SSSA Special Publication no. 55.

so I returned to my alma mater, Brigham Young University, to take courses that would be advantageous to me in the future. Few technical courses were being offered at the time because of the shortage of professors imposed by the war. An available course that appealed to me was undergraduate Physical Chemistry. Thus it happened that I spent all of my time for an academic year studying this subject alone. Later, when I became a graduate student under C.A Black at Iowa State University, I enrolled in a graduate course in the same subject. Insofar as I am aware, I was the first soils student of that university to enroll in any graduate course in the Chemistry Department. The course that I took was largely devoted to thermodynamics and it initiated an interest in the subject that has lasted until the present time.

Soon after joining the Agronomy Department of Purduc University, I wrote a thermodynamic article entitled "Movement and Equilibrium of Water in Heterogeneous Systems with Special Reference to Soils" (Low & Deming, 1953). This article advanced the concept that the convective movement of water in a soil was along a free energy gradient. Prior to its publication the article was supposed to be reviewed locally by a competent reviewer. But no one with sufficient knowledge of thermodynamics could be found in the Agricultural Experiment Station. Therefore, J.B. Peterson, Head of the Agronomy Department, and N.J. Volk, Director of the Agricultural Experiment Station, jointly decided to send the article to a senior professor in the Chemistry Department for review. This professor was noted for being caustic, negative, and critical but he was the only professor in that department who had any acquaintance with agriculture. In addition to teaching physical chemistry, he operated a farm. His review was devastating. Without being specific in his criticisms, he expressed his opinion that the article "was of no value to either agronomists or chemists." Naturally, Peterson and Volk were at a loss to know what to do. But they decided to place their trust in me, a young, newly-hired staff member, and authorized me to go ahead and publish anyway; however, rather than risk the possibility of bringing embarrassment to them, I decided to solicit comments on the article from several eminent soil scientists, namely, M.B. Russell, L.A. Richards, Don Kirkam, and R.K. Schofield. Even though these scientists were not known for having any special competence in thermodynamics, they were highly recognized for their fundamental contributions and so their advice could be trusted. All of them responded positively and, without addressing any details, suggested that the article be published because it presented a new approach that was worthy of consideration. Consequently, the article was published. Just suppose, however, that Peterson and Volk had been afraid to authorize publication, or that the scientists cited above had not been open minded. Then it is quite possible that my career would have taken a different direction and that someone else would have had the assignment of talking about thermodynamics at this symposium. Although the foregoing account was given to illustrate how uncommon it was for agricultural scientists to have a knowledge of thermodynamics, it also illustrates how careful we should be in disparaging the work of young scientists at the beginning of their careers.

A few years after joining the staff at Purdue University, I was invited to spend a month as Visiting Professor at the University of California, Berkeley. The purpose of the visit was to discuss thermodynamics with Kenneth Babcock and

Roy Overstreet. Evidently, they felt the need to engage in discussions on the thermodynamics of soils with a knowledgeable scientist from a different environment. The result was one of the most scientifically exhilarating experiences I ever enjoyed. All day long, every day, the discussions continued with great fervor and enthusiasm. A different opinion existed on whether or not thermodynamics could be applied to microsystems in which the variables were not experimentally determinable. I claimed that it could and they claimed that it could not. But, despite this difference, cordial relations and mutual respect were maintained and substantial benefits accrued to everyone. In particular, we felt more confident about our thermodynamic knowledge because it had withstood the scrutiny of others who had some competence in the subject.

Throughout my career, I have published many articles involving thermodynamics. All of these were pertinent to soil science and some of them dealt with problems over which there were lively controversies. These articles could have helped resolve the controversies. For example, a article on ionic activity measurements in heterogeneous systems (Low, 1954) could have helped resolve a controversy on the suspension effect in the measurement of pH. But since the articles were written in thermodynamic terms, they were given little attention. The net result was that they did not fully serve the purpose for which they were written and their benefit to soil science was minimal. The same could be said about the thermodynamic articles of other soil scientists.

Now what is the reason for the lack of interest in thermodynamics among soil chemists and other soil scientists? It isn't that thermodynamics is an obsolete scientific discipline. Reference to modem chemical journals shows that they are replete with thermodynamic studies. The reason must be that, in general, soil scientists have not been adequately trained in the subject. Probably, this inadequacy stems from insufficient training in the basic sciences in both the undergraduate and graduate curricula of soil science.

The author's emphasis on the limited number of soil scientists trained in thermodynamics should not be construed to mean that soil scientists are generally poorly trained. History does indicate, however, that it is not uncommon for their training to be deficient in the basic sciences. In the generation just before World War II, the formal chemical training of many soil chemists did not extend beyond the sophomore year. It is interesting to note, however, that this deficiency was recognized, at least by a few. As far back as the early 1940s a committee of the Soil Science Society of America recommended that the undergraduate curriculum of students in soil science include two semesters of each of the following subjects: calculus, physics, and physical chemistry (e.g., see the report of the Committee on Undergraduate Training and Graduate Training in Soil Science, Soil Sci. Soc. Am. Proc. 6:507–510, 1941). Fifty years later, a curriculum with the same emphasis on the basic sciences has yet to be adopted.

3–3 UTILITY OF THERMODYNAMICS IN SOIL SCIENCE

A reasonable question to ask is whether or not the utility of thermodynamics in soil science justifies the rigorous training required. In answer to this question I will tell how thermodynamics has helped me. But first it should be noted that, although thermodynamics can be regarded as another scientific tool to unlock the secrets of the universe, it is more fundamental than most other tools. Like mathematics, it applies to every branch of science. It is relevant to physics, chemistry, biology, the earth sciences, engineering, and others. Processes in the earth and its atmosphere cannot be fully understood without it. Hence, any scientific education is incomplete unless thermodynamics is included.

C.A. Black had a great interest in phosphorus fixation. It was natural, therefore, that my thesis research at Iowa State University was on this subject. At the time, it was felt that phosphate anions were adsorbed by clay minerals. With reference to kaolinite, one idea was that phosphate anions replaced hydroxyl ions exposed at the surface of the mineral. But, being imbued with thermodynamic ideas, I hypothesized, with the wholehearted support of Black, that kaolinite was slightly soluble in the soil solution and that it was in equilibrium with the ions that dissolved from it. Accordingly, the following thermodynamic equations were supposed to apply

$$MX \Leftrightarrow M^+ + X^-$$

$$K_{sp} = (M^+)(X^-) \qquad [1]$$

where M^+ is a complex Al cation, X^- is a complex silicate anion, and K_{sp} is the thermodynamic equilibrium constant, i.e., the solubility product constant. Therefore, if enough phosphate were added to the soil solution, M^+ should be precipitated as some kind of phosphate (i.e., phosphate should be fixed) and the kaolinite should decompose to restore equilibrium. Experimental evidence lent support to this hypothesis (Low & Black, 1947).

Although it was generally recognized that soil minerals dissolve in the soil solution, no one thought that a complex soil mineral like kaolinite would behave like a simple, partially soluble salt and have a solubility product constant. Our hypothesis was entirely unique in this regard; however, it excited the imagination of soil chemists and soon the nature of phosphate compounds resulting from phosphate fixation was being investigated extensively. Much was learned as a result. Simple thermodynamic theory had evoked an idea that changed the way we thought about phosphate fixation.

When I arrived at Purdue University in the late autumn of 1949, 1 assumed responsibility for three graduate students and a graduate course in soil chemistry that was to be taught the following semester. Moreover, I was actively involved in preparing the results of my Ph.D. dissertation for publication. Nevertheless, I found time to think and read about thermodynamics. During the course of my reading I became acquainted with a volume entitled *The Collected Works of J. Willard Gibbs* (Gibbs, 1948). Included in the volume was the development of Gibb's famous Phase Rule in which it was shown that μ, the chemical potential of a component, is the same in every phase of a system at equilibrium unless the component is affected by a gravitational field. Then μ + *gh* is the same in every phase of the system. Here *g* is the acceleration of gravity and *h* is the height above a reference. The rigor and beauty of this development impressed me profoundly and I could see that it could be extended to apply to any force field, e.g., a force field emanating from the surface of a clay particle. Therefore, I wrote an article entitled "Force Fields and Chemical Equilibrium in Heterogeneous Systems with Special Reference to Soils" (Low, 1951) in which I used the methods of Willard Gibbs to show that μ + θ is the same in every phase of the system at equilibrium and that

$$\bar{G} = \mu + \theta \qquad [2]$$

where $\bar{G}$ is the partial molar free energy of the component and θ is its molar potential energy. Later, D.M. Anderson and I wrote an article (Low & Anderson, 1958) showing that, when the water in a clay-water phase is in equilibrium with pure, bulk water,

$$\bar{G}_w - G_w^0 = -\bar{V}_w \Pi \qquad [3]$$

where $\bar{G}_w$ and $\bar{V}_w$ are the partial molar free energy and partial molar volume, respectively, of the water in the clay-water phase, G_w^0 is the molar free energy of bulk water and Π is the swelling pressure of the clay. Combination of Eq. [2] and [3] yields

$$\Pi = -(1/\bar{V}_w) \, [(\mu_w - \mu_w^0) + (\theta_w - \theta_w^0)] \qquad [4]$$

where the *w* subscript denotes water, the absence of a superscript denotes the clay-water phase and the zero superscript denotes the phase containing bulk water. The first term on the right-hand-side of Eq. [4] represents the osmotic contribution to Π, i.e., the contribution arising from the difference in ionic concentration in the two phases, and the second term represents the contribution arising from the difference in potential energy of the water. The first term is the only one taken into account in traditional electric double layer theory. Therefore, Eq. [4] called this theory into question and provoked a longstanding controversy between most of my fellow scientists and me. Moreover, it focused attention on the need to determine whether or not clay-water interaction causes $(\theta_w - \theta_w^0)$ to have a significant non-zero value. But how could one make such a determination?

One day I was talking with a graduate student, Jack Hemwall, in the laboratory. On a bench next to us a Bradfield electrodialysis cell was in operation. I noticed that long fingers of clay of a gel-like consistency adhered to the membrane next to the anode compartment and extended into the adjacent clay suspension. This was not unusual. I had seen it happen before. But when I absent-mindedly cut one of the fingers off with a spatula, it floated. Immediately, our attention was drawn to the floating finger of clay. Since clay is heavier than water, and since the finger contained a high concentration of clay, why did it float? The only plausible explanation seemed to be that the water associated with the clay was less dense than normal bulk water. We decided, therefore, to design experiments by which this explanation could be tested. Our idea was that the density of the water could be used as a criterion of the degree of clay–water interaction. If the density of the water in the clay differed significantly from that of bulk water, we would have reason to believe that the interaction of the water with the clay was strong enough to cause θ_w, to differ from θ_w^0 and, thereby, contribute to the magnitude of Π. Other properties of the water in the clay could be used in a similar fashion. Thus, a lifelong study of the water in clay–water systems was initiated by the development of a thermodynamic equation that challenged existing concepts.

Among the water properties that we studied were the thermal expansibility and heat of compression. These two properties were measured by entirely independent experiments. The thermal expansibility was measured in a dilatometer and the heat of compression was measured in a calorimeter. Yet, as shown below, the same results could be extracted from both of them by means of well known thermodynamic relationships.

The thermal expansibility of a clay-water system, $(\partial V/\partial T)_P$, is related to the apparent thermal expansibility of the water within it, $(\partial \varphi_V/\partial T)_P$ by the equation

$$(\partial V/\partial T)_P = m_m (\partial v_m^o/\partial T)_P + m_w (\partial v/\partial T)_P \qquad [5]$$

where V, T, and P are the volume, temperature and pressure of the system, m_m and m_w are the masses of clay and water, respectively, v_m^o is the specific volume of the pure clay and φ_v is the apparent specific volume of the water. The term on the left-hand-side of Eq. [5] was measured by means of the dilatometer. Hence, it was possible to calculate $(\partial \varphi/\partial T)_P$ by using Eq. [5] and the known value of $(\partial v_m^o/\partial T)_P$, the thermal expansibility of the pure clay.

The heat of compression is the heat, Q, released by the application of a pressure, P, to a clay suspension. We measured Q calorimetrically at constant T and several different values of P and then obtained $(\partial Q/\partial P)_T$ from the slope of the plot of Q vs. P. Since Q was released reversibly, we could write $Q = TdS$ and

$$(\partial Q/\partial P)_T = T(\partial S/\partial P)_T \qquad [6]$$

where S is the entropy of the system. Now, one of the Maxwell equations of thermodynamics is

$$(\partial S/\partial P)_T = -(\partial V/\partial T)_P \qquad [7]$$

Combination of Eq. [6] and [7] gives

$$(\partial Q/\partial P)_T = -T(\partial V/\partial T)_P \qquad [8]$$

It was possible, therefore, to use the calorimetric data to calculate $(\partial S/\partial P)_T$, $(\partial V/\partial T)_P$, and, with the help of Eq. [5], $(\partial \varphi_v/\partial T)_P$.

In Fig. 3–1 the values of $(\partial \varphi_v/\partial T)_P$ determined dilatometrically and calorimetrically are compared. This comparison was accomplished by combining graphs published by Clementz and Low (1976). Note how well the two sets of data agree. The slight difference that does exist is constant at all values of m_w/m_c, the mass ratio of water to clay, and is probably attributable to minor errors in calibrating the dilatometer and/or calorimeter. Thus, we see that thermodynamics allowed extra information to be gleaned from the calorimetric experiment. In addition, it allowed the results of the calorimetric experiment to be compared with those of the dilatometric experiment and, consequently, enhanced the reliability of both experiments.

Now, let us return to Eq. [3]. The mathematical definition of $\bar{G}_w$ is

$$\bar{G}_w = (\partial G/\partial n_w)_{P,T} \qquad [9]$$

where G is the free energy of the entire system and n_w is the number of moles of water. Thus $\bar{G}_w$ includes every effect on G produced by the addition of water to the system. This concept has interesting ramifications. Consider, for example, the infrared spectrum of montmorillonite. It includes four peaks representing different Si–O stretching vibrations in the tetrahedral sheet. These vibrations are altered by any change in the crystal structure of the montmorillonite, e.g., a change in the angle of rotation of the silica tetrahedra. Recently, we determined how ν_{Si-O}, the Si–O stretching frequency, changes with m_w/m_c, the mass ratio of water to montmorillonite, for each of these peaks (Yan et al., 1996). Some of our results are presented in Fig. 3–2. Note that ν_{Si-O} changes with m_w/m_c, across a wide range of m_w/m_c. According to spectroscopic theory, the vibrational energy of any vibrator, i, is equal to $hc\nu$ where h is Planck's constant, c is the velocity of light and ν is the frequency of vibration expressed in wavenumbers. Consequently, E_{vib}, the vibrational energy of a clay crystal, is given by

$$E_{vib} = Nhc\Sigma n_i \nu_i \qquad [10]$$

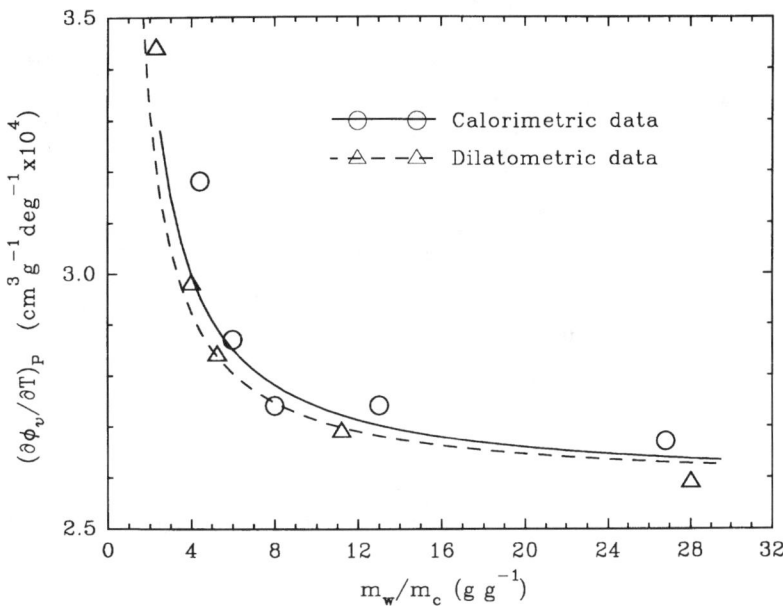

Fig. 3–1. Comparison of the values of $(\partial \varphi_v/\partial T)_P$, the apparent thermal expansibility of the interlayer water in montmorillonite, obtained dilatometrically and calorimetrically.

where N is Avogadro's number and n is the number of moles of the vibrator in the crystal. Now, since ν_{Si-O} changes with m_w/m_c up to very high values of m_w/m_c, we are obliged to conclude that there is a corresponding change in the vibrational energy of the montmorillonite. Further, the vibrational energy is a component of G and so the change in vibrational energy with water content must contribute to $\bar{G}_w$ and, thereby, to Π. It appears, therefore, that Π is affected by changes in the

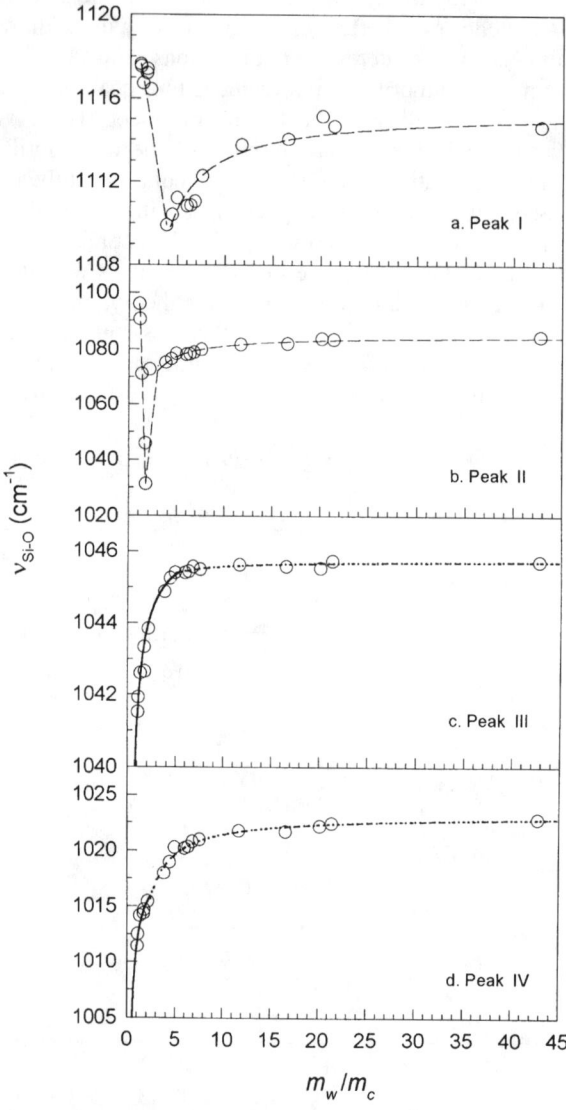

Fig. 3–2. Relation between ν_{Si-O}, the Si–O stretching frequency, and m_w/m_c, the mass ratio of water to montmorillonite, for the four peaks characteristic of Si–O stretching in the infrared spectrum of montmorillonite.

structure of the montmorillonite that occur during hydration. This is a new concept of swelling. Note that it was evoked by a thermodynamic equation.

Recently, we thought it worthwhile to study the adsorption by montmorillonite of simple, uncharged organic compounds with different structures and functional groups (Zhang et al., 1990). As a part of this study, a calorimeter was used to measure the heat of adsorption, Q_a, of each of the compounds as a function of its concentration in the solution in which the montmorillonite was suspended. The amount of each compound adsorbed at each concentration also was measured. Consequently, we were able to construct the graphs in Fig. 3–3, which was taken from the article cited above. In this figure $\Gamma_2^{(n)}$ is the reduced surface excess of the solute, denoted by the subscript 2, and S is the specific surface area of the montmorillonite. The product of $\Gamma_2^{(n)}$ and S is the difference between the amount of solute actually present in the system per unit mass of montmorillonite and that which would be present if the concentration of the solute in the equilibrium solution persisted up to the solid–solution interface. Now, Q_a was regarded as being positive when heat was released from the system and negative when heat was gained by the system and so we could write

$$[\partial Q_a / \partial (\Gamma_2^{(n)} S)]_{T,P} = (\bar{H}_2^s - \bar{H}_2^b) \qquad [11]$$

where $\bar{H}$ is the partial molar enthalpy of the component and the superscripts s and b denote the interfacial phase and bulk phase, respectively. The left-hand-side of

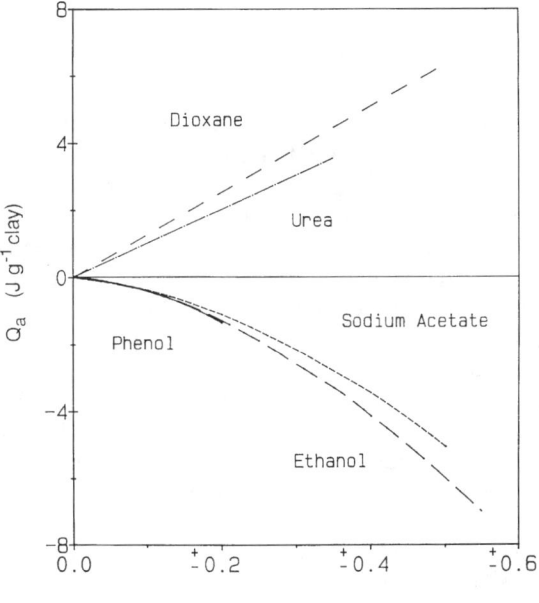

Fig. 3–3. The heat of adsorption, Q_a, at different values of $\Gamma_2^{(n)}S$, the apparent adsorption, for 1,4-dioxane, urea, ethanol, phenol, and the acetate ion (negative values on the abscissa apply to the acetate ion only).

Eq. [11] is determinable at any value of $\Gamma_2^{(n)}S$ from the slope of the relevant plot in Fig. 3–3 at that value of $\Gamma_2^{(n)}S$. Hence, $(\bar{H}_2^s - \bar{H}_2^b)$ was positive for phenol, ethanol and sodium acetate, i.e., the enthalpy of the solute in the interfacial solution was greater than that in the bulk solution. Since the reverse would be true if bonds were formed between the solute and the surface of the montmorillonite, we concluded that such bonds do not form when these three solutes are adsorbed. Moreover we know that

$$\Delta G = \Delta H - T\Delta S \qquad [12]$$

where ΔG, ΔH, and ΔS are the changes in free energy, enthalpy, and entropy of the system, respectively. For a process like adsorption to occur spontaneously at constant T and P, ΔG has to be negative. But we know that ΔH was positive for the adsorption of ethanol, phenol, and sodium acetate. Therefore, we also concluded that ΔS had to be relatively large and positive. A positive ΔS means that adsorption of the organic solute produces a higher degree of randomness or disorganization. This would not have been the case if adsorption caused the solute to be localized at adsorption sites on the solid surface. Evidently, the positive ΔS was caused by disorganization of water molecules that accompanied adsorption. Either the hydration shells of the clay layers or the hydration shells of the organic solutes were disorganized when the solutes entered the interfacial region. The foregoing conclusions are quite unique in the soils literature and warrant further consideration. They could not have been reached without the benefit of thermodynamics.

3–4 SUMMARY AND FUTURE OF THERMODYNAMICS IN SOIL SCIENCE

A brief, personalized history of thermodynamics in soil science is presented. Then the utility of thermodynamics in soil science is discussed and illustrated by examples from the author's experience. Finally, the future of thermodynamics in soil science is considered. It is proposed that thermodynamics will not assume its proper role in soil science unless more emphasis is placed on the basic sciences in the training of soil scientists.

My assignment for the present symposium was to discuss the future of thermodynamics in soil chemistry. This is an impossible task. The principles of thermodynamics are well established and the applications depend on the needs of specific research projects. These needs become apparent only as the research unfolds and problems arise that are amenable to thermodynamic solutions. Had I been asked at the beginning of my scientific career how I was going to apply thermodynamics in the future, I could not have given a satisfactory answer. The same is true now. I cannot predict how thermodynamics will be used in soil chemistry in the future because I cannot predict the problems that different soil chemists will investigate. The only safe prediction that I can make is that thermodynamics will be used, and used profitably, if soil chemists are prepared to use it. Nevertheless, a few general comments about the future may be in order.

The chemistry of the soil is too complex to be understood with elementary chemistry. A thorough understanding of this important subject requires a comprehensive knowledge of chemistry, including thermodynamics. As we have seen, thermodynamics has great utility. In general, it can be used to: (i) inspire new concepts; (ii) extract more information from experimental results; (iii) facilitate the interpretation of experimental results; (iv) indicate which reactions are possible; (v) establish the conditions for equilibrium; and (vi) facilitate understanding of the literature

My emphasis on good training in chemistry, including thermodynamics, should not necessarily discourage any soil chemist who does not presently have that training. I have watched several fellow scientists improve their training after leaving graduate school. To do this they have taken formal courses, read the relevant literature, taken sabbatical leaves at appropriate places, and consulted or collaborated with knowledgeable colleagues. In one or more of these ways Joe White and Charles Roth, my colleagues in the Agronomy Department at Purdue University, learned infrared spectroscopy and another colleague, John Cushman, learned molecular dynamics. All of these colleagues have been generous in sharing their knowledge with me.

When I began my career, I had little practical knowledge of x-ray diffraction and infrared spectroscopy. But I received training in these techniques from my students and, especially, from White and Roth. I also have conducted collaborative research with both of them. The result has been that, although I still cannot pretend to be highly proficient in these techniques, I know when and how to use them and they have proved to be invaluable to me on many occasions. A scientist cannot be truly successful unless he continues to learn throughout his entire professional career.

If we predict the future of thermodynamics in soil chemistry on the basis of past performance, it is not very promising. Thermodynamics cannot assume its proper role in soil chemistry unless the training of soil chemists improves. But we are not necessarily captives of the past. As the demands of society change, soil chemists must improve their training so that they can meet these demands. The future belongs to those who prepare for it. If we don't prepare ourselves to use the best tools of chemistry, others with better training will preempt our roles as soil chemists. The world of science will not wait for us to use inferior techniques to discover principles that must be understood before pressing practical problems can be solved. Many of these problems are related to the soil. The soil is involved in the production of our food, the disposal of our wastes, the construction of our highways and the protection of our drinking water. As one of our most important natural resources, it will increasingly be the focus of national attention. If we are adequately prepared, we will be fully accepted in the scientific community and our expertise will be in demand. Then we will have many opportunities to serve, even in ways that are not yet anticipated. From my perspective, the future of soil chemistry, including soil thermodynamics, is potentially very exciting. I wish that I could be a part of it.

REFERENCES

Clementz, D.M., and P.F Low. 1976. Thermal expansion of interlayer water in clay systems: 1. Effect of water content. p. 485–502. *In* M. Kerker (ed.) Colloid and Interface Science. Vol.3. Academic Press, New York.

Edlefsen, N.E., and A.B.C. Anderson. 1943. Thermodynamics of soil moisture. Hilgardia 15:31–298.

Gibbs, J.W. 1948. The collected works of J. Willard Gibbs. Vol. 1. Yale Univ. Press, New Haven, CT.

Low, P.F. 1951. Force fields and chemical equilibrium in heterogeneous systems with special reference to soils. Soil Sci. 71:409–418.

Low, P.F. 1954. Ionic activity measurements in heterogeneous systems. Soil Sci. 77:29–41.

Low, P.F., and D.M. Anderson. 1958. Osmotic pressure equations for determining thermodynamic properties of soil water. Soil Sci. 86:251–253.

Low, P.F., and C.A. Black. 1947. Phosphate-induced decomposition of kaolinite. Soil Sci. Soc. Am. Proc. 12:180–184.

Low, P.F., and J.M. Deming. 1953. Movement and equilibrium of water in heterogeneous systems with special reference to soils. Soil Sci. 75:187–202.

Yan, L., C.B. Roth, and P.F. Low. 1996. Changes in the $Si\text{-}O$ vibrations of smectite layers accompanying the sorption of interlayer water. Langmuir 12:4421–4429.

Zhang, Z.Z., P.F. Low, J.H. Cushman, and C.B. Roth. 1990. Adsorption and heat of adsorption of organic compounds on montmorillonite from aqueous solutions. Soil Sci. Soc. Am. J. 54:59–66.

4 Kinetics of Soil Chemical Phenomena: Future Directions

Donald L. Sparks

Department of Plant and Soil Sciences
University of Delaware
Newark, Delaware

4–1 INTRODUCTION

Historically, the emphasis in soil chemistry teaching and research has been on equilibrium processes and reactions. While these studies have been useful, they are not often relevant to field settings where chemical reactions are time-dependent.

In the late 1970s and certainly in the 1980s and 1990s the kinetics of environmentally important reactions on natural materials has become and will continue to be a major leitmotif in soil and environmental chemistry. This intense interest is in large part due to the recognition that to accurately predict the fate of contaminants in the subsurface environment, a knowledge of the reaction kinetics is imperative.

While major advances have been made in understanding time-dependent reactions on natural materials such as soils and sediments, there are still many unknowns that are complicated by the complex, heterogeneous nature of natural materials. This review will focus on future research needs, particularly in the area of sorption–release processes in the soil environment, and not on past findings and accomplishments. To learn about the latter, the reader can consult a number of books and reviews (Sparks, 1989, 1991, 1995; Sparks & Suarez, 1991). Future research needs that will be discussed in this review include: models that accurately describe both chemical kinetics and transport processes in multiple site, heterogeneous systems; more extensive studies on the effect of residence time or aging on contaminant retention–release; and mechanistic studies that employ both kinetic and in situ (drying and high vacuums are not employed; aqueous suspensions can be examined) microscopic and spectroscopic techniques.

4–2 KINETIC MODELS

While first-order models have been used widely to describe the kinetics of chemical reactions on natural materials, a number of other simple kinetic models also have been employed. These include various ordered equations such as zero-

---

Copyright © 1998. Soil Science Society of America, 677 S. Segoe Rd., Madison, WI 53711, USA. *Future Prospects for Soil Chemistry.* SSSA Special Publication no. 55.

order, second-order, and fractional-order, and Elovich, power function or fractional power, and parabolic diffusion models. For more complete details and applications of these models one should consult Sparks (1989).

4–2.1 $Z(t)$ and Diffusion Models

In a number of studies it has been shown that several simple kinetic models, as listed above, describe rate data well, based on correlation coefficients and standard errors of the estimate (Chien & Clayton, 1980; Onken & Matheson, 1982; Sparks & Jardine, 1984); however, some of the above kinetic models are empirical and no meaningful rate parameters can be obtained.

Aharoni and Ungarish (1976) and Aharoni (1984) noted that some simple kinetic models are approximations to which more general expressions reduce in certain limited time ranges. They suggested a generalized empirical equation by examining the applicability of power- function, Elovich, and first-order equations to experimental data. By writing these as the explicit functions of the reciprocal of the rate, Z, which is $(dq/dt)^{-1}$ where q is amount of sorbate sorbed and t is time, one can show that a plot of Z vs. t should be convex if the power-function equation is operational, linear if the Elovich equation is appropriate, and concave if the first-order equation is appropriate; however, Z vs. t plots for soil systems are usually S-shaped: convex at small t, concave at large t, and linear at some intermediate t. These findings suggest that the reaction rate can best be described by the power-function equation at small t, by the Elovich equation at an intermediate t, and by a first-order equation at large t. Thus, the S-shaped curve indicates that the above equations may be applicable, each at some limited time range.

One of the reasons a particular kinetic model appears to be applicable may be that the study is conducted during the time range when the model is most appropriate. While sorption, for example, decreases over many orders of magnitude before equilibrium is approached, with most methods and experiments, only a portion of the entire reaction is measured and over this time range the assumptions associated with a simple kinetic model (power function, Elovich, and first-order) are valid. Aharoni and Suzin (1982a,b) showed that the S-shaped curves could be well described using homogeneous and heterogeneous diffusion models. In homogeneous diffusion situations, the final and initial portions of the S-shaped curves (conforming to the power-function and first-order equations, respectively) predominated, whereas, in instances where the heterogeneous diffusion model was operational, the linear portion of the S-shaped curve, that conformed to the Elovich equation, predominated.

4–2.2 Implications of Diffusion Models

The finding that slower reactions at the soil particle–liquid interface can be described by diffusional models indicates that the kinetics of chemical processes cannot be considered separately from transport phenomena. Thus, such a combination of processes cannot be treated using first-order or other-order chemical kinetics equations. When one states that a reaction between the molecular species A and B is of first-order with respect to A, one assumes that the molecules of A have equal chances of participating in the reaction and therefore the rate is proportional to the concentration C_A. This reasoning can be extended to a reaction between an adsorbing surface and an adsorptive solute. The concentration C_A, in this case, refers to the number of reactive sites per unit area, which corresponds to the number of unoccupied sites per unit area $(1 - \theta_A)$; however, by using first-order kinetics (or other-order kinetics) one tacitly assumes that all of the surface sites are potential reactants at any time, and they have an opportunity of participating in the sorption process. If one assumes that there are sites that cannot be reached directly from the fluid phase, but can be reached after the sorbate has undergone sorption and desorption at other sites, one cannot separate chemical kinetics from transport kinetics. The overall kinetic process obeys a diffusion equation (Aharoni & Sparks, 1991).

4–2.3 Multiple Site Models

Based on the previous discussion, it is evident that simple chemical kinetics models such as ordered reaction models and the power function and Elovich models may not be appropriate to describe reactions in heterogeneous systems such as soils, sediments, and soil components. In these systems, where there is a range of particle sizes and multiple retention sites, both chemical kinetics and transport phenomena are occurring simultaneously, and a fast reaction is often followed by a slower reaction(s).

In such systems, nonequilibrium models that describe both chemical and physical nonequilibrium and that consider multiple components and sites are more appropriate. Physical nonequilibrium is ascribed to some rate-limiting transport mechanism such as film diffusion (FD) or particle diffusion (PD) while chemical nonequilibrium is due to a rate-limiting mechanism at the particle surface (i.e., the chemical reaction, CR). Nonequilibrium models include two-site, multiple site, radial diffusion (pore diffusion), surface diffusion, and multiprocess models (Table 4–1).

The term *sites* can have a number of meanings (Brusseau & Rao, 1989): (i) specific, molecular scale reaction sites; (ii) sites of differing degrees of accessibility (external, internal); (iii) sites of differing sorbent type (organic matter and inorganic mineral surfaces); and (iv) sites with different sorption mechanisms. With chemical nonequilibrium sorption processes, the sorbate may undergo two or more types of sorption reactions, one of which is rate-limiting. For example, a metal cation may sorb to organic matter by one mechanism and to mineral surfaces by another mechanism with one of the mechanisms being rate-limiting.

4–2.4 Chemical Nonequilibrium Models

Chemical nonequilibrium models describe time-dependent reactions at sorbent surfaces. The one-site model is a first-order approach that assumes that the reaction rate is limited by only one process or mechanism on a single class of sorbing sites and that all sites are of the time-dependent type (Table 4–1). In many cases this model appears to describe soil chemical reactions quite well; however, often it does not. This model would seem not appropriate for most heterogeneous

Table 4-1. Comparison of sorption kinetic models.†‡

Conceptual model	Fitting parameter(s)	Model limitations
One-site model $$S \xrightarrow{k_d} C$$	k_d	Cannot describe biphasic sorption–desorption
Two-site model $$S_1 \xrightarrow{X_1 K_p} C \xrightarrow{k_d} S_2$$	k_d, K_p, X_1 §	Cannot describe the bleeding or slow, reversible, nonequilibrium desorption for residual sorbed compounds (Karickhoff, 1980)
Radial diffusion penetration retardation (pore diffusion) model (Wu & Gschwend, 1986) $$S' \xrightarrow{K_p} C' \xrightarrow{D_{eff}} C$$	$D_{eff} = f(n,t) D_m n/(1-n)\rho_s K_p$ ¶	Cannot describe instantaneous uptake without additional correction factor (Ball & Roberts, 1991); did not describe kinetic data for times >10^3 min (Wu & Gschwend, 1986)
Dual-resistance surface diffusion model (Miller & Pedit, 1992) $$S' \xrightarrow{D_s} C'_s \xrightarrow{k_b} C$$	D_s, k_b	Model calibrated with sorption data predicted more desorption than occurred in the desorption experiments (Miller & Pedit, 1992)
Multi-site continuum compartment model (Connaughton et al., 1993) $$F(t) = 1 - [M(t)/M] = 1 - (\beta/\beta + t)^\alpha$$	α, β	Model needs comparison to other models, α and β may vary with sorbent properties and model should be applied to sorption
Pore space diffusion model (Fuller et al., 1993) $$\{\varepsilon + [S_a K_s C(r)^{(1-1/n)}]/n\}[\partial C(r)/\partial t] = D_e[\partial^2 C(r)/\partial r^2] + [2\partial C(r)/r\partial r]$$	$D_e/a^2, \varepsilon, K_s, 1/n, F_{eq}$	Assumption of homogeneous, spherical particles and diffusion only in aqueous phase
Multiple Particle Class Pore Diffusion Model (Pecit & Miller, 1995) $$\{\theta^i_P + \rho^i_a [\partial q^i_r(r,t)/\partial C^i_P(r,t)]\}[\partial C^i_P(r,t)/\partial t] = (\theta^i_P D^i_P/r^2)(\partial/\partial r)[r^2 \partial C^i_P(r,t)/\partial r] - \theta^i_P \lambda^i_P C^i_P(r,t) - \rho^i_a \lambda^i_r q^i_r(r,t)$$	$\theta^i_P, \rho^i_a, D^i_P, \lambda^i_P, \lambda^i_r$	Multiple fitting parameters; variations in sorption equilibrium and rates that might occur within a particle class or an individual particle grain are not addressed

† Partially adapted from Connaughton et al. (1993).
‡ Abbreviations used are as follows: S, concentration of the bulk sorbed contaminant (g g^{-1}); C, concentration of the bulk aqueous-phase contaminant (g mL^{-1}); k_d, first-order desorption rate coefficient (min^{-1}); S_2, concentration of the sorbed contaminant that is rate limited (g g^{-1}); S_1, concentration of the contaminant that is in equilibrium with the bulk aqueous concentration (g g^{-1}); X_1, fraction of the bulk sorbed contaminant that is in equilibrium with the aqueous concentration; K_p, sorption equilibrium partition coefficient (mL g^{-1}); D_{eff}, effective diffusivity of sorbate molecules or ions in the particles (cm^{2}s^{-1}); S', concentration of contaminant in immobile bound state (mol g^{-1}); C', concentration of contaminant free in the pore fluid (mol cm^{-3}); n, porosity of the sorbent (cm^3 of fluid cm^{-3} total); D_m, pore fluid diffusivity of the sorbate (cm^2 s^{-1}); ρ_s, specific gravity of the sorbent (g cm^{-3}); $f(n,t)$, pore geometry factor; k_b, boundary layer mass transfer coefficient (m s^{-1}); R, radius of the spherical solid particle, assumed constant (m); ρ, macroscopic particle density of the solid phase (g m^{-3}); C'_s, solution-phase solute concentration corresponding to an equilibrium with the solid-phase solute concentration at the exterior of the particle (g L^{-1}); D_s, surface diffusion coefficient (m s^{-1}).
§ K_p can be determined independently.
¶ K_p, D_m, and ρ_s can be determined independently; $F(t)$, fraction of mass released through time t; $M(t)$, mass remaining after time t; M, total initial mass; β, scale parameter necessary for determination of mean and standard deviation of k_s; α, shape parameter; ε, internal porosity of sorbent; $C(r)$, concentration of sorptive in the aqueous phase in the pore fluid at radial distance r; S_a is the surface of sorbent per unit volume of solid; $1/n$, the adsorption isotherm slope; K_s, adsorption isotherm intercept; D_e, effective diffusion coefficient; a, radius of the aggregate, F_{eq}, equilibrium fraction of adsorption sites; θ^i_p, intraparticle porosity of particle class i; ρ^i_a, apparent particle density of particle class i; r, radial distance; $C^i_p(r,t)$, intraparticle fluid-phase solute concentration of the particle class i; D^i_p, pure diffusion coefficient for particle class i; λ^i_p, intraparticle fluid-phase first-order reaction rate coefficient for particle class i; λ^i_r, intraparticle solid-phase first-order reaction rate coefficient for particle class i; $q^i_r(r,t)$, intraparticle solid-phase solute concentration of particle class i.

systems since multiple sorption sites exist. The two-site (two compartment, two box) or bicontinuum model has been widely used to describe chemical nonequilibrium (Leenheer & Ahlrichs, 1971; Hamaker & Thompson, 1972; Karickhoff, 1980; Karickhoff & Morris, 1985; McCall & Agin, 1985; Jardine et al., 1992) and physical nonequilibrium (Nkedi-Kizza et al., 1984; Lee et al., 1988; van Genuchten & Wagenet, 1989). This model assumes that there are two reactions occurring, one that is fast and reaches equilibrium quickly and a slower reaction that can continue for long time periods (Table 4–1). The reactions can occur in either series or in parallel (Brusseau & Rao, 1989).

In describing chemical nonequilibrium with the two-site model it is assumed that the sorbent has two types of sites. One site involves an instantaneous equilibrium reaction and the other site involves the time-dependent reaction. The instantaneous equilibrium reaction is described by an equilibrium isotherm equation while a first-order equation is usually employed to describe the time-dependent reaction.

Jardine et al. (1985) modeled the transport of Al through Ca-saturated kaolinite columns using a two-site nonequilibrium transport process. They assumed type-1 sites were in local equilibrium with the solution phase and involved an instantaneous Ca–Al exchange mechanism. Type-2 sites involved a time-dependent Al polymerization reaction mechanism and were described by first-order kinetics.

The polymerization mechanism was indirectly confirmed by investigating the effect of influent pH on Al transport. When influent pH was lowered, the slower kinetic reaction was eliminated and the Al breakthrough curve was described with a one-site equilibrium model.

With the two-site model there are two adjustable or fitting parameters, the fraction of sites at local equilibrium (X_1) and the rate constant (k). A distribution (K_d) or partition coefficient (K_p) is determined independently from a sorption–desorption isotherm. Connaughton et al. (1993) used a two-site model to describe naphthalene desorption from contaminated soil (Fig. 4–1a). The model did not describe the data well and fitted and estimated desorption rate coefficients (k_d) did not agree, with the estimated k_d values being higher than the values obtained from fitting. The estimated k_d values were based on the relation log k_d = 0.301 −0.688 log K_p (Brusseau & Rao, 1989). Connaughton et al. (1993) related this discrepancy to the greater desorption times of their experiments and to the use of the two-site model to describe the entire desorption process. The two-site model, that the $K_p - k_d$ relationship was based on, assumes that the initial desorption is instantaneous, which is not the case for naphthalene desorption. One major disadvantage in using the two-site model to describe heterogeneous systems such as soil is the assumption that only two sorptive sites are present. Thus, the fitting parameters in the two-site model probably do not conform to actual reaction rates on multiple sites in soils. Moreover, it is difficult to relate the fitting parameters to known properties of the sorbent. For example Wu and Gschwend (1986) found two different sets of fitting parameters described tetrachlorobenzene sorption on a Charles River sediment if different sediment mean aggregate sizes were used. Consequently, the parameters had to be experimentally determined for each sediment size.

To account for the multiple sites that may exist in heterogeneous systems Connaughton et al. (1993) developed a multi-site compartment model (Γ) that incorporates a continuum of sites or compartments with a distribution of rate coefficients that can be described by a gamma density function. A fraction of the sorbed mass in each compartment is at equilibrium and there is a desorption rate coefficient or distribution coefficient for each compartment or site (Table 4–1). The multisite model has two fitting parameters α, a shape parameter, and 1/β, which is a scale parameter that determines the mean standard deviation of the rate coefficients. Figure (4–1B) shows application of the Γ model to desorption of

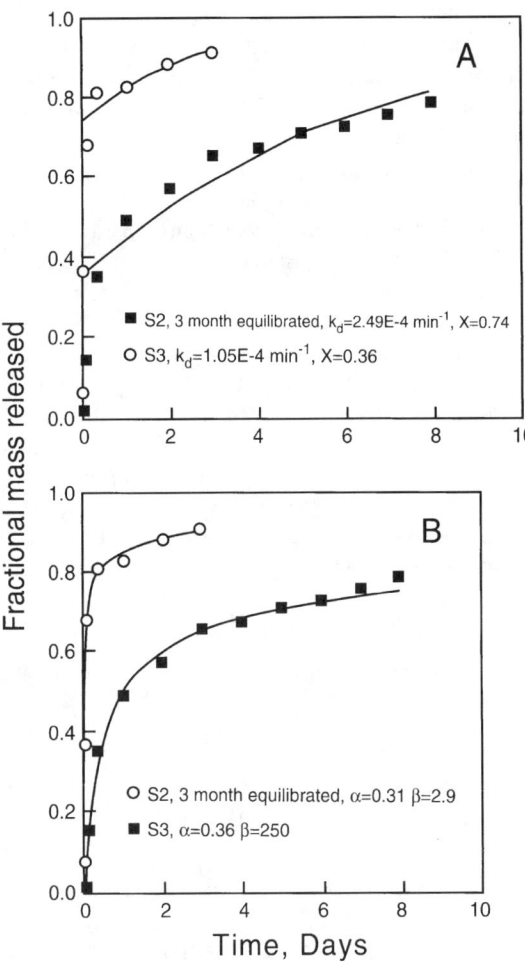

Fig. 4–1. (A) Fitted two-site model with release profiles of naphthalene from two soils, S2 and S3; R^2 = 0.88 and 0.91 for S2 and S3 regression fits, respectively, where X is the fraction of sites at instantaneous equilibrium and k_d is the desorption rate coefficient. (B) Mass fractional release of naphthalene from S2 and S3 soils with fitted multisite continuum compartment (Γ) model. (from Connaughton et al., 1993).

naphthalene from contaminated soils. The entire desorption process was described well with this model.

4–2.5 Physical Nonequilibrium Models

There are a number of physical models that can be used to describe nonequilibrium reactions. Since transport processes in the mobile phase are not usually rate-limiting, physical nonequilibrium models focus on diffusion in the immobile phase or intra-aggregate-diffusion processes (e.g., pore and/or surface diffusion). The transport between mobile and immobile regions is accounted for in physical nonequilibrium models in three ways (Brusseau & Rao, 1989): (i) explicitly with Fick's law to describe the physical mechanism of diffusive transfer; (ii) explicitly by using an empirical first-order mass-transfer expression to approximate solute transfer; and (iii) implicitly by using an effective or lumped dispersion coefficient that includes the effects of sink–source differences and hydrodynamic dispersion and axial diffusion.

A pore diffusion model (Table 4–1) has been used by a number of investigators to study sorption processes using batch systems (Wu & Gschwend, 1986; Steinberg et al., 1987; Ball & Roberts, 1991; Harmon et al., 1992; Pignatello et al., 1993). Wu and Gschwend (1986) successfully used the pore diffusion model to describe chlorobenzene congener sorption–desorption on soils and sediments. Figure 4–2 shows experimental and model fits for tetrachlorobenzene and pentachlorobenzene sorption on soils. The sole fitting parameter in this model is the effective diffusion coefficient (D_e). This parameter can be estimated a priori from chemical and colloidal properties; however, this estimation is only valid if the sorbent material has a narrow particle size distribution such that an accurate, average particle size can be defined. Moreover, in the pore diffusion model it is assumed that there is an average representative D_e, which means there is a continuum in properties across an entire pore size spectrum. This is not a valid assumption for micropores (<2.0 nm) since there are higher adsorption energies

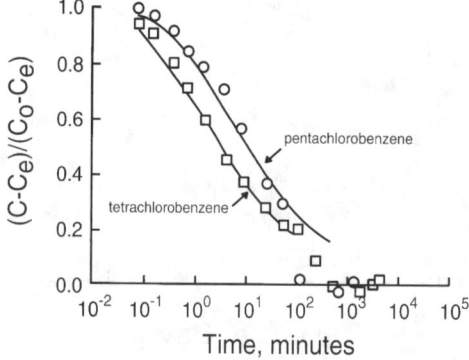

Fig. 4–2. Experimental and model-fitting results for pentachlorobenzene and tetrachlorobenzene sorption on Iowa soils where C is the dissolved concentration of organic chemical in the bulk solution, C_o is the initial concentration and C_e is the equilibrium concentration. The points represent experimental data and the solid lines represent fit of the data to the radial diffusion (pore diffusion) model (from Wu & Gschwend, 1986).

of sorbates in micropores, which causes increased sorption. The increased sorption causes reduced diffusive transport rates and nonlinear isotherms for sorbents with pores < several sorbate diameters in size. Other factors can cause reduced transport rates in micropores including steric hindrance that increases as the pore size approaches the solute size and greatly increased surface area to pore volume ratios, which occurs as pore size decreases. Another problem with the pore diffusion model is that sorption and desorption kinetics may have been measured across a narrow concentration range. This is a problem since a sorption–desorption mechanism in micropores at one concentration may be insignificant at another concentration.

Fuller et al. (1993) used a pore space diffusion model (Table 4–1) to describe arsenate adsorption on ferrihydrate that included a subset of sites whereby sorption was at equilibrium. A Freundlich model was used to describe sorption on these sites. Diffusion into the particle was described by Fick's second law of diffusion and homogeneous, spherical aggregates, and diffusion only in the aqueous phase were assumed.

Figure 4–3 shows the fit of the model when sorption at all sites was controlled by intra-aggregate diffusion. The fit was better when sites that had attained sorption equilibrium were included (Fig. 4–3). The latter model assumed that there was an initial rapid sorption on external surface sites before intra-aggregate diffusion.

Pedit and Miller (1995) have developed a general multiple-particle class pore diffusion model that accounts for differences in physical and sorptive properties for each particle class (Table 4–1). The model includes both instantaneous equilibrium sorption and time-dependent pore diffusion for each particle class. The pore diffusion portion of the model assumes that solute transfer between the intra-particle fluid and the solid phases is fast vis a' vis intra-particle pore diffusion processes.

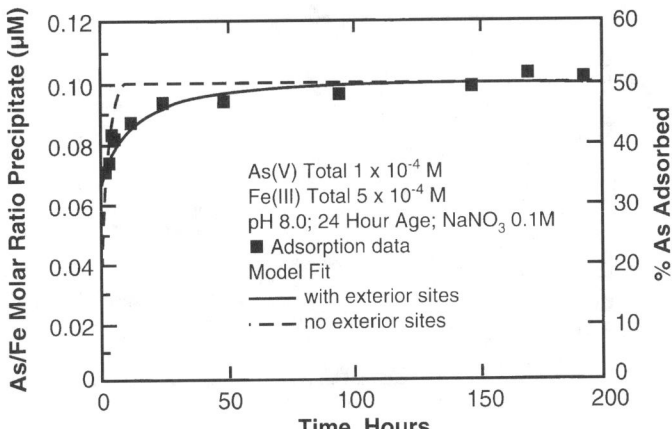

Fig. 4–3. Comparison of pore space diffusion model fits of As (V) sorption with experimental data (dashed curve represents sorption where all surface sites are diffusion-limited and the solid curve represents sorption on equilibrium sites plus diffusion-limited sites; from Fuller et al., 1993).

Surface diffusion models, assuming a constant surface diffusion coefficient, have been used by a number of researchers (Weber & Miller, 1988; Miller & Pedit, 1992). The dual resistance model (Table 4–1) combines both pore and surface diffusion.

4–3 RESIDENCE TIME EFFECTS ON REACTION RATES

While some soil chemical reactions are rapid, such as some ion exchange and adsorption reactions, it is well known that many reactions are quite slow. There also is good evidence that as the time of contact between the sorbate and sorbent (residence time) increases, the rate of many soil chemical reactions become increasingly retarded. The slow kinetics have been attributed to intraparticle and interparticle diffusion processes, sites of differing reactivity and surface precipitation phenomena; however, the mechanisms are not well understood and mechanistic conclusions have primarily been based on macroscopic and kinetic studies. To definitively understand the mechanism(s), in-situ spectroscopic and microscopic studies, along with kinetic investigations, are needed.

4–3.1 Metal Reactions

Ainsworth et al. (1994) studied the adsorption–desorption of Co^{2+}, Cd^{2+}, and Pb^{2b+} on hydrous ferric oxide (HFO) as a function of oxide aging and metal-oxide residence time (aging). Oxide aging did not cause hysteresis of metal cation sorption–desorption. Aging the oxide with the metal cations resulted in hysteresis with Cd^{2+} and Co^{2+} but little hysteresis was observed with Pb^{2+}. With Pb^{2+}, between pH 3 and 5.5 there was slight hysteresis during a 21-wk aging process (hysteresis varied from <2% difference between sorption and desorption to ≈10%). At pH 2.5 Pb desorption was complete within a 16 h desorption period

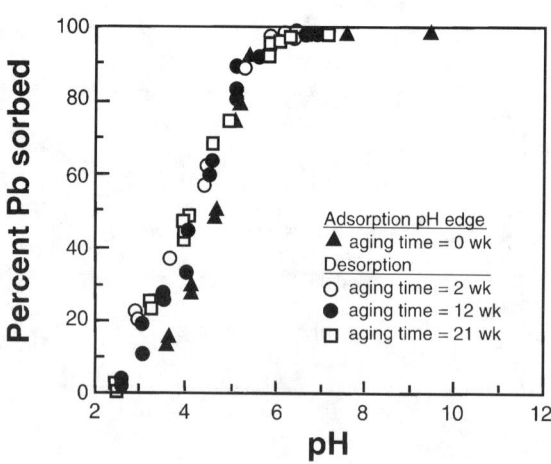

Fig. 4–4. Fractional sorption–desorption of Pb^{2+} to hydrous Fe-oxide (HFO) as a function of pH and HFO-Pb^{2+} aging time (from Ainsworth et al., 1994).

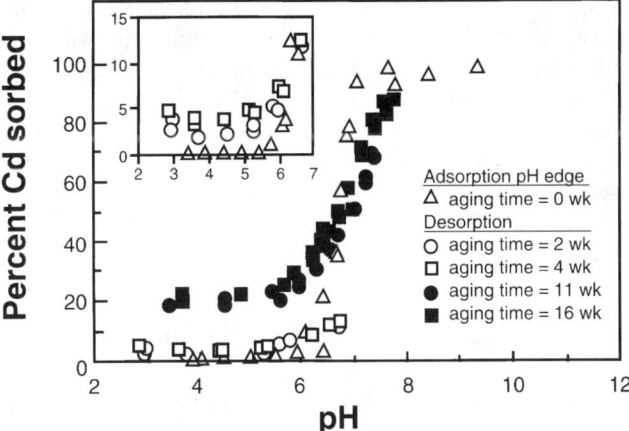

Fig. 4–5. Fractional sorption–desorption of Cd^{2+} to hydrous Fe-oxide (HFO) as a function of pH and HFO-Cd^{2+} aging time; insert shows adsorption–desorption of Cd^{2+} to HFO at 2- and 4-wk aging times (from Ainsworth et al., 1994).

and was not affected by aging time (Fig. 4–4); however, with Cd and Co, extensive hysteresis was observed during a 16 wk aging period and the hysteresis increased with aging time (Fig. 4–5 and 4–6). After 16 wk of aging 20% of the Cd and 53% of the Co was not desorbed, and even at pH 2.5, hysteresis was observed. Ainsworth et al. (1994) attributed the hysteresis to Co and Cd incorporation into a recrystallizing solid (probably goethite) via isomorphic substitution and not to micropore diffusion.

Bruemmer et al. (1988) studied Ni^{2+}, Zn^{2+}, and Cd^{2+} adsorption on goethite, and found at pH 6 that as reaction time increased from 2 h to 42 d (at

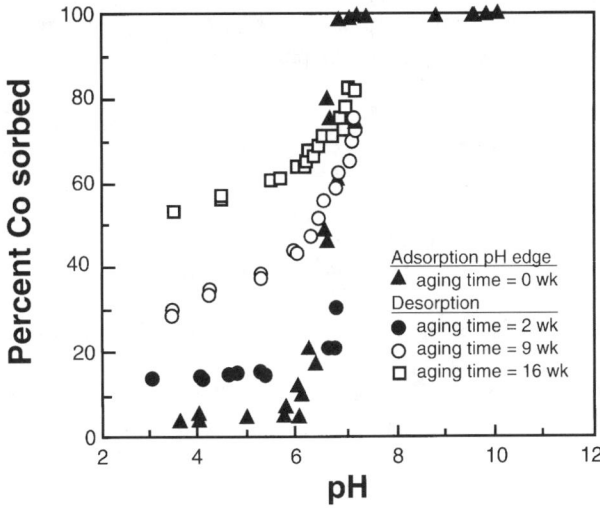

Fig. 4–6. Fractional adsorption of Co^{2+} to hydrous Fe-oxide (HFO) as a function of pH and HFO-Co^{2+} aging time (from Ainsworth et al., 1994).

293 K), adsorbed Ni^{2+} increased from 12 to 70% of total adsorption, and total increases in Zn^{2+} and Cd^{2+} adsorption over this time increased 33 and 21%, respectively. Bruemmer et al. (1988) attributed the slow kinetics to diffusion-controlled reactions on external and internal surface sites.

Backes et al. (1995) did not observe an increase in Co^{2+} or Cd^{2+} adsorption on Fe- and Mn-oxides as the sorption period increased from 12 to 15 wk; however, increased sorption time did result in a decrease in the rate and amount of desorption of Co^{2+} and Cd^{2+} from goethite, Co^{2+} from hausmennite and Cd^{2+} from crystomelane.

4–3.2 Organic Chemical Reactions

There have been a number of studies on the kinetics of organic chemical sorption–desorption with soils and soil components. Many of these investigations have shown that sorption–desorption is characterized by a rapid, reversible stage followed by a much slower, nonreversible stage (Karickhoff et al., 1979; DiToro & Horzempa, 1982; Pavlostathis & Mathavan, 1992) or biphasic kinetics. The rapid phase has been ascribed to retention of the organic chemical in a labile form that is easily desorbed; however, the much slower reaction phase involves the entrapment of the chemical in a nonlabile form that is difficult to desorb. This slower sorption–desorption reaction has been described to diffusion of the chemical into micropores of organic matter and inorganic soil components (Wu & Gschwend, 1986; Steinberg et al., 1987; Ball & Roberts, 1991). The labile form of the chemical is available for microbial attack while the nonlabile portion is resistant to biodegradation.

An example of the biphasic kinetics that is observed for many organic chemical reactions in soils–sediments is shown in Fig. 4–7. In this study 55% of the labile polychlorinated biphenyls (PCBs) were desorbed from sediments in a 24-h period, while little of the remaining 45% nonlabile fraction was desorbed in 170 h (Fig. 4–7a). During another 1-yr period about 50% of the remaining nonlabile fraction was desorbed (Fig. 4–7b).

In another study with volatile organic compounds (VOCS), Pavlostathis and Mathavan (1992) observed a biphasic desorption process for field soils contaminated with trichloroethylene (TCE), tetrachloroethylene (PCE), toluene (TOL), and xylene (XYL). A fast desorption reaction occurred in 24 h followed by a much slower desorption reaction beyond 24 h. In 24 h, 9–29%, 14–48%, 9–40%, and 4–37% of the TCE, PCE, TOL, and XYL, respectively, were released.

A number of studies have also shown that with aging the nonlabile portion of the organic chemical in the soil–sediment becomes more resistant to release (McCall & Agin, 1985; Steinberg et al., 1987; Pavlostathis & Mathavan, 1992; Scribner et al., 1992; Pignatello et al., 1993); however, Connaughton et al. (1993) did not observe the nonlabile fraction increasing with age for naphthalene contaminated soils.

One way to gauge the effect of time on organic contaminant retention in soils is to compare K_d (sorption distribution coefficient) values for freshly aged and aged soil samples. In most studies, K_d values are measured based on a 24 h equilibration between the soil and the organic chemical. When these values are compared to K_d values for field soils previously reacted with the organic chemical (aged samples) the latter have much higher K_d values, indicating that much more of the organic chemical is in a sorbed state. For example, Pignatello and Huang (1991) measured K_d values in freshly aged (K_d) and aged soils (K_{app}, apparent sorption distribution coefficient) reacted with atrazine (2-chloro-4-ethylamino-6-isopropylamino-1,2,5-triazine) and metolachlor [2-chloro-N-(2-thyl-6-methylphenyl)-N-(2-methoxy-1-methylethyl) acetamide], two widely used herbicides. The aged soils had been treated with the herbicides 15 to 62 mo before sampling. The K_{app} values ranged from 2.3 to 42 times higher than the K_d values (Table 4–2).

Scribner et al. (1992), studying simazine (a widely used triazine herbicide for broadleaf and grass control in crops) desorption and bioavailability in aged

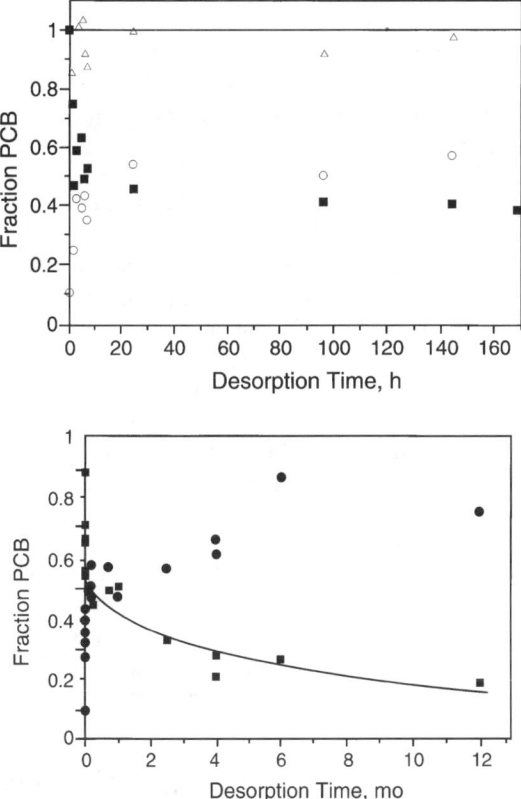

Fig. 4–7. (A) Short-term polychlorinated biphenyl (PCB) desorption in hours from Hudson River sediment contaminated with 25 mg kg^{-1} PCB. Distribution of the PCB between the sediment (■) and XAD-4 resin (○) is shown, as well as the overall mass balance (△). The resin acts as a sink to retain the PCB that is desorbed (from Carroll et al., 1994). (B) Long-term PCB desorption in months (mo) from Hudson River sediment contaminated with 25 mg kg^{-1} PCB. Distribution of the PCB between the sediment (■) and XAD-4 resin (○) is shown. The line represents a nonlinear regression of the data by a two-site model (from Carroll et al., 1994).

Table 4–2. Sorption distribution coefficients for herbicides in freshly aged and aged soils.†

Herbicide	Soil	K_d‡	K_{app}§
		L kg^{-1}	
Metolachlor	CVa	2.96	39
	CVb	1.46	27
	W1	1.28	49
	W2	0.77	33
Atrazine	CVa	2.17	28
	Cvb	1.32	29
	W3	1.75	4

† Adapted from Pignatello and Huang (1991); herbicides had been added to soils 31 mo prior to sampling for CVa and CVb soils, 15 mo for the W1 and W2 soils, and 62 mo for the W3 soil.
‡ Sorption distribution coefficient of freshly aged soil based on a 24-h equilibration period.
§ Apparent sorption distribution coefficient in contaminated soil (aged soil) determined using a 24-h equilibration period.

soils, found that K_{app} values were 15 times higher than K_d values. Scribner et al. (1992) also showed that 48% of the simazine added to the freshly aged soils was biodegradable during a 34-d incubation period while none of the simazine in the aged soil was biodegraded.

4–3.3 Implications of Aging Studies

One of the implications of the above studies on the effects of residence time on metal and organic chemical reactions on natural materials is that while many transport and degradation models for contaminants in soils and waters assume that the sorption process is an equilibrium process, the studies clearly show that kinetic reactions must be considered when making predictions about the mobility and fate of organic pollutants. Moreover, calculation of K_d values based on a 24-h equilibration period, which are commonly used in fate and risk assessment models, can be inaccurate since 24-h K_d values often overestimate the amount of contaminant in the solution phase.

The findings that many contaminants are quite persistent in the soil environment have both good and bad features. The beneficial aspect is that the contaminants are less mobile and may not be readily transported in groundwater supplies. The negative aspect is that their persistence and inaccessibility to microbes may make decontamination more difficult, particularly if in-situ remediation techniques such as biodegradation are employed.

4–4 USE OF MOLECULAR APPROACHES TO CONFIRM REACTION MECHANISMS

Kinetic studies, particularly if energies of activation are calculated and stopped-flow or interruption techniques are employed (Sparks, 1989), can reveal information about reaction mechanisms at the soil particle–solution interface; however, to definitively confirm the mechanism, molecular and/or atomic resolution surface techniques should be employed to corroborate the proposed mechanism hypothesized from kinetic studies. These techniques can be either used separately or preferably, simultaneously with the kinetic investigations. While the latter approach is preferable, only limited studies have been reported in the literature. Examples of both approaches will be cited in the following discussions.

There are a plethora of surface probe techniques—both spectroscopic and microscopic. These will not be reviewed in detail here. The reader should consult a number of monographs and textbooks for extensive details on these techniques (Hawthorne, 1988; Hochella & White, 1990; Charlet & Manceau, 1993; Fendorf et al., 1994; Schulze & Bertsch, 1995). Many of the surface techniques are non-in situ, meaning that the sorbent must be examined dry and/or under vacuum. The effects of these treatments on the surface chemistry of the sorbent and of the contrast between such a system and what would be found in the natural environment are not well understood.

Magnetic [nuclear magnetic resonance (NMR) and electron paramagnetic resonance (EPR)] and vibrational [Fourier transform infrared (FTIR) and Raman] spectroscopies can be used to study reactions in situ. While these methods provide atomic level information, they do not always provide precise information on the local structural environment of a species. X-ray absorption fine structure (XAFS) spectroscopy is an in situ technique that can provide direct information on the local structural environment, e.g., surface complexed and/or surface precipitated metals, and oxidation states of metals (Charlet & Manceau, 1993; Fendorf et al., 1994; Schulze & Bertsch, 1995; Sparks, 1995).

While XAFS provides local chemical information, it yields no spatial resolution information. Transmission electron microscopy can provide spatial resolution of surface changes and can be combined with electron spectroscopics to determine elemental analysis; however, electron microscopies require a vacuum environment that may cause an alteration in the solid sample. Also, bombardment of the sample with electrons could damage the sample. To avoid these problems, atomic force microscopy (AFM) or other surface probing microscopies (SPM) can be used.

4–4.1 Use of Kinetic and Spectroscopic Approaches

There are a few examples in the literature of studies where mechanisms of metal reactions on soil components have been hypothesized via kinetic experiments and verified in separate spectroscopic investigations (Fuller et al., 1993; Waychunas et al., 1993). An example of this approach can be found in the recent research of Fuller et al. (1993) and Waychunas et al. (1993), who studied the kinetics and mechanisms of As(V) sorption on ferrihydrite. Adsorption was investigated during coprecipitation, in which As(V) was present in solution during the hydrolysis and precipitation of Fe, and after coprecipitation (post-synthesis adsorption). In the post-synthesis adsorption studies, As(V) uptake was initially rapid and then slowly increased for up to 8 d. The rapid uptake was ascribed to adsorption on surface sites near the outside of aggregates, while the slower adsorption was attributed to diffusion of As(V) to adsorption sites on ferrihydrite surfaces within aggregates of colloidal particles. The latter were caused by coagulation and crystallite growth processes. These processes resulted in a decrease in the number of adsorption sites, and as aggregates formed, adsorption sites were buried in large clusters of the particles (Fuller et al., 1993). In the coprecipitation studies, initial As(V) uptake was much greater than observed for the post-synthesis adsorption studies, and the uptake rate was not diffusion-controlled as As(V) was coordinated by surface sites before crystallite growth.

The mechanistic hypotheses, based on the kinetic studies, were verified with companion XAFS studies (Waychunas et al., 1993). Analyses of the XAFS data provided no evidence for surface precipitation, one possible mechanism that has been proposed for slow metal sorption processes. Arsenate retention in both the coprecipitation and post-synthesis adsorption studies involved primarily an inner-sphere bidentate and monodentate binding on sites initially adsorbing arsenate. Waychunas et al. (1993) hypothesized that these defect sites probably adsorb As(V) as a bidentate complex first, and then sorb as a monodentate complex. Monodentate complexes accounted for about 30% of all As–Fe correlations and occurred at only low As loading levels. Recently, Manceau (1995) has questioned some of the As–Fe distances and pointed out the importance of edge linkages between Fe octahedra and arsenate tetrahedra.

Ideally, one would prefer to study reaction mechanisms by following reaction rates spectroscopically (time-resolved or real-time studies) using in-situ approaches. Such studies are scarce in the literature. The quick EXAFS (extended x-ray fine structure spectroscopy) technique, abbreviated QEXAFS, depends on a constant monochromator scan rate and fast array defectors to obtain a full EXAFS spectrum in a fraction of a second compared to tens of minutes in a traditional EXAFS method. Thus, ms or μs reactions could be spectroscopically monitored (Frahm, 1991; Lytle & Greegar, 1991; Dobson, 1994; Schulze & Bertsch, 1995). Such studies would be very useful for many soil particle–solution reactions that are rapid. Energy dispersive EXAFS, abbreviated DEXAFS, also can be used to determine a full EXAFS spectrum in a fraction of a second (Baker et al., 1991); however, detection can only be determined in the transmission mode. For slower reactions, it should be possible to follow the reaction mechanisms over time using standard XAFS techniques.

A recent example of time-resolved in-situ spectroscopic analyses is the research of Hunter and Bertsch (1994). They employed attenuated total reflectance Fourier transform infrared spectroscopy (ATR-FTIR) to quantitatively measure the degradation kinetics of tetraphenylboron (TPB) on clay minerals. The mechanisms of degradation were ascribed to surface-facilitated oxidation at Lewis acid and Bronsted acid sites. First-order models, based on these mechanisms, described the time-dependent data quite well.

4–4.2 Use of Kinetic and Microscopic Approaches

Scanning force microscopy (SFM) is being used increasingly as an in situ technique for imaging mineral surfaces immersed in aqueous solution and studying the kinetics of dissolution, precipitation, and heterogeneous nucleation reactions (Dove & Hochella, 1993; Gratz & Hillner, 1993; Bosbach & Rammensee, 1994; Junta & Hochella, 1994; Stipp et al., 1994; Maurice et al., 1995; Fendorf

Table 4–3. Studies on the kinetics of mineral reactions using scanning force microscopy (SFM).†

Dove et al., 1992	Calcite precipitation
Hellman et al., 1992	Albite dissolution
Hillner et al., 1992a,b	Calcite growth and dissolution
Johnsson et al., 1992	Muscovite dissolution
Dove & Hochella, 1993	Calcite precipitation mechanisms and inhibition by orthophosphate
Gratz & Hillner, 1993	Step dynamics and spiral growth on calcite
Bosbach & Rammensee, 1994	Gypsum growth and dissolution
Junta & Hochella, 1994	Mn(II) oxidation on hematite, goethite and albite
Stipp et al., 1994	Calcite surface structure
Maurice et al., 1995	Dissolution of hematite in organic acids
Fendorf et al., 1996	Precipitation kinetics of chromium hydroxide on goethite and silica

† Studies are listed in chronological order.

et al., 1996). SFM permits a direct measure of surface-controlled growth and dissolution rates by providing three-dimensional data on changes in microtopography. In situ SFM has the perhaps unique ability to detect different processes, such as dissolution and secondary phase formation, occurring simultaneously on a mineral surface (Maurice, 1998). Some of the studies in which SFM have been used to study the kinetics of mineral reactions are reported in Table 4–3.

Recently, Fendorf et al. (1996) studied the kinetics of Cr(III) reactions on single goethite and silica particles using a flow-cell mounted in a SFM. This procedure enabled one to study the reactions in an aqueous environment (in situ) and to react the surface while imaging (real-time measurements). Figure 4–8 shows an image of the unreacted silica in an aqueous environment. The surface is mostly flat and smooth with no island outcroppings. One hour after a 1 mM Cr(III) solution at pH 6 was introduced into the flow cell one sees that the Cr(III) solution caused a dramatic change in the surface morphology of silica (Fig. 4–8). Surface clusters have formed on the surface and within 2 h (figure not shown) the clusters have expanded in width and girth. The precipitates form as discrete surface clusters on the silica surface rather than distributing across the surface (Fendorf et al., 1996).

4–5 CONCLUSIONS AND FUTURE RESEARCH NEEDS

Research on the kinetics and mechanisms of reactions at the soil particle-solution interface will be a common theme in soil and environmental sciences for decades to come. This research emphasis is in large part due to the recognition that reactions controlling the fate and transport of contaminants in the subsurface environment are time-dependent. To accurately predict contaminant behavior in soils over time, one must understand the kinetics of the reactions. For further advancement to occur in this area the following research is needed: (i) more accurate kinetic models that describe reactions on multireactive, heterogeneous particle surfaces; (ii) long-term sorption–desorption rate studies; (iii) a better understanding of residence time effects on nutrient, radionuclide, metal, and organic retention–release mechanisms on soils and other natural materials, (iv) an increased knowledge of nucleation–precipitation and dissolution reaction rates at the mineral–water interface and their effect on nutrient–contaminant mobility in the soil environment; (v) increased studies on the kinetics and mechanisms of redox processes in soils, particularly the role that soil components such as Mn-oxides have on oxidation-reduction of inorganic and organic pollutants; and (vi) increased use of time-resolved, in situ spectroscopic and microscopic techniques to confirm reaction mechanisms.

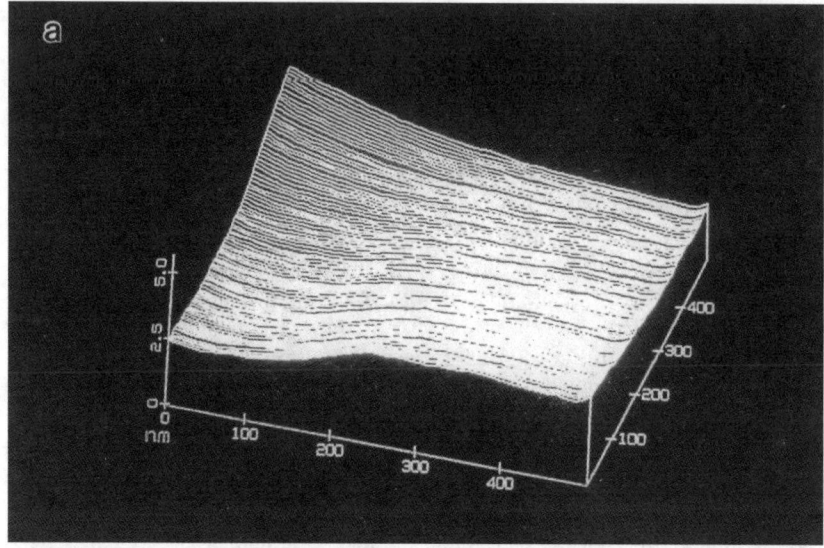

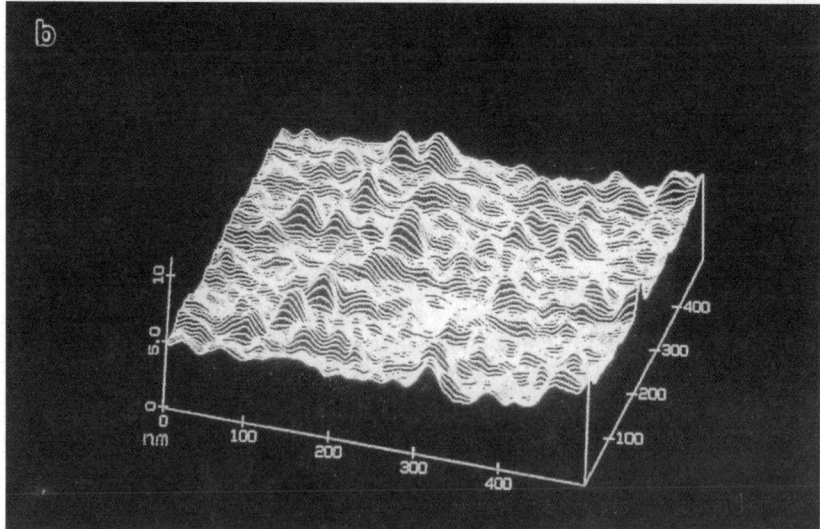

Fig. 4–8. Using a flow cell a single-particle of silica was imaged in an aqueous environment. The unreacted silica (a) is relatively flat and smooth across the 500 × 500 nm scan region; no pronounced outcroppings from the surface are observed. After reacting with 1 mM Cr(III) for 1 h at pH 6.0 (b) a different surface morphology is apparent: distinct surface clusters have formed that protrude away from the silica surface. These clusters continue to expand away from the surface and fuse together after 2 h (not shown) of reacting with Cr(III) (from Fendorf et al., 1996).

REFERENCES

Aharoni, C. 1984. Kinetics of adsorption: The S-shaped $Z(t)$ plot. Adsorpt. Sci. Technol. 1:1–29.

Aharoni, C., and D.L. Sparks. 1991. Kinetics of soil chemical reactions: A theoretical treatment. p. 1–18. *In* D.L. Sparks and D.L. Suarez (ed.) Rates of soil chemical processes. SSSA Spec. Publ. 27. SSSA, Madison, WI.

Aharoni, C., and Y. Suzin. 1982a. Application of the Elovich equation to the kinetics of occlusion: 1. Homogenous microporosity. J. Chem. Soc., Faraday Trans. 178: 2313–2320.

Aharoni, C. and Y. Suzin. 1982b. Application of the Elovich equation to the kinetics of occlusion: 3. Heterogenous microporosity. J. Chem. Soc., Faraday Trans. 178: 2329–2336.

Aharoni, C., and M. Ungarish. 1976. Kinetics of activated chemisorption. 1. The non-Elovichian part of the isotherm. J. Chem. Soc., Faraday Trans. 172:400–408.

Ainsworth, C.C., J.L. Pilou, P.L. Gassman, and W.G. Van Der Sluys. 1994. Cobalt, cadmium, and lead sorption to hydrous iron oxide: Residence time effect. Soil Sci. Soc. Am. J. 58:1615–1623.

Backes, C.A., R.G. McLaren, A.W. Rate, and R.S. Swift. 1995. Kinetics of cadmium and cobalt desorption from iron and manganese oxides. Soil Sci. Soc. Am. J. 59:778–785.

Baker, G., A.J. Dent, G. Derbyshire, G.N. Greaves, C.R.A. Catlow, J.W. Couves, and J.M. Thomas. 1991. Time resolved structural studies of nickel exchanged zeolite and nickel oxide using energy dispersive EXAFS. p. 738–741. *In* S.S. Hasnain (ed.) X-ray absorption fine structure. Ellis Harwood, New York.

Ball, W.P., and P.V. Roberts. 1991. Long-term sorption of halogenated organic chemicals by aquifer material: 1. Equilibrium. Environ. Sci. Technol. 25(7):1223–1236.

Bernasconi, C.F. 1976. Relaxation kinetics. Academic Press, New York.

Bosbach, D., and W. Rammensee. 1994. In situ investigation of growth and dissolution on the (010) surface of gypsum by scanning force microscopy. Geochim. Cosmochim. Acta. 58:843–849.

Bruemmer, G.W., J. Gerth, and K.G. Tiller. 1988. Reaction kinetics of the adsorption and desorption of nickel, zinc and cadmium by goethite: I. Adsorption and diffusion of metals. J. Soil Sci. 39:37–52.

Brusseau, M.L., and P.S.C. Rao. 1989. Sorption nonideality during organic contaminant transport in porous media. CRC Crit. Rev. Environ. Control 19:33–99.

Carroll, K.M., M.R. Harkness, A.A. Bracco, and R.B. Balcarel. 1994. Application of a permeant/polymer diffusion model to the desorption of polychlorinated biphenyls from Hudson River sediments. Environ. Sci. Technol. 28:253–258.

Charlet, L., and A. Manceau. 1993. Structure, formation, and reactivity of hydrous oxide particles: Insights from x-ray absorption spectroscopy. p. 117–164. *In* J. Buffle and H.P. van Leeuwen (ed.) Environmental particles. Lewis Publ., Boca Raton, FL.

Chien, S.H., and W.R. Clayton. 1980. Application of Elovich equation to the kinetics of phosphate release and sorption in soils. Soil Sci. Soc. Am. J. 44:265–268.

Connaughton, D.F., J.R. Stedinger, L.W. Lion, and M.L. Shuler. 1993. Description of time-varying desorption kinetics: Release of naphthalene from contaminated soils. Environ. Sci. Technol. 27:2397–2403.

DiToro, D.M., and L.M. Horzempa. 1982. Reversible and resistant components of PCB adsorption–desorption: Isotherms. Environ. Sci. Technol. 16:594–602.

Dobson, B.R. 1994. Quick scanning EXAFS facilities at Daresbury. SRS Synchrotron Radiat. News 7(1):21–24.

Dove, P.M., and M.F. Hochella, Jr. 1993. Calcite precipitation mechanisms and inhibition by orthophosphate: In situ observations by scanning force microscopy. Geochim. Cosmochim. Acta. 57:705–714.

Dove, P.M., M.F. Hochella, Jr., and R.J. Reeder. 1992. In situ investigation of near-equilibrium calcite precipitation by atomic force microscopy. p. 141–144. *In* Y.K. Kharaka and A.S. Maest (ed.) Water–rock interaction VII. A.A. Balkema, Rotterdam.

Fendorf, S.E., G. Li, and M.E. Gunter. 1996. Micromorphologies and stabilities of chromium (III) surface precipitates elucidated by scanning force microscopy. Soil Sci. Soc. Am. J. 60:99–106.

Fendorf, S.E., D.L. Sparks, G.M. Lamble, and M.J. Kelley. 1994. Applications of x-ray absorption fine structure spectroscopy to soils. Soil Sci. Soc. Am. J. 58:1583–1595.

Frahm, R. 1991. Quick XAFS: Potentials and practical applications in materials science. p. 731–737. *In* S.S. Hasnain (ed.) X-ray absorption fine structure. Ellis Harwood, New York.

Fuller, C.C., J.A. Davis, and G.A. Waychunas. 1993. Surface chemistry of ferrihydride: 2. Kinetics of arsenate adsorption and coprecipitation. Geochim Cosmochim. Acta 57:2271–2282.

Gratz, A.J., and P.E. Hillner. 1993. Poisoning of calcite growth viewed in the atomic force microscope (AFM). J. Cryst. Growth 129:789–793.

Hamaker, J.W., and J.M. Thompson. 1972. Adsorption. p. 39–151. *In* C.A.I. Goring and J.W. Hamaker (ed.) Organic chemicals in the environment. Marcel Dekker, New York.

Harmon, T.C., L. Semprini, and P.V. Roberts. 1992. Simulating solute transport using laboratory-based sorption parameters. J. Environ. Eng. 118:666–689.

Hawthorne, F.C. (ed.) 1988. Spectroscopic methods in mineralogy and geology. Rev. Mineral. Vol. 18. Mineral. Soc. Am., Washington, DC.

Hellman, R., B. Drake, and K. Kjoller. 1992. Using atomic force microscopy to study the structure, topography and dissolution of albite surfaces. p. 149–152. *In* Y.K. Kharaka and A.S. Maest (ed.) Water–rock interaction VII. A.A. Balkema, Rotterdam.

Hillner, P.E., A.J. Gratz, S. Manne, and P.K. Hansma. 1992a. Atomic-scale imaging of calcite growth and dissolution in real-time. Geology 20:359–362.

Hillner, P.E., S. Manne, A.J. Gratz, and P.K. Hansma. 1992b. AFM images of dissolution and growth on a calcite crystal. Ultramicroscopy 44:1387–1393.

Hochella, M.F., Jr., and A.F. White (ed.) 1990. Mineral–water interface geochemistry. Rev. Mineral. Vol. 23. Ann Arbor Sci., Ann Arbor, MI.

Hunter, D.B., and P.M. Bertsch. 1994. *In situ* measurements of tetraphenylboron degradation kinetics on clay mineral surfaces by FTIR. Environ. Sci. Technol. 28:686–691.

Jardine, P.M., F.M. Dunnivant, H.M. Selim, and J.F. McCarthy. 1992. Comparison of models for describing the transport of dissolved organic carbon in aquifer columns. Soil Sci. Soc. Am. J. 56:393–401.

Jardine, P.M., J.C. Parker, and L.W. Zelazny. 1985. Kinetics and mechanisms of aluminum adsorption on kaolinite using a two-site nonequilibrium transport model. Soil Sci. Soc. Am. J. 49:867–873.

Johnsson, P.A., M.F. Hochella, Jr., G.A. Parks, A.E. Blum, and G. Sposito. 1992. Direct observation of muscovite basal-plane dissolution and secondary phase formation: An XPS, LEED, and SFM study. p. 159–162. *In* Y.K. Kharaka and A.S. Maest (ed.) Water–rock interaction VII. A.A. Balkema, Rotterdam.

Junta, J.L., and M.F. Hochella, Jr. 1994. Manganese (II) oxidation at mineral surfaces: A microscopic and spectroscopic study. Geochim. Cosmochim. Acta 58:4985–4999.

Karickhoff, S.W. 1980. Sorption kinetics of hydrophobic pollutants in natural sediments. p. 193–205. *In* R.A. Baker (ed.) Contaminants and sediments. 2. Ann Arbor Sci., Ann Arbor, MI.

Karickhoff, S.W., D.S. Brown, and T.A. Scott. 1979. Sorption of hydrophobic pollutants on natural sediments. Water Res. 13:241–248.

Karickhoff, S.W., and K.R. Morris. 1985. Sorption dynamics of hydrophobic pollutants in sediment suspensions. Environ. Toxicol. Chem. 4:469–479.

Lee, L.S., P.S.C. Rao, M.L. Brusseau, and R.A. Ogwada. 1988. Nonequilibrium sorption of organic contaminants during flow through columns of aquifer materials. Environ. Toxicol. Chem. 7:779–793.

Leenheer, J.A., and J.L. Ahlrichs. 1971. A kinetic and equilibrium study of the adsorption of carbaryl and parathion upon soil organic matter surfaces. Soil Sci. Soc. Am. Proc. 35:700–704.

Lytle, F.W., and R.B. Greeger. 1991. New developments in XAS experiments. p. 625–633. *In* S.S. Hasnain (ed.) X-ray absorption fine structure. Ellis Harwood, New York.

Manceau, A. 1995. The mechanism of anion adsorption on iron oxides: Evidence of the bonding of arsenate tetrahedra on free $Fe(O,OH)_6$ edge. Geochim. Cosmochim. Acta 59:3647–3653.

Maurice, P.A. 1998. Scanning probe microscopy of environmental surfaces. p. 109–153. *In* P.M. Huang et al. (ed.) Structure and surface reactions of soil particles. JohnWiley & Sons, New York.

Maurice, P.A., M.F. Hochella, Jr., G.A. Parks, G. Sposito, and U. Schwertmann. 1995. Evolution of hematite surface microtopography upon dissolution by simple organic acids. Clays Clay Miner. 43(1):29–38.

McCall, P.J., and G.L. Agin. 1985. Desorption kinetics of picloram as affected by residence time in the soil. Environ. Toxicol. Chem. 4:37–44.

Miller, C.T., and J. Pedit. 1992. Use of a reactive surface-diffusion model to describe apparent sorption–desorption hysteresis and abiotic degradation of lindane in a subsurface material. Environ. Sci. Technol. 26(7):1417–1427.

Nkedi-Kizza, P., J.W. Biggar, H.M. Selim, M. Th. van Genuchten, P.J. Wierenga, J.M. Davidson, and D.R. Nielsen. 1984. On the equivalence of two conceptual models for describing ion exchange during transport through an aggregated Oxisol. Water Resour. Res. 20:1123–1130.

Onken, A.B., and R.L. Matheson. 1982. Dissolution rate of EDTA-extractable phosphate from soils. Soil Sci. Soc. Am. J. 46:276–279.

Pavlostathis, S.G., and G.N. Mathavan. 1992. Desorption kinetics of selected volatile organic compounds from field contaminated soils. Environ. Sci. Technol. 26:532–538.

Pedit, J.A., and C.T. Miller. 1995. Heterogenous sorption processes in subsurface systems: 2. Diffusion modeling approaches. Environ. Sci. Technol. 29(7):1766–1772.

Pignatello, J.J., and L.Q. Huang. 1991. Sorptive reversibility of atrazine and metolachlor residues in field soil samples. J. Environ. Qual. 20:222–228.

Pignatello, J.J., F.J. Ferrandino, and L.Q. Huang. 1993. Elution of aged and freshly added herbicides from a soil. Environ. Sci. Technol. 27:1563–1571.

Schulze, D.G., and P.M. Bertsch. 1995. Synchrotron x-ray techniques in soil, plant, and environmental research. Adv. Agron. 55:1–66.

Scribner, S.L., T.R. Benzing, S. Sun, and S.A. Boyd. 1992. Desorption and bioavailability of aged simazine residues in soil from a continuous corn field. J. Environ. Qual. 21:115–120.

Sparks, D.L. 1989. Kinetics of soil chemical processes. Academic Press, San Diego, CA.

Sparks, D.L. 1991. Chemical kinetics and mass transfer processes in soils and soil constituents. p. 585–637. *In* J. Bear and M.Y. Corapcioglu (ed.) Transport processes in porous media. Kluwer Academic Publ., Dordrecht, the Netherlands.

Sparks, D.L. 1995. Environmental soil chemistry. Academic Press, San Diego, CA.

Sparks, D.L., and P.M. Jardine. 1984. Comparison of kinetic equations to describe K–Ca exchange in pure and in mixed systems. Soil Sci. 138:115–122.

Sparks, D.L., and D.L. Suarez (ed.). 1991. Rates of soil chemical processes. SSSA Spec. Publ. 27. SSSA, Madison, WI.

Steinberg, S.M., J.J. Pignatello, and B.L. Sawhney. 1987. Persistence of 1,2 dibromoethane in soils: Entrapment in intra particle micropores. Environ. Sci. Technol. 21:1201–1208.

Stipp, S.L.S., C.M. Eggleston, and B.S. Nielsen. 1994. Calcite surface structure observed at microtopographic and molecular scales with atomic force microscopy (AFM). Geochim. Cosmochim. Acta 58:3023–3033.

van Genuchten, M.Th., and R.J. Wagenet. 1989. Two-site/two-region models for pesticide transport and degradation: Theoretical development and analytical solutions. Soil Sci. Soc. Am. J. 53:1303–1310.

Waychunas, G.A., B.A. Rea, C.C. Fuller, and J.A. Davis. 1993. Surface chemistry of ferrihydrite: 1. EXAFS studies of the geometry of coprecipitated and adsorbed arsenate. Geochim. Cosmochim. Acta 57:2251–2269.

Weber, W.J., Jr., and C.T. Miller. 1988. Modeling the sorption of hydrophobic contaminants by aquifer materials: 1. Rates and equilibria. Water Res. 22:457–464.

Wu, S., and P.M. Gschwend. 1986. Sorption kinetics of hydrophobic organic compounds to natural sediments and soils. Environ. Sci. Technol. 20:717–725.

5 Elucidating Fundamental Mechanisms in Soil and Environmental Chemistry: The Role of Advanced Analytical, Spectroscopic, and Microscopic Methods

Paul M. Bertsch and Douglas B. Hunter

Advanced Analytical Center for Environmental Sciences
University of Georgia
Aiken, South Carolina

5–1 INTRODUCTION

Soil chemists have a rich history of using advanced analytical and spectroscopic methods for the characterization of soil constituents and for the examination of solute interactions with mineral surfaces and humic macromolecules. The term advanced is, of course, arbitrary as well as relative. Whereas x-ray diffractometers and atomic absorption (AA) spectrometers (Table 5–1) were once thought of as sophisticated instrumentation that, at the time, provided important insights into the nature of soil constituents and soil solution components, they now are quite routine and standard. In fact, the sophisticated replacements of the AA, i.e., the ICP and then ICP-MS spectrometer have become routine in most soil chemistry laboratories. Other analytical methods for chemical analyses, such as HPLC, IC, GC–GC-MS, and a host of the so-called hyphenated techniques also have become common, yet powerful tools that are used by soil chemists to provide unparalleled characterization data of soil and soil solution constituents. The detailed characterization provided by these analytical methods is leading to a more complete understanding of the complex processes that control the fate and behavior of nutrients and contaminants in soil systems.

It is, however, the explosion in the types and capabilities of spectroscopic, analytical, and imaging techniques over the past decade that has had the most significant impact on advancing our molecular level understanding of soil chemical processes. It is noteworthy that, based on the published abstracts (ASA, 1995), ≈25% of the papers being presented in the Division of Soil Chemistry during this national meeting use what we consider advanced spectroscopic or imaging tech-

---

Copyright © 1998. Soil Science Society of America, 677 S. Segoe Rd., Madison, WI 53711, USA. *Future Prospects for Soil Chemistry.* SSSA Special Publication no. 55.

Table 5–1. Spectroscopic, analytical, and imaging techniques and their acronyms.

AA	Atomic absorption
AES	Auger electron spectroscopy
AFM	Atomic force microscopy
ATR-IR	Attenuated total reflectance infrared spectroscopy
CFM	Chemical force microscopy
DAFS	Diffraction anomalous fine structure spectroscopy
EDX	Energy dispersive x-ray analysis
EELFS	Extended energy loss fine structure
EELS	Electron energy loss spectroscopy
EPR	Electron paramagnetic resonance spectroscopy
EPR-SF	Electron paramagnetic resonance-stopped flow spectroscopy
ESR	Electron spin (paramagnetic) resonance spectroscopy
EXAFS	Extended x-ray absorption fine structure spectroscopy
FT-Raman	Fourier transform raman spectroscopy
FTIR	Fourier transform infrared spectroscopy
GC–GC-MS	Gas chromatography–GC-mass spectrometry
GI-EXAFS	Grazing incidence-extended x-ray absorption fine structure spectroscopy
HPLC	High performance liquid chromatography
HR-TGA	High resolution thermogravimetric analysis
IC	Ion chromatography
ICP	Inductively coupled plasma emission
ICP-MS	Inductively coupled plasma emission-mass spectrometry
IR	Infrared spectroscopy
LFM	Lateral force microscopy
LSCM	Laser scanning confocal microscopy
MAS NMR	Magic angle spinning nuclear magnetic resonance spectroscopy
MFM	Magnetic force microscopy
MRI	Magnetic resonance imaging
NMR	Nuclear magnetic resonance spectroscopy
PAS	Photoacoustic spectroscopy
PIXE	Proton induced x-ray emission
QEXAFS	Quick extended x-ray absorption fine structure spectroscopy
ReflEXAFS	Reflection extended x-ray absorption fine structure spectroscopy
SEXAFS	Surface extended x-ray absorption fine structure spectroscopy
SFM	Scanning force microscopy
SIMS	Secondary ion mass spectroscopy (spectrometry)
SPINOE	Spin polarized-induced nuclear overhauser effect
SPM	Scanning probe microscopy
STM	Scanning tunneling microscopy
SXRF	Synchrotron x-ray fluorescence spectroscopy
TEM	Transmission electron microscopy
XAFS	X-ray absorption fine structure
XANES	X-ray absorption near edge structure spectroscopy
XAS	X-ray absorption spectroscopy
XPS	X-ray photoelectron spectroscopy

niques. This can be contrasted to only ≈6% of the papers presented in 1990 (ASA, 1990). Furthermore, the recent *Methods in Soil Analysis* (Sparks et al., 1996) contains chapters on a host of advanced analytical and spectroscopic techniques that were not considered in the second edition of this publication (Page et al., 1982). Combining a number of these advanced spectroscopic techniques with the power of the analytical capabilities mentioned previously will undoubtedly lead to unprecedented information that will help us unravel many of the complex mysteries that have hampered a more complete understanding of soil chemical processes.

Because of the rapid advances in electronics and computing power, most techniques, even many common ones, have been totally transformed in terms of overall capabilities. For example, the introduction of high resolution thermogravimetric analysis (HR–TGA) in recent years has greatly enhanced the capabilities of this traditional routine method (*vide infra*). Other newer techniques, such as scanning probe microscopy (SPM), continue to experience rapid development, with new innovations becoming available on a regular basis (Eggleston, 1994; Dove & Chermak, 1994). Thus, investigators should not let past experiences influence their opinion about the capabilities (or lack of) of a given technique until they investigate the most recent capabilities.

It is an intractable task indeed to attempt to provide a cogent discussion of advanced analytical, spectroscopic, and microscopic methods and their application to soil and environmental chemistry in a short presentation. Most of all the techniques could hardly be dealt with in individual chapters and many have entire books dedicated to their applications. It is not our intention to provide a general description of spectroscopic and microscopic techniques or their theoretical background. Many comprehensive reviews can be consulted to provide the appropriate background to many of the techniques that will be touched on herein. Rather, our focus will be on how, in light of current applications, we envision some of the advanced techniques will be applied to complex problems in soil and environmental problems over the next decade.

5–2 SPECTROSCOPIC METHODS AND THE ELECTROMAGNETIC SPECTRUM

Spectroscopy deals with the interaction of electromagnetic radiation with matter and the shear number of spectroscopic techniques is a function of both the large frequency or energy range of the electromagnetic radiation involved and the specific approach for probing the interaction across a given frequency range (Plate 5–1; Table 5–1). Thus, while techniques such as IR, NMR, and EXAFS cover diverse regions of the electromagnetic spectrum, MAS NMR, FTIR, SEXAFS, and ReflEXAFS are specific experimental approaches within a specific energy or frequency range. Many other techniques are very similar in principle but probe matter in different energy ranges. For example, EXAFS and EELFS are quite similar in principle, differing in the incident energy of the electromagnetic radiation, i.e., x-rays vs. electrons. Because of the explosion in novel approaches for excitation or detection using a given spectroscopic method, the number of acronyms describing these techniques has become overwhelming and sometimes confusing. It is easier to sort through this collage by focusing on the common aspects of the methods rather than the seemingly complex diversity.

5–3 ENVIRONMENTAL SAMPLES AND THE SPECTROMETER ENVIRONMENT

A very important consideration when applying advanced spectroscopic techniques to environmental samples is the appropriateness of the spectrometer environment for addressing the specific scientific question. Many spectroscopic tools traditionally employed by surface chemists for interrogating chemical species at interfaces are invasive methods that require the sample to be desiccated and placed in high or ultrahigh vacuum environments. These invasive methods include familiar techniques such as XPS, AES, EELS, SIMS, and EDX. While these high vacuum methods can be used effectively in combination with other techniques to characterize *stable* phases in soil systems, they must be employed cautiously when used to investigate the coordination or bonding environments or chemical forms (oxidation state and others) of nutrients or contaminants at mineral surfaces in an environmental context (Johnston et al., 1993). It is *unlikely* that coordination environments of nutrients or contaminants in desiccated samples placed in a high vacuum environment are relevant to conditions commonly encountered in soils and it is *likely* that these extreme conditions enhance changes in coordination environment and oxidation state and enhance the formation of secondary phases or coprecipitates. In many cases, even properties of surface functional or reporter groups (i.e., functional groups of the sorbent having diagnostic properties perturbed by interfacial reactions) can be expected to be influenced by the absence of solvent molecules (Johnston et al., 1993). Misapplication of data generated by the use of such invasive techniques to probe the coordination environment of chemical species at mineral surfaces has caused and continue to cause, credibility problems with colleagues who prefer an entirely macroscopic approach to soil chemistry. This is unfortunate as the skepticism created often leads to a much more difficult acceptance of new, robust, *in situ* , noninvasive spectroscopic methods when they are initially applied to problems in soil and environmental chemistry.

In situ or noninvasive spectroscopic techniques allow samples to be examined in a fully aquated form with little or no sample manipulation. Among the most common of these techniques are fluorescence, UV-Vis, vibrational (Raman, IR), NMR, ESR, and Mossbauer spectroscopies along with PAS and XAS. While these techniques can, in principle, be used to examine soil constituents in their natural state, samples are sometimes dried or manipulated in some manner that partially or totally negates the potential advantages over the noninvasive methods discussed previously. Furthermore, many of these techniques lack or lacked sufficient sensitivity to examine realistic concentrations of nutrients or contaminants associated with soil minerals. This has led to many studies that were or still are conducted at unrealistic concentrations or surface loadings. Although these studies can still provide useful information, the data are often over-interpreted or over extended in an environmental context. This also has resulted in credibility problems that inhibit the acceptance of studies using newer instruments or experimental apparati that provide sufficient sensitivity to examine constituents associated with soil minerals or humic components at low concentrations and, thus provide meaningful molecular level information.

5–4 FUNDAMENTAL MECHANISMS AND THE ISSUE OF SCALE

A major challenge in choosing suitable spectroscopic techniques to address fundamental questions in soil and environmental chemistry is the issue of scale

(Plate 5–2). Traditionally, advanced spectroscopic methods have been usually employed to examine atomic- and molecular-level interactions with surfaces of isolated monomineralic phases or other surrogates of soil minerals. While atomic and molecular scale interactions at mineral surfaces are important at regulating soil chemical processes, it is at the grain boundary and pore scales and, in some instances, even the landscape scale where these processes are manifested. Thus, not only is it important to investigate molecular-level phenomenon on isolated monomineralic phases but it is also important to probe pore and grain scales within soils at the molecular level for the purposes of generating a mechanistic understanding of critical soil chemical processes in complex heterogenous mineral assemblages.

In recent years, noninvasive microspectroscopic capabilities with sufficient sensitivity to examine soil constituents have become increasingly available. FT-Raman, FT-IR, laser-based fluorescence, and micro-XAS are being used to probe chemical heterogeneity on a scale ranging from less than one to several hundred micrometers (e.g., Guilment et al., 1994; Schaeberle et al., 1995; Kim et al., 1996; Schulze & Bertsch, 1995). These techniques pose tremendous potential to examine such micromorphological features as grain and ped faces, pore matrix boundaries, heterogeneous mosaics within soil thin sections, and regions within the rhizosphere. In addition to providing direct molecular information on the chemical speciation of nutrients and contaminants in these microregions, many of these techniques can follow dynamic processes related to chemical species transformations. Advanced three dimensional imaging capabilities using such techniques as x-rays, lasers, and NMR also have allowed for the examination of important processes occurring at the pore scale, including, in some instances, the tracking of dynamic processes. The ability to probe fundamental soil chemical processes across such a wide range of spatial scales will continue to make spectroscopic techniques critical tools for a larger number of researchers in soil and environmental chemistry.

5–5 APPLICATIONS OF ADVANCED ANALYTICAL, SPECTROSCOPIC, AND MICROSCOPIC TECHNIQUES TO PROBLEMS IN SOIL AND ENVIRONMENTAL CHEMISTRY

An example of an established technique that has been transformed by advances is thermogravimetric analysis (TGA). The introduction of the high-resolution capability (HR-TGA) is transforming this rather simple, routine method into a much more powerful analytical tool for probing structures and components of complex mineral and biological systems (Bertsch & Hunter, 1998, unpublished data). For example, studies in our laboratory have demonstrated that the cellulose and pectin type components of isolated cell walls can be easily resolved, as can the variations in pyrolysis temperature as a function of both genetic diversity and Al loading (Fig. 5–1). Results of these investigations coupled with FTIR and ^{27}Al NMR results suggest that Al may be interacting with nonionizable functional groups, contrary to previous beliefs (Hunter & Bertsch, 1998, unpublished data). Another HR-TGA investigation of goethites suggests that individual populations of surface OH groups, i.e., A-, C-, and B-type hydroxyls, can be identified and quantified by distinct dehydroxylation temperatures (Ford, 1996). As the technique can be easily interfaced with FTIR, whereby the off-gases are swept directly into an IR cell, there may be substantial potential of this technique to examine organic-mineral interactions. Although there have been few applications of HR-TGA in the soil chemistry literature to date it is a very simple and accessible technique that should find many applications in future studies.

There also has been significant progress made in the capabilities of more advanced spectroscopic techniques that have pushed the envelope on detection limits that were considered inherently unachievable a decade ago. A recent study using ^{27}Al NMR demonstrated the capability to provide information on the chemical speciation of Al in solutions in the 1 to 10 mM concentration range (Faust et al., 1995). Additional recent advances in NMR have greatly enhanced the sensitivity and capabilities of the technique and these promise to become integrated into soil and environmental chemistry research in the future. Notably, the availability of a near gigahertz magnetic fields and the recent advances in hyperpolarization of Xe via optical pumping methods have introduced unprecedented imaging capabilities that were, until recently, thought to be impossible by NMR or MRI techniques (Albert et al., 1994). Additionally, the recent demonstration that hyperpolarized Xe can transfer polarization to protons, a process referred to as the spin polarization-induced nuclear Overhauser effect (SPINOE), allows for the polarization of protons without the need for radiofrequency excitation and has

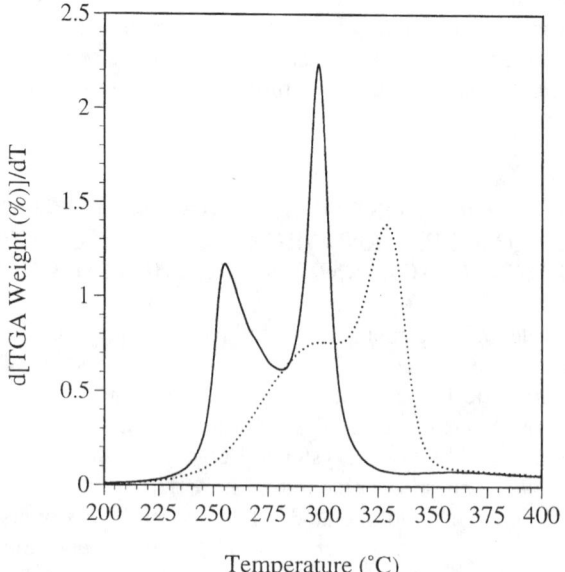

Fig. 5–1. First derivative conventional (. . . .) and HR-TGA (_____) patterns for isolated cell wall materials. The distinct peaks in the HR-TGA are related to the pectic type components (low temperature weight loss) and the cellulose (high temperature weight loss) fraction. The integrated areas of the peaks can be used to quantify the relative amounts of each present.

Future Prospects for Soil Chemistry

© 1998 SSSA

Future Prospects for Soil Chemistry

© 1998 SSSA

Chapter 5—Elucidating Fundamental Mechanisms in Soil and Environmental Chemistry: The Role of Advanced Analytical, Spectroscopic, and Microscopic Methods
Paul M. Bertsch and Douglas B. Hunter
Color Plates 1 through 4

Chapter 8—New Ideas on the Chemical Make-Up of Soil Humic and Fulvic Acids
Morris Schnitzer and Hans-Rolf Schulten
Color Plates 1 through 4

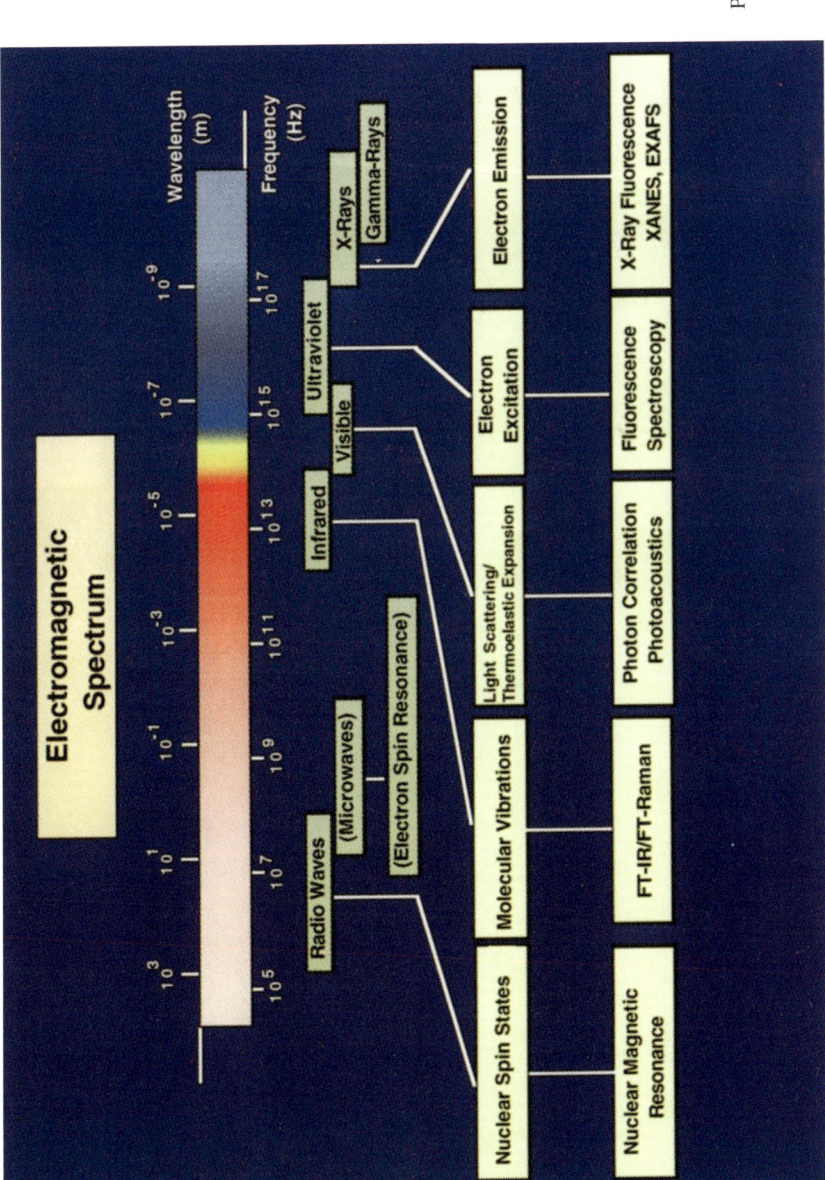

Plate 5–1. Electromagnetic spectrum and the approximate energy ranges of various spectroscopic techniques.

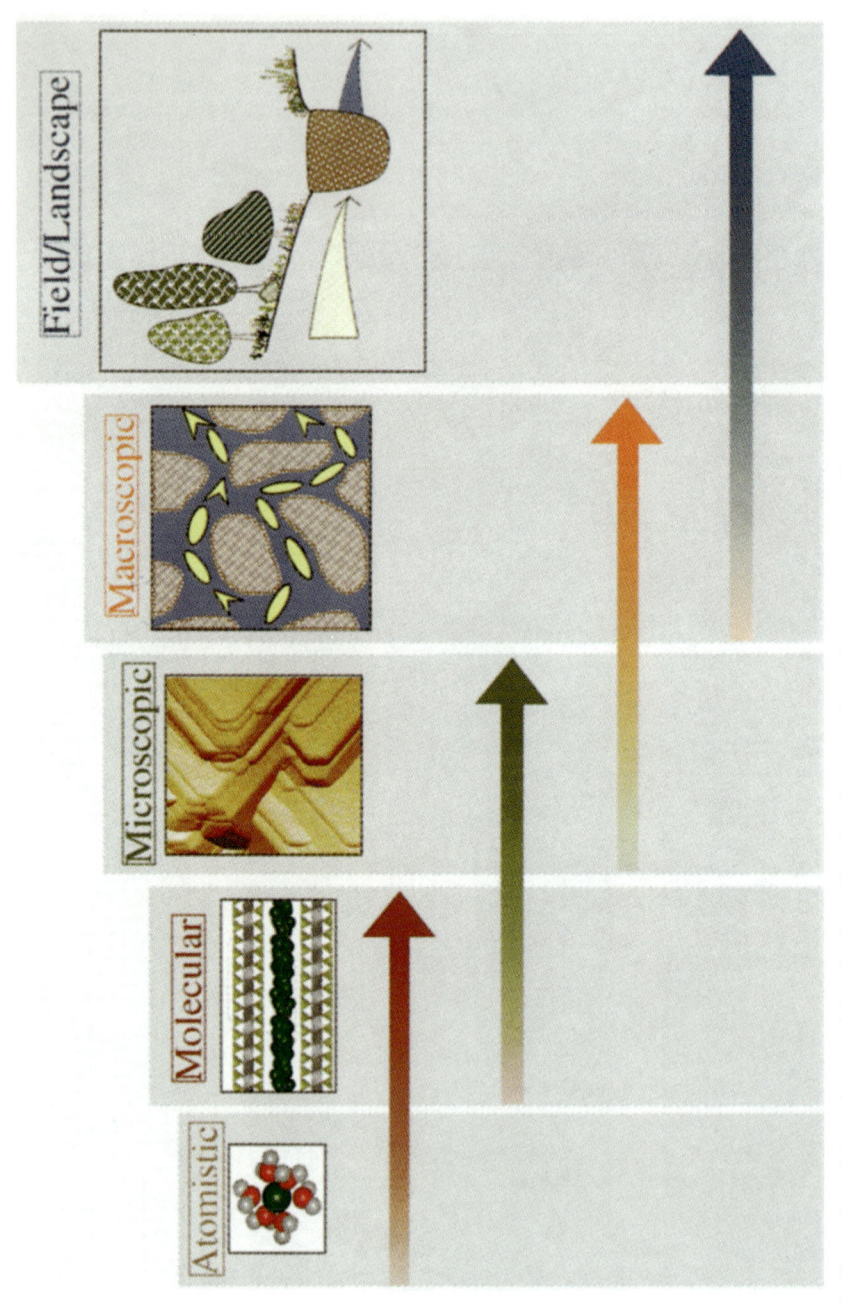

Plate 5–2. Illustration of the various spatial scales that soil and environmental chemists are concerned with.

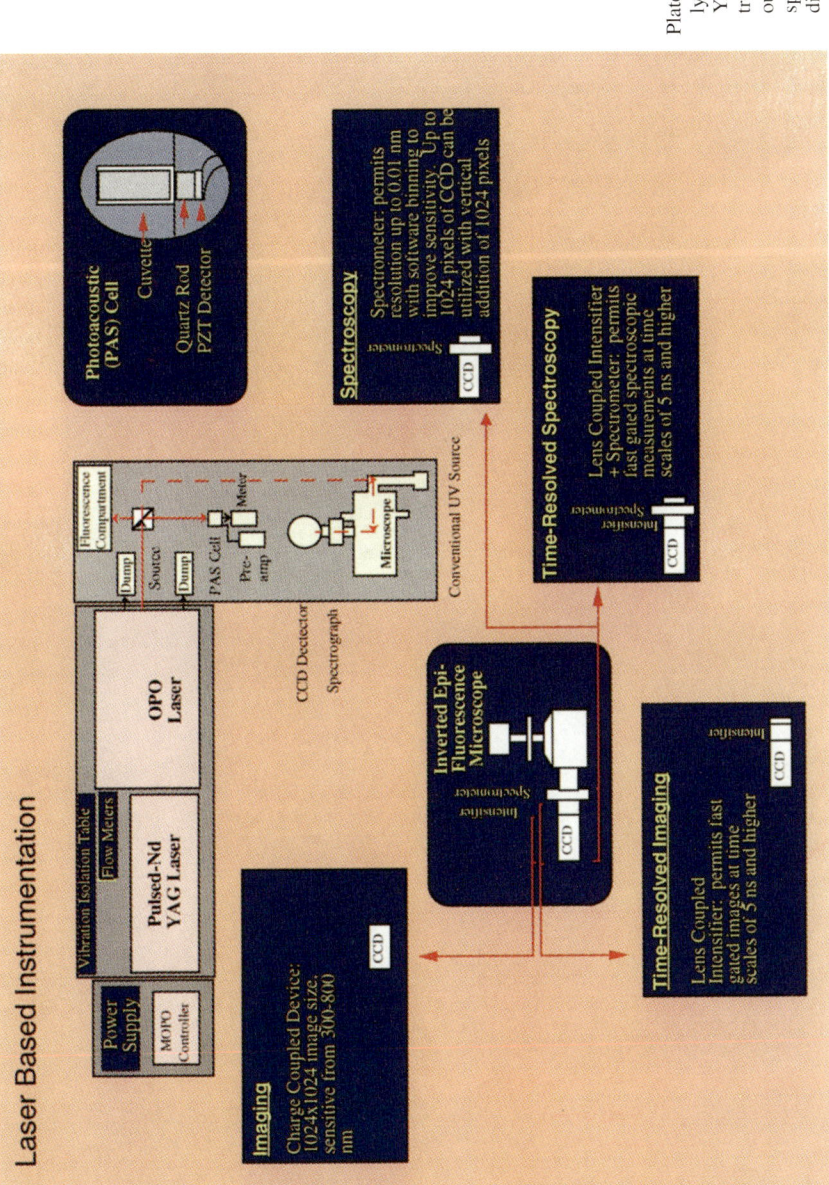

Plate 5–3. Illustration of a newly commissioned tunable Nd-YAG–OPO laser-based spectroscopy unit with the various spectroscopic and microspectroscopic capabilities displayed.

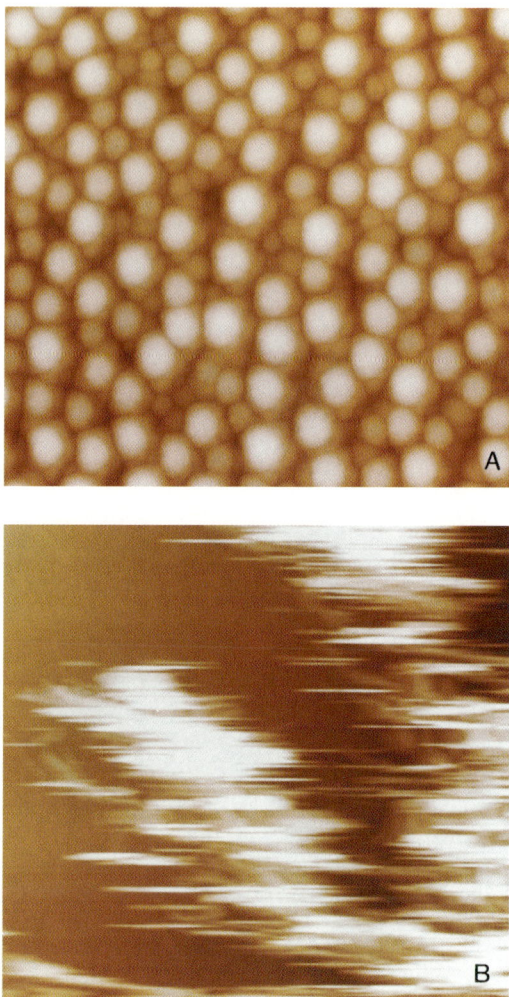

Plate 5–4. SFM images of monodisperse polystyrene beads (120 nm) (a) collected using the Fluid Tapping Mode (image 3 × 3, μm), (b) collected using conventional contact mode (image 3 × 3, μm), and (c) collected using the tapping mode following the generation of Image b (image 7 × 7, μm; Courtesy of Digital Instruments).

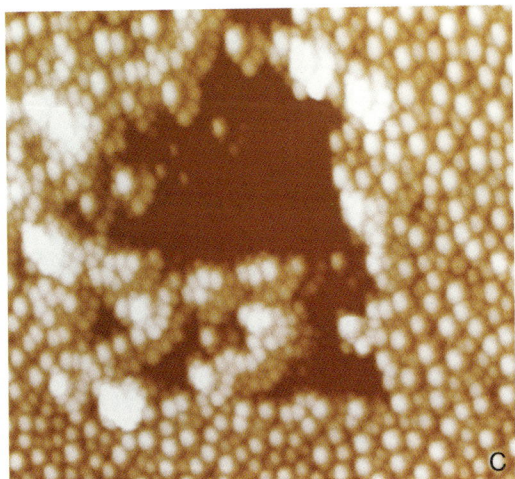

Plate 5–4. Continued.

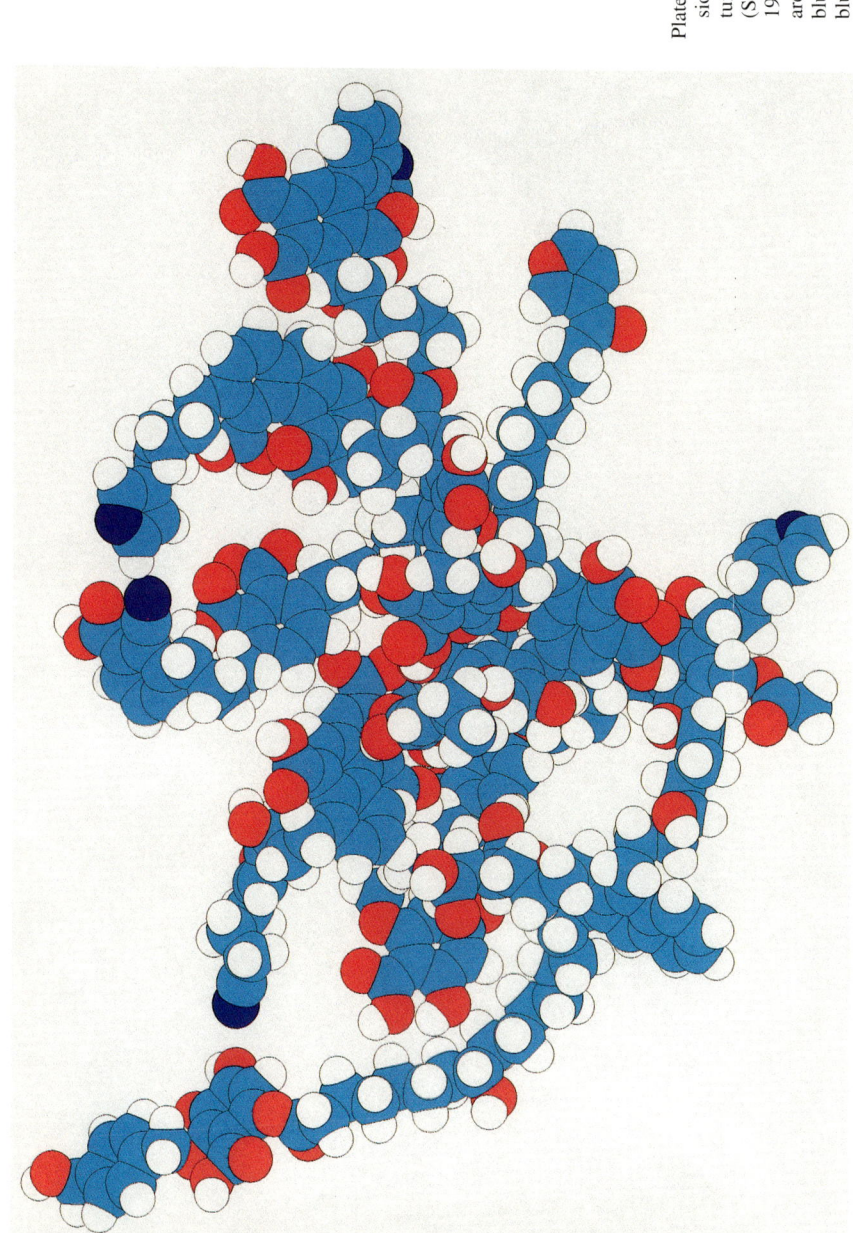

Plate 8–1. Three-dimensional chemical structure of humic acids (Schnitzer & Schulten, 1995). Element colors are: H (white); C (light blue); O (red); N (dark blue).

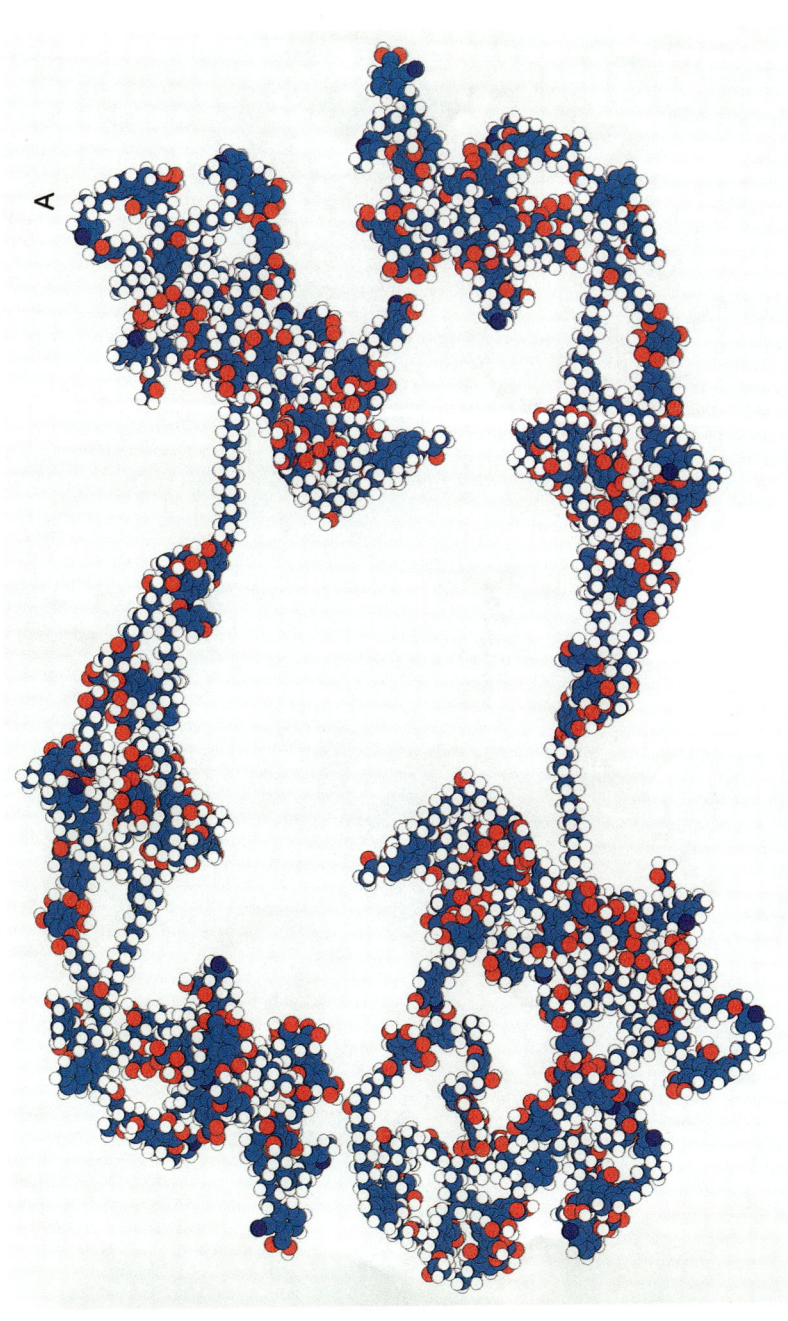

Plate 8–2. Three-dimensional structures of a humic acid colloid: (a) color plot of an open aggregate of nine humic acid subunits (6821 atoms; *Disks*). The element colors are: H (white), C (light blue), O (red), and N (dark blue); and (b) black–white display of the voids in the structural skeleton and links by two intermolecular H bonds (*Sticks*) (Schulten, 1997, unpublished data).

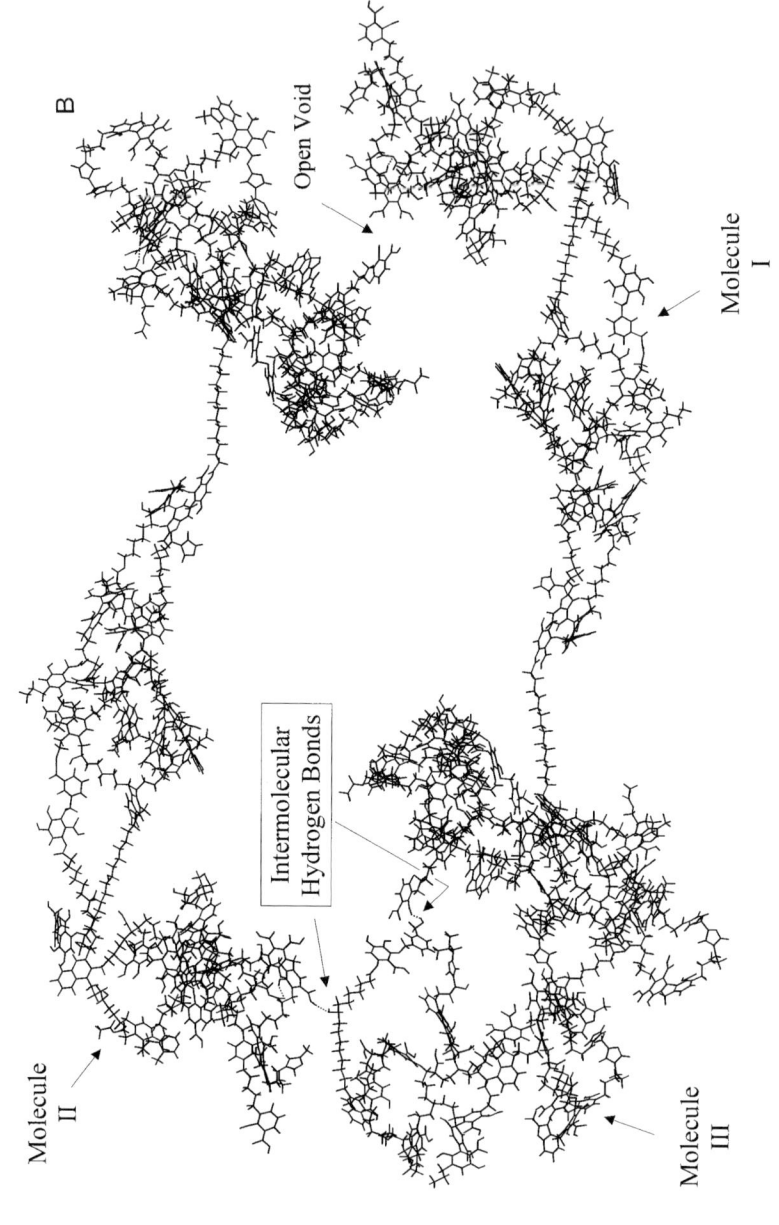

Plate 8–2. Continued.

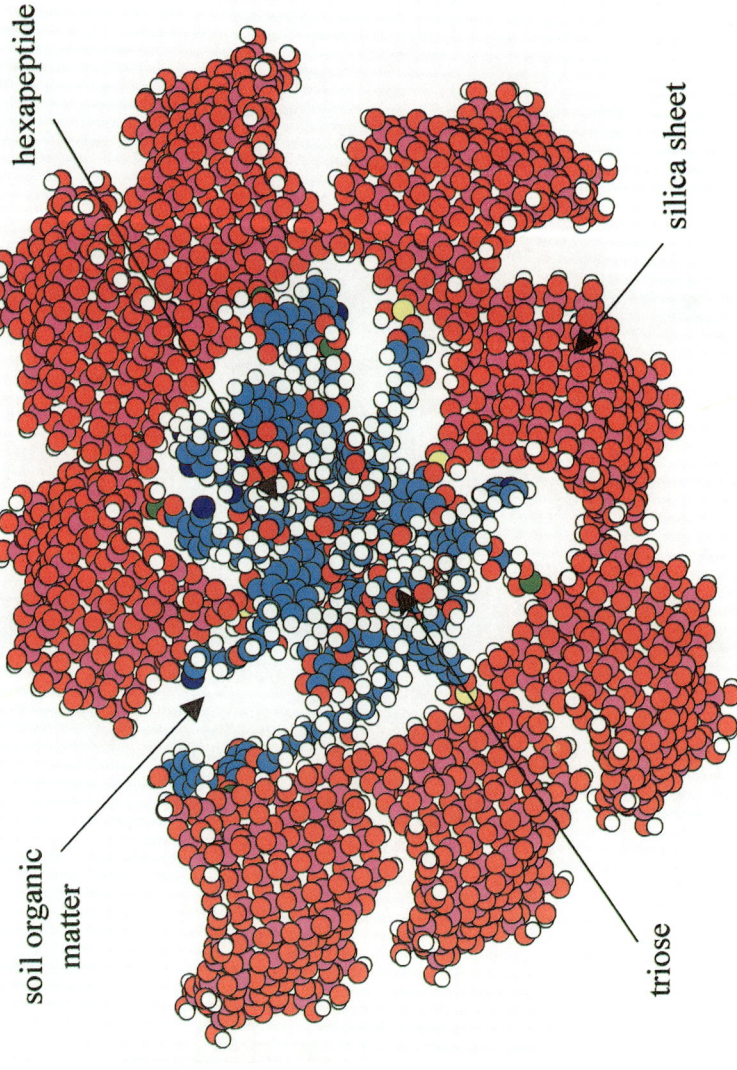

Plate 8–3. Three-dimensional structure of a model soil particle: Color plot of the soil surface consisting of the humic acids (cf. Plate 8–1), a triose, a hexapeptide, five Fe^{3+} and four Al^{3+} cations, and eight silica sheets. The humic–mineral complex (3196 atoms) is bound to the silica particles by four trivalent Fe bridges and four Al^{3+} bridges. The fifth Fe cation is still available for mineral surface attachment and bonding. For element colors see Plates 8–1 and 8–2, plus Si (violet), Al (yellow), and Fe (green) (Schulten, 1997, unpublished data).

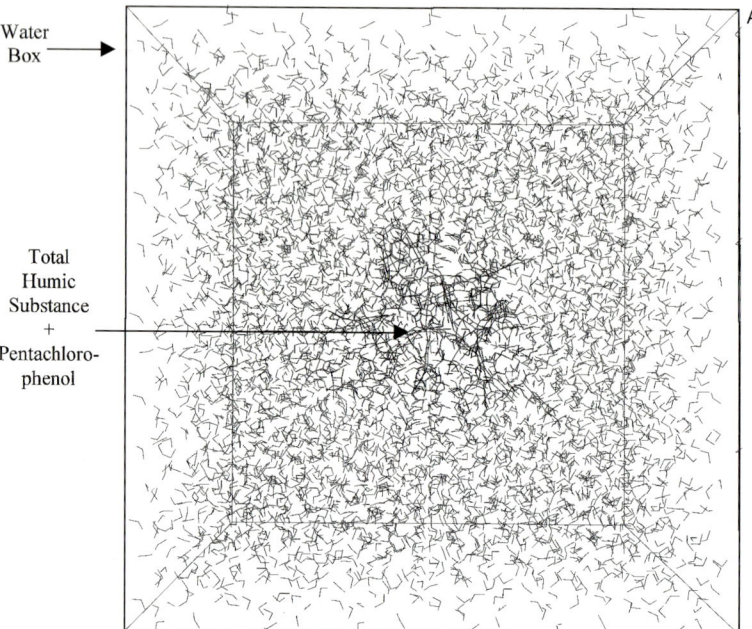
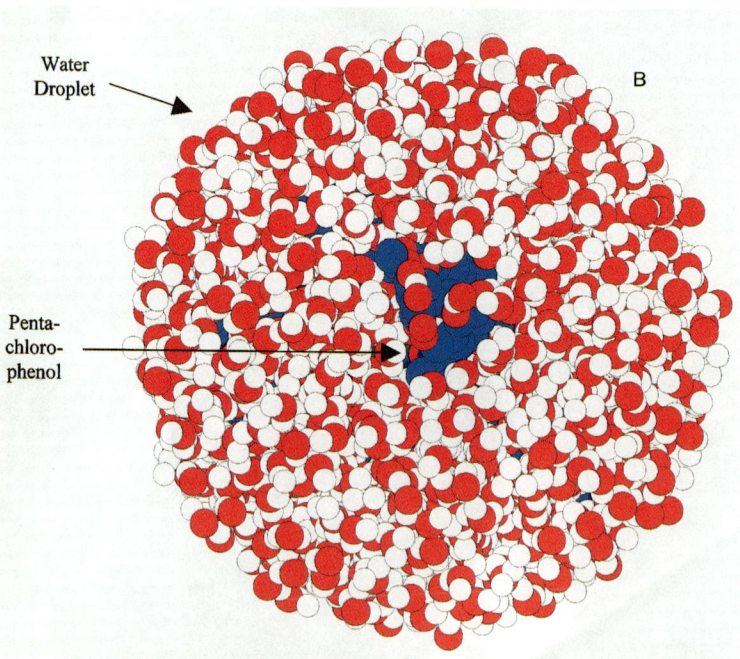

Plate 8–4. Three-dimensional structures of a humic pentachlorophenol complex in water (17205 atoms): (a) water box of 5479 water molecules is formed with the dissolved complex; and (b) spheric droplet is cut out of the water box for closer investigations of the complex–water interactions. For element colors see Plate 8–2, with the exception that chlorine atoms are in green color (Schulten, 1997, unpublished data).

great potential for probing the local nature of surfaces and humic macromolecules (Navon et al., 1996).

A novel application of EPR to investigate the kinetics of Mn(II) sorption and desorption on the mineral birnessite using a stopped flow apparatus has recently been reported (Fendorf et al., 1993). This study demonstrated the ability to probe sorption reactions in colloidal suspensions on a ms time scale at mM concentrations and the utility of EPR-SF for providing direct measurements of surface reactions involving EPR-active ions. Other applications of this powerful technique to investigate sorption–desorption of EPR-active metals and organics will undoubtedly be integrated into future investigations.

Other investigations using modified ATR-IR methods have demonstrated the capability to examine the interactions and complexation environments of organic solutes at the mineral-water interface by FTIR (Bibler & Stumm, 1994; Hug & Sulzberger, 1994; Hunter & Bertsch, 1994; Sun & Doner, 1996). The study by Hug and Sulberger (1994) investigated oxalate surface complexes on TiO_2 as a function of solution parameters (ionic strength, pH, and others), where several surface complexes were observable. The investigation by Hunter and Bertsch (1994) examined the surface facilitated degradation of tetraphenylboron (TPB) on clay minerals, where the kinetics of TPB degradation to diphenylboron, triphenylboron, and biphenyl was followed in real time (Fig. 5–2). This study demonstrated that high quality kinetics data could be collected using in situ FTIR spectroscopy at environmentally relevant concentrations. Both studies indicated that subtraction of background absorption in aqueous suspensions using conventional ATR cells was unreliable and, thus demonstrate the need for innovations specifically designed for examining environmentally relevant systems.

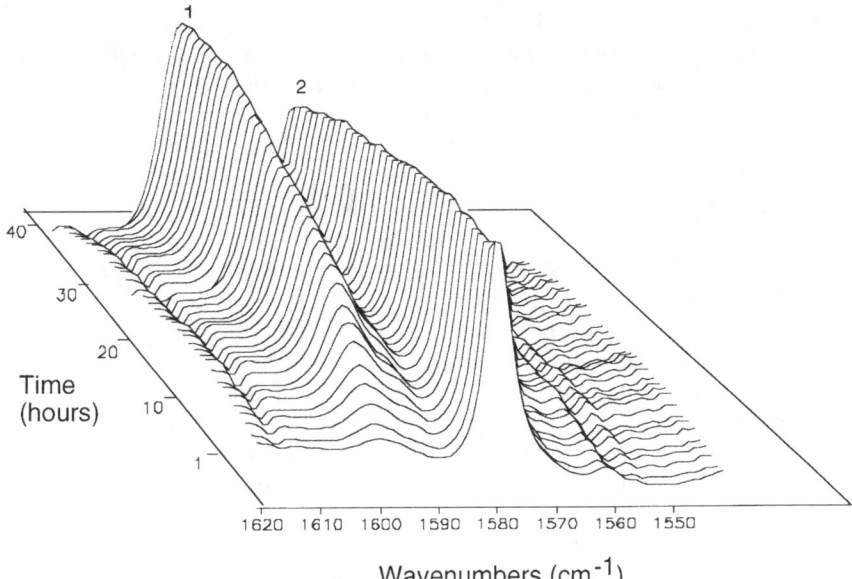

Fig. 5–2. Real-time FTIR spectra showing the degradation of tetraphenylboron (2) to diphenylboric (1) acid and biphenyl on bentonite surfaces (from Hunter & Bertsch, 1994).

Another technique that has been greatly enhanced in the past several years is fluorescence spectroscopy and microspectroscopy. The introduction of extremely low dark current charge coupled device (CCD) detectors and rapid image processing capabilities have enhanced the sensitivity of these techniques to single or near single molecule detection (Moerner, 1994; Kohler et al., 1995). We have recently commissioned a fluorescence microscopy–spectroscopy system that interfaces to a novel Nd-YAG-optical parametric oscillator (OPO) laser system that provides tunable wavelength laser irradiation from 220 to 2000 nm (Plate 5–3). The system is capable of collecting synchronous spectra and pulse-gated measurements of fluorescence lifetimes that can be gated to 5 ns. The system was originally configured to be used as a photoacoustic spectrometer, but the addition of an inverted epifluorescence microscope with a two way imaging light path and a high sensitivity liquid N_2-cooled CCD detector has greatly expanded its capabilities (Plate 5–3). As currently configured, the system can be used as a fluorescence–luminescence spectrometer or, when used in conjunction with the epifluorescence scope, as a microspectrometer providing both fluorescence and luminescence imaging and spectroscopy with near single molecule detection under ideal conditions. Thus, the system can be used to probe the local environment of surface sorbed constituents, or be used to examine processes at the pore and grain boundary scales. The system is currently being used to probe microscale heterogeneities in the distributions of organic and inorganic fluorophores on thin sections of environmental samples and the initial data collected provides evidence for the potential of the technique (Fig. 5–3).

Other advances in optical and electronic technology have allowed laser scanning confocal microscopy (LSCM) to become a powerful technique that has recently been applied to the imaging of pore structure of geomaterials (Fredrich et al., 1995). The reconstruction of three- dimensional images have demonstrated the ability to examine the geometric complexity of pore structure within quartz sandstone at a resolution of 200 nm. The potential of combining such outstanding image resolution with a range of microspectroscopic techniques indicates that tremendous progress in probing soil chemical reactions at grain and pore boundaries will be made in the coming years.

Arguably, it has been the introduction and application of scanning probe microscopy and synchrotron-based x-ray techniques over the past decade that promises to have the greatest impact on advancing research in soil and environmental chemistry. Scanning probe microscopy (SPM) represents an ever increasing class of microscopic techniques that provide high resolution (atomic or near atomic scale) multidimensional images of solid surfaces by monitoring the interactions between sharp tips and the surface. The class of SPM is based on the specific nature of the tip-surface interaction ranging from tip-sample interactions represented by electrical currents, as in the classic scanning tunneling microscope (STM), to the specific forces between the tip and the sample as in atomic force microscopy (AFM; Eggleston, 1994). Other subsets of SPM include lateral force microscopy (LFM), magnetic force microscopy (MFM), and chemical force microscopy (CFM), where tips are chemically modified to interrogate specific chemical features associated with a surface. This latter area is one that is currently experiencing significant advances. Although there exists some confusion in the literature concerning the terminology, AFM, LFM, MFM, and CFM all can be collectively grouped into the scanning force microscopy (SFM) umbrella. Since many studies actually probe surfaces at a scale that includes groups of atoms or individual molecules rather than at the atomic-scale as in STM, SFM is becoming the generally accepted term. The distinct advantage of SFM compared with SEM and TEM is the relatively low cost and that the samples do not have to be coated and subjected to ultrahigh vacuum conditions, thus providing a relatively easy yet extremely powerful technique for interrogating environmentally relevant samples at the atomic or molecular scale.

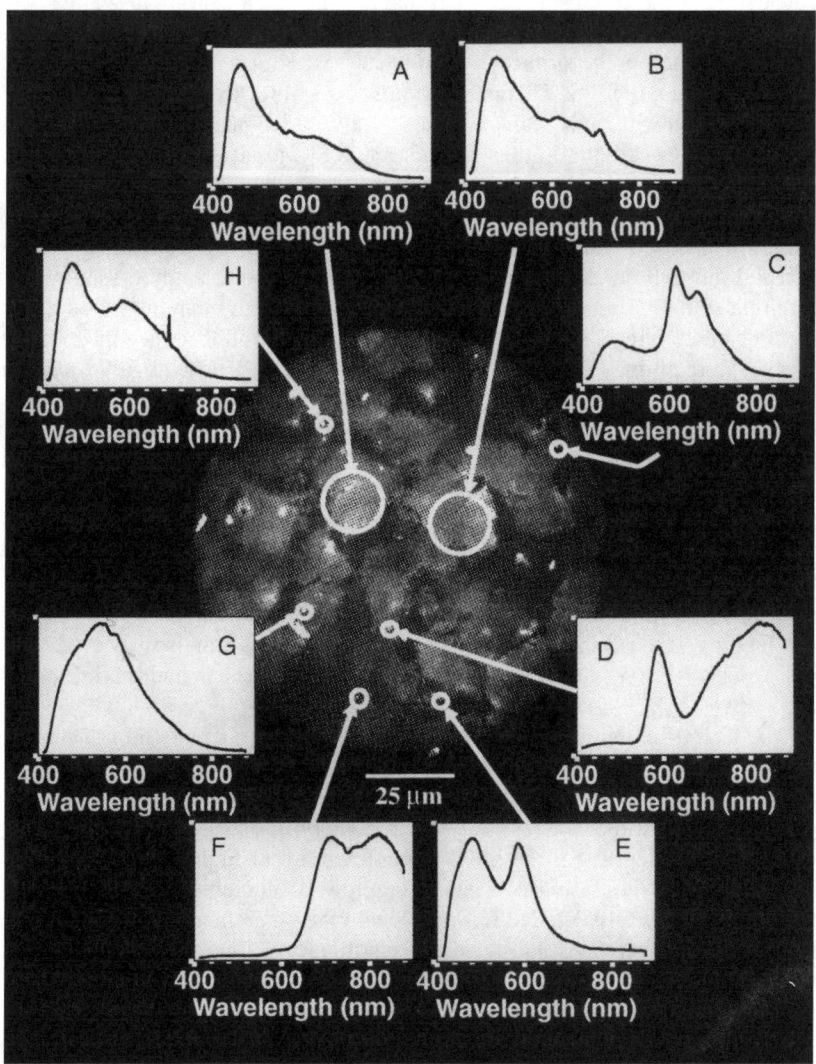

Fig. 5–3. Thin section of hydrothermal dolomite displaying microscale variations in luminescence spectra collected with the epifluorescence microscope and CCD detector described in Plate 5–3 (Romanek & Hunter, 1996, unpublished data).

There have been numerous applications of SFM in recent years for examining mineral surfaces (see Nagy & Blum, 1994, for extensive reviews). The SFM has been used extensively to investigate dissolution and precipitation reactions of mineral surfaces and micromorphological and microtopological aspects of either well characterized mineral surfaces or complex mineral assemblages, as well as to examine sorbed inorganic and organic solutes (e.g., Dove & Hochella, 1993; Eggleston & Stumm, 1993; Heaton & Engstrom, 1994; Nagy & Blum, 1994; Putnis et al., 1995; Fendorf et al., 1996; Sullivan et al., 1997). Two advances that have greatly enhanced the capabilities of SFM for the examination of reactions at the solid–solution interface are the so called tapping-mode SFM and the fluid cell apparatus. Unlike the traditional contact mode SFM where the tip is rastered across the surface, the tip contacts the surface at a known frequency in tapping-mode SFM. This innovation is critical for examining many fragile surfaces and most organic sorbates associated with surfaces, since it reduces the disturbance caused by traditional contact mode SFM that can impart severe artifacts to the images generated (Plate 5–4). The fluid cell apparatus is another important innovation since it allows the sample to be imaged in contact with water (or some other fluid) and to follow dynamic processes as a bathing solution is passed through the cell. The technique has been used to study crystal growth inhibition in the presence of foreign ions, oxidation-precipitation and other surface precipitation reactions, as well as organic solute sorption at the mineral water interface in real time (Dove & Hochella, 1993; Junta & Hochella, 1994; Fendorf et al., 1996; Sullivan et al., 1997). These studies are demonstrating that surface defects are far more important as surface reactive sites than previously thought. Another study reported the ability to apply AFM to a polished rock surface during exposure to nitric acid using a fluid cell attachment (Heaton & Engstrom, 1994). This study demonstrated the ability to observe nanometer-scale changes in topographic features within micrometer-scale regions. Changes in the texture of the rock surface related to differential dissolution of magnetite and fayalite and in grain boundary morphologies also were discernible. Most recently, a flow cell with tapping mode capability has been introduced that promises to expand the application of SFM to a wide variety of problems in soil and environmental chemistry.

Although extremely powerful and in many ways quite simple, deriving meaningful images with SPM techniques is not trivial, especially when applied to complex surfaces (Eggleston, 1994). Thus, while many applications of SPM are appearing in the soil and environmental chemistry literature, one must carefully examine the evidence that the images presented are of high quality and free of artifacts. Despite these caveats, SPM is a rapidly advancing family of techniques that provide extremely detailed information on the nature and structure of mineral and biological surfaces and they will continue to be powerful tools, complimenting other wet chemical and spectroscopic data, for examining a range of complex environmental samples.

Synchrotron-based x-ray techniques are also revolutionizing soil and environmental chemistry research. It is only within the past several years that synchrotron-based x-ray techniques have found application in soil and environmental sciences, but the impact has been significant (e.g., Fendorf & Sparks, 1996;

Schulze & Bertsch, 1995). The most used synchrotron-based x-ray technique in soil and environmental chemistry to date has been x-ray absorption spectroscopy, including XANES and EXAFS. While XANES spectroscopy is useful for probing local geometrical arrangement of first and more distant neighbors surrounding the central absorbers and for determining its oxidation state, EXAFS provides specific information on the identity and distance of surrounding atoms (Schulze & Bertsch, 1995). There are several attributes that make these techniques so appealing for investigating soil chemical processes. In addition to being predominately noninvasive, there are many different synchrotron-based x-ray techniques that can be applied across a wide range of spatial scales. For example, EXAFS has been applied to determine the coordination environment of metals and metalloids at mineral surfaces (±0.002 nm; *vide infra*), x-ray microscopy has been used to examine colloids and clay aggregates ($\approx$30–100 nm; Niemeyer et al., 1994; Thieme et al., 1994), while spatially resolved x-ray absorption spectroscopy (XAS) techniques have been combined with x-ray imaging to probe chemical heterogeneity along grain boundaries, within soil pores, and in the rhizosphere of growing plants ($\approx$1–500 µm; Bertsch et al., 1994; Schulze et al., 1995; Tokunaga et al., 1996).

The XAS is a powerful technique that can provide detailed chemical and structural information about a specific absorbing element, whether it is a major component of a bulk solid phase (both crystalline and noncrystalline), a trace component of the bulk phase, a soluble species, or a surface sorbed component. For most elements of the periodic table, XAS can provide specific information on the local environment of an absorber, including coordination number, identity of and distances to, nearest and sometimes next nearest neighboring atoms. This type of information is not generally available via other techniques, particularly for poorly ordered mineral phases. Additionally, the energy of specific features around the absorption edge of an element often provides information on the oxidation state, coordination geometry, and on transitions between core levels and partially occupied or continuum levels in the excited state. There are a number of comprehensive reviews of the principles and applications of XAS to complex systems (Brown et al., 1988, 1989; Brown, 1990; Schulze & Bertsch, 1995; Fendorf & Sparks, 1996; Hayes & Katz, 1996).

Applications of both XANES and EXAFS spectroscopy to elucidating cation environments in minerals and glasses began appearing in the earth sciences literature in the early 1980s and this pioneering work has been reviewed extensively (e.g., Calas et al., 1987; Brown et al., 1988, 1989; Brown, 1990; Charlet & Manceau, 1993). More recently, numerous studies have demonstrated the usefulness of XANES and EXAFS for providing structural information on important noncrystalline components in soils (Combes et al., 1986, 1989, 1990; Manceau et al, 1992a,b; Drits et al., 1994; Ildefonse et al., 1994) and specific chemical speciation information on contaminants associated with sorptive phases in soils (e.g., Hayes et al.,1987; Chisholm-Brause et al., 1989a,b; Chisholm-Brause et al., 1990a,b; Roe et al., 1991; Charlet & Manceau, 1992a,b; Dent, et al., 1992; Manceau & Charlet, 1992; Bidoglio et al., 1993; Waychunus et al., 1993; Fendorf et al., 1994; Chisholm-Brause et al., 1994; Bertsch et al., 1994; Fendorf & Sparks,

1994; O'Day et al., 1994a,b, 1995, 1996; Papelis et al., 1996; Papelis & Hayes, 1996; Scheidegger et al., 1996).

While a number of spectroscopic methods can be used to interrogate the surface environment of monomineralic systems via probe molecules and reporter groups (Motschi, 1987; Johnston et al., 1993), few are as versatile nor provide the specific molecular level information available from a well planned and executed XAS experiment. Recent investigations have demonstrated the ability of XAS to elucidate removal mechanisms, i.e., distinguish adsorption from precipitation for a wide range of metals and metalloids sorbed to important mineral phases commonly found in soils (Hayes et al., 1987; Brown et al., 1989; Chisholm-Brause et al., 1990a,b; Roe et al., 1991; Charlet & Manceau, 1992a,b; Combes et al., 1992; Dent et al., 1992; Manceau & Charlet, 1992; Manceau et al., 1992a,b; Bidoglio et al., 1992, 1993; Waychunas et al., 1993; Fendorf et al., 1994; O'Day et al., 1994a,b). Many of these studies also have demonstrated the ability of EXAFS to provide specific microscopic information on the adsorption mechanism, i.e., distinguish between inner- and outer-sphere surface association to provide direct evidence on surface orientation, i.e., to suggest mono- or bidentate surface complexation, and to provide information on the chemical speciation of the sorbed component, including oxidation state, multinuclearity, and ternary complexation. A major finding in many of these studies has been the ability to detect evidence for multinuclear clusters for a range of metals and metalloids sorbed to diverse mineral phases at quite low surface coverages, generally ≤5 to 10%. Thus, these studies are demonstrating that surface clustering and surface-induced polynuclear formation are important mechanisms contributing to metal sorption reactions that have been traditionally ignored in surface complexation modeling. Recently, it has been demonstrated that XAS also can be used to distinguish between interlayer and external sorption sites of 2:1 clay minerals (Papelis & Hayes, 1996). An active area of research involves applying XAFS techniques to soils, humic materials, and other complex environmental samples to extract specific information on the coordination and bonding environments of contaminants and nutrients. Many of these initial studies have demonstrated the great potential of XAFS in this regard (Manceau, 1995; O'Day, et al., 1995; Xia et al., 1997).

There are a number of other developments in XAS and spatially resolved XAS that will greatly facilitate future applications in soil and environmental sciences. The quick EXAFS technique or QEXAFS relies on a constant monochromator scan rate and fast array detectors to collect a full EXAFS spectrum in a few seconds versus tens of minutes in a conventional experiment. In principle, msec or μsec time scales are achievable (Lytle & Greeger, 1991; Frahm, 1991). This technique will greatly facilitate the collection of data for elements that are present in low concentrations and will allow the design of experiments where dynamic processes can be examined in situ. Another technique used for time resolved XAS is energy dispersive EXAFS, which employs a bent triangular monochromator crystal and a large area photodiode array detector that will allow for parallel acquisition of a full EXAFS spectrum in a small fraction of a second (Baker et al., 1991). The disadvantage of this approach compared with QEXAFS is that detection is limited only to transmission mode, i.e., fluorescence experiments on low concentration samples are not possible at the current time.

A surface sensitive EXAFS technique is grazing incidence EXAFS (GI-EXAFS). In this technique, the sample is oriented in the incident beam at an angle that is smaller than the critical angle for total reflection. By varying the incidence angle, EXAFS spectra can be obtained from a few nanometers in depth to a total depth approaching the reciprocal of the absorption coefficient, which can be on the order of several millimeters (Greaves et al., 1991). A recent investigation, using a well characterized single crystal of α-alumina and GI-XAFS, has provided evidence for the presence of a well structured outer-sphere complex of Pb(II) within defined interstices on a specific crystal face (0001; Bargar et al., 1996; Fig. 5–4). In contrast, Pb(II) was found to be sorbed primarily as an inner-sphere complex on the α-alumina (1102) surface. This study demonstrates that innovative applications of experimental approaches using synchrotron-based techniques have the potential to interrogate the coordination environments of surface sorbed species at low surface coverages with a remarkable degree of structural detail. Another technique that may hold promise to applications in soil and environmental sciences is diffraction anomalous fine structure (DAFS; Mizuki, 1996). In

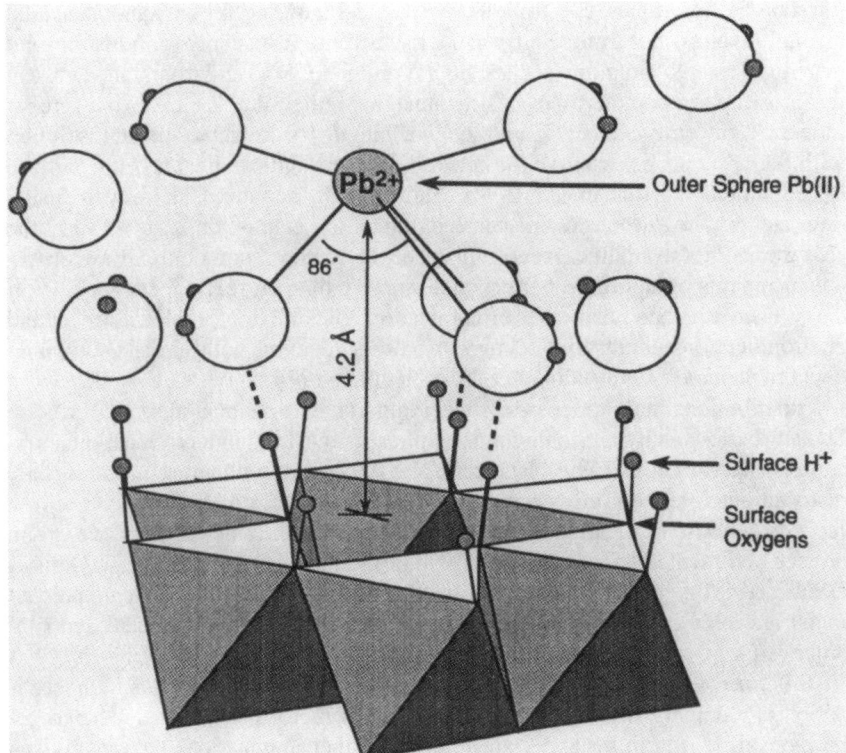

Fig. 5–4. Schematic representation of a highly ordered outer-sphere complex of Pb(II) on the α-Al$_2$O$_3$ (0001) face derived from GIXAFS measurements. Hydrogen bonds are indicated by the dashed lines (from Bargar et al., 1996).

this technique, XANES and EXAFS are collected at the Bragg diffraction angle of a given phase. This measurement scheme may be particularly useful when operating in the microprobe mode especially if the heterogeneous sample contains discrete, yet multiple phases of a given element. Although this is an active area of research in the materials science area, there are no published reports of its use in soil or environmental chemistry.

Perhaps the most powerful technique for examining heterogeneous soil and environmental samples is spatially resolved XAS, where discrete regions within a complex mosaic can be interrogated. Previous investigations have demonstrated that spatially-resolved synchrotron x-ray fluorescence spectroscopy (SXRF) is a very powerful technique for measuring elemental concentrations and distributions of both bulk and contaminant elements within complex matrices (Sutton et al., 1993; Smith & Rivers, 1995; Sutton et al., 1994; Smith, 1995). SXRF spectroscopy is a rapid nondestructive technique for the quantitative determination of elements ($Z > 20$) in a wide variety of solid phases in the parts-per-million to parts-per-billion concentration range under ambient conditions. The high intensity, linear polarization, and natural collimation of synchrotron radiation, contribute to the high sensitivity and achievable spatial resolution (≤ 10 μm) of SXRF. There are a number of advantages of SXRF over other commonly employed microprobe techniques. Electron microprobe techniques offer outstanding spatial resolution, but suffer from relatively poor sensitivity (especially for transition and heavier elements), require a ultra high vacuum (UHV) environment, and can result in appreciably greater beam damage to the sample. Likewise, proton induced x-ray emission (PIXE), although achieving comparable spatial resolution with SXRF, is not as sensitive for heavier elements and can impart beam damage to the sample (Sutton et al., 1994). Furthermore, advances in x-ray focusing devices and the increased brilliance of the third generation sources (e.g, the Advanced Photon Source, APS) will result in spatial resolution and sensitivity enhancements of an order-of-magnitude or more than currently achievable.

There has been a number of applications of SXRF to problems in soil and environmental chemistry including applications dealing with U distributions in contaminated soils and sediments (Bertsch et al., 1994, 1997; Duff et al., 1997), Mn distributions in the rhizosphere of living plants (Schulz et al., 1995), Cr and Se distributions in hyperaccumulator plants employed in phytoremediation applications (Bertsch et al., 1997; Hunter et al., 1997), contaminant (Ni, As, and Se) distributions in annuli of bone fragments and protein coatings from turtle shells, Se distributions surrounding anaerobic micro sites in soil micro aggregates and in ponded sediments (Tokunaga et al., 1994a,b), contaminant metal concentrations associated with groundwater colloids (Kaplan et al., 1994), and contaminant metal and elemental distributions and associations in soil and in coal and coal combustion waste products (Bertsch et al, 1994; Torok et al., 1994).

Some of these applications involve simple nondestructive in situ determination of elements in small samples or when elemental distributions in heterogeneous regions within a sample are required. Other applications have employed two-dimensional scans of samples or regions within a sample that have provided complete images comprised of elemental maps. The spatial resolution of the images of one-half the beam size (≈ 4 μm) is easily achievable by acquiring data at points separated by a commensurate step size. Active research and development efforts in the area of spatially resolved EXAFS at the µm scale are currently under way but will require additional advances in a number of areas if high quality data is to be collected. Continuing advances in monochromators, x-ray focusing devices, and detectors, coupled with the availability of the increased brilliance of APS, will undoubtedly result in novel applications of XAS in soil and environmental sciences that, when combined with other wet chemical and advanced techniques, will ultimately define new standards in the determination of chemical speciation in complex environmental samples.

5–6 ADVANCED ANALYTICAL, SPECTROSCOPIC, AND MICROSCOPIC TECHNIQUES AND THE FUTURE OF SOIL CHEMISTRY

Although advanced analytical, spectroscopic, and microscopic methods have long been important tools in soil and environmental chemistry, recent advances in instrumentation have created unprecedented opportunities for researchers. The ability to examine complex systems at the atomic and molecular levels has allowed soil chemists to explore historically hypothesized mechanisms that have sculptured the discipline. Rapid advances in almost all advanced analytical, spectroscopic, and microscopic techniques are greatly enhancing sensitivity and applicability to complex systems. Reports of exciting new applications, such as single molecule EPR measurements (Kohler et al., 1995), laser enhanced NMR (Buckingham & Parlett, 1994; Tycko et al., 1995), near field optical microspectroscopy (Hess et al., 1994; Zenhausern et al., 1995), ion cyclotron (Fourrier transform) mass spectrometry (Struyf et al., 1996), and many others are common, and indicate that many more powerful tools will be added to our analytical–imaging–spectroscopic toolbox over the next several years. For many applications, soil chemists have been leaders in the application of advanced analytical, spectroscopic, and microscopic techniques to complex environmental samples and this should continue. Additionally, the coupling of clever, more traditional wet chemical experiments with advanced techniques as well as their adaptation to be more compatible with complex samples has become a hallmark of soil chemistry research. There is little question that the future of soil chemistry will be impacted significantly by advances in technology as it relates to the improvement and continued development of advanced analytical, spectroscopic, and microscopic techniques and their application to complex environmental samples.

ACKNOWLEDGMENTS

Both authors were partially funded during preparation of this manuscript by Cooperative Agreement DE-F609-96SR18546 between the U.S. Department of Energy and the University of Georgia Research Foundation.

REFERENCES

Albert, M.S., G.D. Cates, B. Driehuys, W. Happer, B. Saam, C.S. Springer, Jr., and A. Wishnia. 1994. Biological magnetic resonance imaging using laser-polarized ^{129}Xe. Nature (London) 370:199–201.

ASA. 1990. Agronomy abstracts. ASA, Madison, WI.

ASA. 1995. Agronomy abstracts. ASA, Madison, WI.

Baker, G., A.J. Dent, G.N. Derbyshire, C.R. Greaves, A. Catlow, J.W. Couves, and J.M. Thomas. 1991. Time resolved structural studies of nickel exchanged zeolite and nickel oxide using energy dispersive EXAFS. p. 738–741. In S.S. Hasnain (ed.) X-ray absorption fine structure. Ellis Harwood, New York.

Bargar, J.R., S.N. Towle, G.E. Brown, Jr., and G.A. Parks. 1996. Outer-sphere Pb(II) adsorbed at specific surface sites on single crystal-alumina. Geochim. Cosmochim. Acta 60:3541–3547.

Bertsch, P.M., D.B Hunter, P. Nuessle, and S.B. Clark. 1997. Molecular characterization of contaminants in soils and waste-forms by spatially resolved XRF and XANES spectroscopy. J. Physique. 7(C2):817–818.

Bertsch, P.M., D.B. Hunter, S.R. Sutton, S. Bajt, and M.L. Rivers. 1994. In situ chemical speciation of uranium in soils and sediments by micro x-ray absorption spectroscopy. Environ. Sci. Technol. 28(5):980–984.

Bibler, M.V., and W. Stumm. 1994. An in-situ ATR-FTIR study: The surface coordination of salicylic acid on aluminum and iron(III) oxides. Environ. Sci. Technol. 28:763–768.

Bidoglio, G., P.N. Gibson, E. Haltier, N. Omenetto, and M. Lipponen. 1992. XANES and laser fluorescence spectroscopy for rare earth speciation at mineral water interface. Radiochim. Acta 58–59:191–197.

Bidoglio, G., P.N. Gibson, M.O. O'Gorman, and K.J. Roberts. 1993. X-ray absorption spectroscopy investigation of surface redox transformation of thallium and chromium on colloidal mineral oxides. Geochim. Cosmochim. Acta 57:2389–2394.

Brown, G.E., Jr. 1990. Spectroscopic studies of chemisorption reaction mechanisms at oxide-water interfaces. Rev. Mineral. 23:309–363.

Brown, G.E., Jr., G. Calas, G.A. Waychunas, and J. Petiau. 1988. X-ray absorption spectroscopy and its application in mineralogy and geochemistry. Rev. Mineral. 18:431–512.

Brown, G.E., Jr., G.A. Parks, and C.J. Chisholm-Brause. 1989. In situ x-ray absorption spectroscopic studies of ions at oxide-water interfaces. Chimia 43:248–256.

Buckingham, A.D., and L.C. Parlett. 1994. High-resolution nuclear magnetic resonance spectroscopy in a circularly polarized laser beam. Science (Washington, DC) 264:1748–1750.

Calas, C., G.E. Brown, G.A. Waychunas, and J. Petiau. 1987. X-ray absorption spectroscopic studies of silicate glasses and minerals. Phys. Chem. Minerals 15:19–29.

Charlet, L., and A.A. Manceau. 1992a. X-ray absorption spectroscopic study of the sorption of Cr(III) at the oxide-water interface: I. Molecular mechanism of Cr(III) oxidation on Mn oxides. J. Colloid Interface Sci. 148:425–442.

Charlet, L., and A.A. Manceau. 1992b. X-ray absorption spectroscopic study of the sorption of Cr(III) at the oxide-water interface: II. Adsorption, coprecipitation, and surface precipitation on hydrous ferric oxide. J. Colloid Interface Sci. 148:443–458.

Charlet, L., and A. Manceau. 1993. Structure, formation, and reactivity of hydrous oxide particles: Insights from x-ray absorption spectroscopy. p. 117–164. In J. Buffle and H.P. van Leeuwan (ed.) Environmental particles. Vol. 2. Lewis Publ., Ann Arbor, MI.

Chisholm-Brause, C.J., G.E. Brown, Jr., and G.A. Parks. 1989a. EXAFS investigation of aqueous Co(II) adsorbed on oxide surfaces in-situ. Physica B+C 158:646–648.

Chisholm-Brause, C.J., S.D. Conradson, C.T. Buscher, P.G. Eller, and D.E. Morris. 1994. Speciation of uranyl sorbed at multiple binding sites on montmorillonite. Geochim. Cosmochim. Acta 58:3625–3631.

Chisholm-Brause, C.J., K.F. Hayes, A.L. Roe, G.E. Brown, Jr., G.A. Parks, and J.O. Leckie. 1990a. Spectroscopic investigation of Pb(II) complexes at the Al_2O_3–water interface. Geochim. Cosmochim. Acta 54:1897–1909.

Chisholm-Brause, C.J., P.A. O'Day, G.E. Brown, Jr., and G.A. Parks. 1990b. Evidence for multinuclear metal-ion complexes at solid-solution interfaces from x-ray absorption spectroscopy. Nature (London) 348:528–530.

Chisholm-Brause, C.J., A.L. Row, K.F. Hayes, G.E. Brown, Jr., G.A. Parks, and J.O. Leckie. 1989b. XANES and EXAFS study of aqueous Pb(II) absorbed on oxide surfaces. Physica B 158:674–675.

Combes, J.M., C.J. Chisholm-Brause, G.E. Brown, Jr., G.A. Parks, S.D. Conradson, P.G. Eller, R. Triay, II, D.E. Hobart, and A. Meijer. 1992. EXAFS spectroscopic study of neptunium(V) sorption at the a-FeOOH–water interface. Environ. Sci. Technol. 26:376–382.

Combes, J.M., A. Manceau, and G. Calas. 1986. Study of the local structure in poorly-ordered precursors of iron oxi-hydroxides. J. Phys. C. 8:697–701.

Combes, J.M., A. Manceau, and G. Calas. 1990. Formation of ferric oxides from aqueous solutions: A polyhedral approach by x-ray absorption spectroscopy. II. Hematite formation from ferric gels. Geochim. Cosmochim. Acta 54:1084–1094.

Combes, J.M., A. Manceau, G. Calas, and J.Y. Bottero. 1989. Formation of ferric oxides from aqueous solutions: A polyhedral approach by x-ray absorption spectroscopy: I. Hydrolysis and formation of ferric gels. Geochim. Cosmochim. Acta 53:583–594.

Dent, A.J., J.D.F. Ramsay, and W. Swanton. 1992. An EXAFS study of uranyl ion in solution and sorbed onto silica and montmorillonite clay colloids. J. Colloid Interface Sci. 150:45–60.

Dove, P., and J. Chermak. 1994. Mineral–water interactions: Fluid cell applications of scanning force microscopy. p. 139–170. In K.L. Nagy and A.E. Blum (ed.) Scanning probe microscopy of clay minerals. The Clay Minerals Soc., Boulder, CO.

Dove, P.M., and M.F. Hochella, Jr. 1993. Calcite precipitation mechanisms and inhibition by orthophosphate: In situ observations by scanning force microscopy. Geochim. Cosmochim. Acta 57:705–714.

Drits, V.A., B.A. Sakharov, A.L. Salyn, and A. Manceau. 1994. Structural model for ferrihydrite. Clay Miner. 28:185–207.

Duff, M.C., C.A. Amrhein, P.M. Bertsch, and D.B. Hunter. 1997. The chemistry of uranium in evaporation pond sediment in the San Joaquin Valley, CA, using x-ray fluorescence and XANES techniques. Geochim. Cosmochim. Acta 61:73–81.

Eggleston, C.M. 1994. High-resolution scanning probe microscopy: Tip–surface interaction, artifacts, and applications in mineralogy and geochemistry. p. 1–90. In K.L. Nagy and A.E. Blum (ed.) Scanning probe microscopy of clay minerals, The Clay Minerals Soc., Boulder, CO.

Eggleston, C. M., and W. Stumm. 1993. Scanning tunneling microscopy of Cr(III) chemisorbed on a-Fe2O3 (001) surfaces from aqueous solutions: Direct observation of surface mobility and clustering. Geochim. Cosmochim. Acta 57:4843–4850.

Faust, B.C., W.B. Labiosa, K'o H. Dia, J.S. MacFall, B.A. Browne, A.A. Ribeiro, and D.D. Richter. 1995. Speciation of aqueous mononuclear Al(III)-hydroxo and other Al-(III) complexes at concentrations of geochemical relevance by aluminum-27 nuclear magnetic resonance spectroscopy. Geochim. Cosmochim. Acta 59:2651–2661.

Fendorf, S.E., G.M. Lamble, M.G. Stapleton, M.J. Kelley, and D.L. Sparks. 1994. Mechanisms of chromium(III) sorption on silica: I. Cr(III) surface structure derived by extended x-ray absorption five structure spectroscopy. Environ. Sci. Technol. 28:284–289.

Fendorf, S.E., G. Li, and M.E. Gunter. 1996. Micromorphologies and stabilities of chromium (III) surface precipitates elucidated by scanning force microscopy. Soil Sci. Soc. Am. J. 60:99–106.

Fendorf, S.E., and D.L. Sparks. 1994. Mechanisms of chromium(III) sorption of silica. II: Effect of reaction conditions. Environ. Sci. Technol. 28:290–297.

Fendorf, S.E., and D.L. Sparks. 1996. X-ray absorption fine structure spectroscopy. p. 377–416. In D.L. Sparks (ed.) Methods of soil analysis. SSSA Book Ser. 3. SSSA, Madison, WI.

Fendorf, S.E., D.L. Sparks, J.A. Franz, and D.M. Camaioni. 1993. Electron paramagnetic resonance stopped-flow kinetic study of manganese(II) sorption-desorption on birnessite. Soil Sci. Soc. Am. J. 57:57–62.

Ford, R.G. 1996. The effects of aging on the partitioning of divalent transition metals coprecipitated with iron oxides. Ph.D. diss. Clemson Univ., Clemson, SC.

Frahm, R. 1991. Quick XAFS: Potentials and practical applications in materials science. p. 731–737. In S.S. Hasnain (ed.) X-ray absorption fine structure. Ellis Harwood, New York.

Fredrich, J.T., B. Menendez, and T.F. Wong. 1995. Imaging the pore structure of geomaterials. Science (Washington, DC) 268:276–279.

Greaves, G.M., S. Pizzini, K.J. Roberts, N.T. Barrett, and S. Kalbitzer. 1991. Glancing angle XAFS for the study of real surfaces. p. 232–237. In S.S. Hasnain (ed.) X-ray absorption fine structure. Ellis Horwood, New York.

Guilment, J., S. Markel, and W. Windig. 1994. Infrared chemical micro-imaging assisted by interactive self-modeling multivariate analysis. Appl. Spectroscopy 48:320–326.

Hayes, K.F., and L.E. Katz. 1996. Application of x-ray absorption spectroscopy for surface complexation modeling of metal ion sorption. p. 147–223. In P.V. Brady (ed.) Physics and chemistry of mineral surfaces. CRC Press, Boca Raton, FL.

Hayes, K.F., A.L. Roe, G.E. Brown, Jr., K.O. Hodgson, J.O. Leckie, and G.A. Parks. 1987. *In situ* x-ray absorption study of surface complexes at oxide–water interfaces: Selenium oxyanions on a-FeOOK. Science (Washington, DC) 238:783–786.

Heaton, J.S., and R.C. Engstrom. 1994. *In situ* atomic force microscopy of the differential dissolution of fayalite and magnetite. Environ. Sci. Technol. 28:1747–1754.

Hess, H.F., E. Betzig, T.D. Harris, L.N. Pfeiffer, and K.W. West. 1994. Near-field spectroscopy of the quantum constituents of a luminescent system. Science (Washington, DC) 264:1740–1745.

Hug, S.J., and B. Sulzberger. 1994. *In situ* fourier transform infrared spectroscopic evidence for the formation of several different surface complexes of oxalate on TiO_2 in the aqueous phase. Langmuir 10:3587–3597.

Hunter, D.B., and P.M. Bertsch. 1994. *In situ* measurements of tetraphenylboron degradation kinetics on clay mineral surface by IR. Environ. Sci. Technol. 28:686–691.

Hunter, D.B., P.M. Bertsch, K.M. Kemner, and S.B. Clark. 1997. Distribution and chemical speciation of metals and metalloids in biota collected from contaminated environments by spatially resolved XRF, XANES, and EXAFS. J. Physique. 7(C2):767–771.

Ildefonse, P., R.J. Kirkpatrick, B. Monetz, G. Calas, A.M. Flank, and P. Lagarde. 1994. ^{27}Al MAS NMR and aluminum x-ray absorption near edge structure study of imogolite and allophanes. Clays Clay Miner. 42:276–287.

Johnston, C.T., G. Sposito, and W.L. Earl. 1993. Surface spectroscopy of environmental particles by fourier-transform infrared and nuclear magnetic resonance spectroscopy. p. 1–36. *In* J. Buffle and H. P. van Leeuwen (ed.) Environmental particles. Lewis Publ., Ann Arbor, MI.

Junta, J., and M.F. Hochella, Jr. 1994. Manganese (II) oxidation at mineral surfaces: A microscopic and spectroscopic study. Geochim. Cosmochim. Acta 58:4985–4999.

Kaplan, D.I., D.B. Hunter, P.M. Bertsch, S. Bajt, and D.C. Adriano. 1994. Application of synchrotron x-ray fluorescence spectroscopy and energy dispersive x-ray analysis to identify contaminant metals on groundwater colloids. Environ. Sci. Technol. 28:1186–1189.

Kim, H-B., M. Hayashi, K. Nakatani, and N. Kitamura. 1996. *In situ* measurements of ion- exchange processes in single polymer particles: Laser trapping microspectroscopy and confocal fluorescence microspectroscopy. Anal. Chem. 68:409–414.

Kohler, J., A.C.J. Brouwer, E.J.J. Groenen, and J. Schmidt. 1995. Single molecule electron paramagnetic resonance spectroscopy: Hyperfine splitting owing to a single nucleus. Science (Washington, DC) 268:1457–1460.

Lytle, F.W., and R.B. Greeger. 1991. New developments in XAS experiments. p. 625–633. *In* S.S. Hasnain (ed.) X-ray absorption fine structure. Ellis Harwood, New York.

Manceau, A. 1995. The mechanism of anion adsorption on iron oxides: Evidence for the bending of arsenate tetrahedra on free $Fe(O,OH)_6$ edges. Geochim. Cosmochim. Acta 59:3647–3653.

Manceau, A., and L. Charlet. 1992. X-ray absorption spectroscopic study of the sorption of Cr(III) at the oxide-water interface: I. Molecular mechanism of Cr(III) oxidation on Mn oxides. J. Colloid Interface Sci. 148:425–442.

Manceau, A., L. Charlet, M.C. Boisset, B. Didier, and L. Spadini. 1992a. Sorption and speciation of heavy metals on hydrous Fe and Mn oxides. From microscopic to macroscopic. Appl. Clay Sci. 7:201–223.

Manceau, A., A.I. Gorshkov, and V. Drits. 1992b. Structural chemistry of Mn, Fe, Co and Ni in Mn hydrous oxides: I. Information from XANES spectroscopy. Am. Miner. 77:1133–1143.

Mizuki, J. 1996. Diffraction anomalous fine structure (DAFS). p. 372–382. *In* Y. Iwasawa (ed.) X-ray absorption fine structure for catalysits and surfaces. Series on Synchrtoron Radiation Techniques and Applications. Vol. 2. World Scientific Publ. Co., Singapore.

Moerner, W.E. 1994. Examining nanoenvironments in solids on the scale of a single, isolated impurity molecule. Science (Washington, DC) 265:46–53.

Motschi, H. 1987. Aspects of the molecular structure in surface complexes: Spectroscopic investigations. p. 111–125. *In* W. Stumm (ed.) Aquatic surface chemistry. Wiley Interscience, New York.

Nagy, K.L., and A.E. Blum. (ed.) 1994. Scanning probe microscopy of clay minerals, The Clay Minerals Soc., Boulder, CO.

Navon, G., Y.-Q. Song, T. Room, S. Appelt, R.E. Taylor, and A. Pines. 1996. Enhancement of solution NMR and MRI with laser-polarized xenon. Science (Washington, DC) 271:1848–1851.

Niemeyer, J., J. Thieme, P. Guttmann, T. Wilhein, D. Rudolph, and G. Schmahl. 1994. Direct imaging of aggregates in aqueous clay-suspensions by x-ray microscopy. Progr. Colloid Polym Sci. 95:139–142.

O'Day, P.A., G.E. Brown, Jr., and G.A. Parks. 1994a. X-ray absorption spectroscopy of cobalt(II) multinuclear surface complexes and surface precipitates on kaoline. J. Colloid Interface Sci. 165:269–289.

O'Day, P.A., S.A. Carroll, G.A. Waychunas, and B. Phillips. 1995. XAS of trace element coordination in natural sediments at ambient and cryogenic temperatures. Physica B. 208–209:309–310.

O'Day, P.A., C.J. Chisholm-Brause, S.N. Towle, G.A. Parks, and G.E. Brown, Jr. 1996. X-ray absorption spectroscopy of Co(II) sorption complexes on quartz (SiO_2). Geochim. Cosmochim. Acta. 60:2515–2532.

O'Day, P.A., G.A. Parks, and G.E. Brown, Jr. 1994b. Molecular structure and binding sites of cobalt(II) surface complexes on kaolinite from x-ray absorption spectroscopy. Clays Clay Miner. 42:337–355.

Page, A.L., R.H. Miller, and D.R. Keeney. (ed.). 1982. Methods of soil analysis. Part 2. Chemical and microbiological properties. 2nd ed. Agron. Monogr.9. ASA and SSSA, Madison, WI.

Papelis, C., G.E. Brown, Jr., G.A. Parks, and J.O. Leckie. 1996. X-ray absorption studies of cadmium and selenate absorption on aluminum oxide. Langmuir 11:2041–2048.

Papelis, C., and K.F. Hayes. 1996. Distinguishing between interlayer and external sorption sites of clay minerals using x-ray absorption spectroscopy. Coll. Surfaces 107:89–96.

Putnis, A., J.L. Junta-Rosso, and M.F. Hochella, Jr. 1995. Dissolution of barite by a chelating ligand: An atomic force microscopy study. Geochim. Cosmochim. Acta 59:4623–4632.

Roe, A.L., K.F. Hayes, C.J. Chisholm-Brause, G.E. Brown, Jr., G.A. Parks, and J.O. Leckie. 1991. X-ray absorption study of lead complexes at a-FeOOH–water interfaces. Langmuir 7:367–373.

Schaeberle, M.D., C.G. Karakatsanis, C.J. Lau, and P.J. Treado. 1995. Raman chemical imaging: Noninvasive visualization of polymer blend architecture. Anal. Chem. 67:4316–4321.

Scheidegger, A.M., G.M. Lamble, and D.L. Sparks. 1996. Investigation of Ni adsorption on pyrophyllite: An XAFS study. Environ. Sci. Technol. 20:548–554.

Schulze, D.G., and P.M. Bertsch. 1995. Synchrotron x-ray techniques in soil, plant, and environmental research. Adv. Agron. 55:1–66.

Schulze, D.G., T. McCay-Buis, S.R. Sutton, and D.M. Huber. 1995. Determination of manganese oxidation states in soils using x-ray absorption near-edge structure (XANES) spectroscopy. Soil Sci. Soc. Am. J. 59:1540–1548.

Smith, J.V. 1995. Synchrotron x-ray sources: Instrumental characteristics. New applications in microanalysis, tomography, absorption spectroscopy and diffraction. Analyst 120:1231–1245.

Smith, J.V., and M.L. Rivers. 1995. Synchrotron x-ray microanalysis. p. 163–233. In P.J. Potts et al. (ed.) Microprobe techniques in the earth sciences. Chapman & Hall, London.

Sparks, D.L., A.L. Page, P.A. Helmke, R.H. Loeppert, P.N. Soltanour, M.A. Tabatabai, C.T. Johnson, and M.E. Sumner (ed.). 1996. Methods of soil analysis. Part 3. Chemical methods. SSSA Book Ser. 5. ASA and SSSA, Madison, WI.

Struyf, H., L. Van Vaeck, P. Kennis, R. Gijbels, and R. Van Grieken. 1996. Chemical characterization of neo-ceramic powders by time-of-flight and fourier transform laser microprobe mass spectrometry. Rapid Commun. Mass Spectrometry 10:699–706.

Sullivan, E.J., D.B. Hunter, and R.S. Bowman. 1997. Topological and thermal properties of surfactant-modified clinoptilolite studied by tapping-mode atomic force microscopy and high-resolution thermogravimetric analysis. Clays Clay Miner. 45:42–53.

Sun, X., and H.E. Doner. 1996. An investigation of arsenate and arsenite bonding structures on goethite by FTIR. Soil Sci. 161:865–872.

Sutton, S.R., M.L. Rivers, S. Bajt, and K.W. Jones. 1993. Synchrotron x-ray fluorescence microprobe analysis with bending magnets and insertion devices. Nucl. Instrum. Methods Phys. Res. B75:553–558.

Sutton, S.R., M.L. Rivers, S. Bajt, K. Jones, and J.V. Smith. 1994. Synchrotron x-ray fluorescence microprobe: A microanalytical instrument for trace element studies in geochemistry, cosmochemistry, and the soil and environmental sciences. Nucl. Instrum. Methods Phys. Res. A347:412–416.

Thieme, J., J. Neimeyer, P. Guttmann, T. Wilhein, D. Rudolph, and G. Schmahl. 1994. X-ray microscopy studies of aqueous colloid systems. Progr. Colloid Polym Sci. 95:135–138.

Tokunaga, T.K., I.J. Pickering, and G.E. Brown, Jr. 1996. Selenium transformations in ponded sediments. Soil Sci. Soc. Am. J. 60:781–790.

Tokunaga, T.K., S.R. Sutton, and S. Bajt. 1994. Mapping of selenium concentrations in soil aggregates with synchrotron x-ray fluorescence microprobe. Soil Sci. 158:421–434.

Torok, S., G. Faigel, K.W. Jones, M.L. Rivers, S.R. Sutton, and S. Bajt. 1994. Chemical characterization of environmental particulate matter using synchrotron radiation. X-Ray Spectrum. 23:3–6.

Tykco, R., S.E. Barrett, G. Dabbagh, L.N. Pfeiffer, and K.W. West. 1995. Electronic states in gallium arsenide quantum wells probed by optically pumped NMR. Science (Washington, DC) 268:1460.

Waychunas, G.A., B.A. Rea, C.C. Fuller, and J.A. Davis. 1993. Surface chemistry of ferrihydrite: 1. EXAFS studies of the geometry of coprecipitated and adsorbed arsenate. Geochim. Cosmochim. Acta 57:2251–2269.

Xia, K., W. Bleam, and P.A. Helmke. 1997. Studies of the nature of binding sites of first row transition elements bound to aquatic and soil humic substances using x-ray absorption spectroscopy. Geochim. Cosmochim. Acta 61:2223–2235.

Zenhausern, F., Y. Martin, and H.K. Wickramasinghe. 1995. Scanning interferometric apertureless microscopy: Optical imaging at 10 angstrom resolution. Science (Washington, DC) 269:1083–1085.

6 Chemistry of the Soil Solution[1]

W. L. Lindsay and K. M. Catlett

Colorado State University
Fort Collins, Colorado

6–1 INTRODUCTION

Soil solution is the liquid phase into which are dissolved some of all the solids and gases that are present in soil. Some reactions that occur between solid phases and soil solution attain equilibrium rapidly, while others occur more slowly, and some may never attain equilibrium. It is folly to talk about the soil solution without considering the solid and gaseous phases that control its composition. The dissolution and precipitation of solid phases in soils largely control the composition of soil solution. Measurement of metal ion activities and calculation of saturation indices of possible solid phases are useful parameters for examining chemical reactions and testing equilibrium relationships.

Increasingly, soil chemists and those who use soil chemical information have shifted their focus from soil fertility and mineralogy to environmental problems dealing with the fate of metal and organic contaminants in soils. There is greater emphasis on understanding the chemical reactions and solubility controls of metals in soils and the breakdown and persistence of organic contaminants. Environmental problems are becoming a major thrust of soil chemistry.

6–2 DYNAMIC EQUILIBRIA IN SOILS

Let us consider for a moment the dynamic equilibria that occur in soils as depicted in Fig. 6–1 (Lindsay, 1979). The soil solution is the focal point of this diagram. Plants take up nutrients from soil solution and lower their activities in the immediate vicinity of absorbing roots. Plants differ in their ability to take up nutrients and to prevent their exudation back into the soil.

Ions are held by surface adsorption and exchange sites in soils. These adsorbed ions tend to maintain equilibrium with the soil solution. Exchange reactions usually occur rapidly and provide an important mechanism for buffering ion activities in solution. The precipitation and dissolution of minerals can ultimately control the activity of ions in solution and in turn the activity of ions on exchange sites. Ion activity gradients are the driving forces that move ions into

---

[1] Contribution from Department of Soil and Crop Sciences, Colorado State University, Fort Collins, Colorado 80523. Supported by the Colorado Agric. Exp. Stn. Regional Project W-184.

Copyright © 1998. Soil Science Society of America, 677 S. Segoe Rd., Madison, WI 53711, USA. *Future Prospects for Soil Chemistry.* SSSA Special Publication no. 55.

and out of exchange sites. The nature of exchange sites and the exchange capacity of soils affect the final equilibrium especially when ion activities are not controlled by solid phases.

Rainfall dilutes the soil solution while evapotranspiration concentrates it. Water draining through soil carries with it dissolved salts, moving them deeper into the soil profile and ultimately into the groundwater. Water-soluble fertilizers applied to soils readily dissolve, and reaction products, more suited to the chemical environment of soils, precipitate.

Organic matter also plays an important role in soils. Microorganisms hasten its decomposition and the synthesis of new biochemical reaction products. Many reactions involving organic matter in soils do not attain equilibrium. If they did, most organic matter would decompose to CO_2, H_2O, and mineral salts. The dashed line leading from organic matter to soil solution as shown in Fig. 6–1 is a reminder that equilibrium for this reaction is generally not attained. Enzymes, catalysts, and temperature changes are important factors that affect the rates of reactions involved in the synthesis and degradation of organic products. Organic matter and particularly organic acids are important in solubilizing metals and facilitating their transport in soils.

Gases continually exchange between soil air and soil solution. Most importantly O_2 is consumed and CO_2 is released in respiration processes that universally occur in soils. As the water content of soils increases, air spaces are diminished and gaseous diffusion is greatly restricted. As a result, oxidized soils become reduced, and $CO_2(g)$ levels increase. Reduction processes modify the composition of soil solution in many important ways.

The dynamic equilibria that occur in soils as shown here continually change as reactions approach nearer and nearer to equilibrium. Many of these equilibria

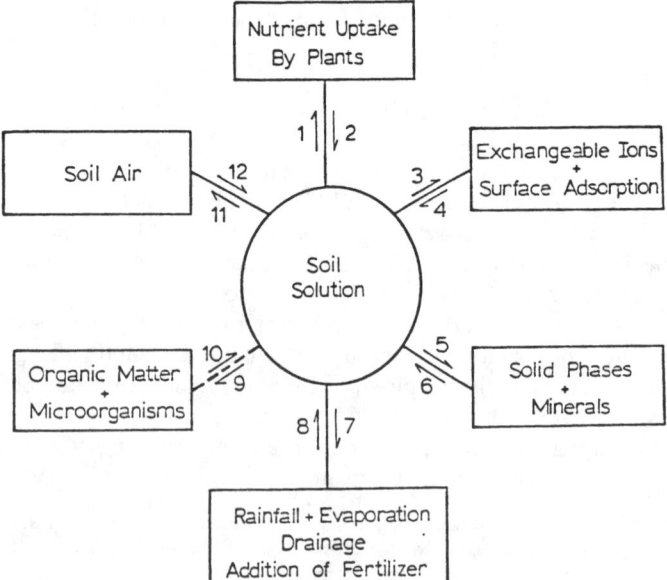

Fig. 6–1. Dynamic equilibria in soil that affect the composition of soil solution (Lindsay, 1979).

are interactive so we need to understand the individual reactions before we can understand the combined reactions. This will be a challenge for the future. The kinetics of chemical reactions in soils are only beginning to be understood. New techniques, such as stop flow methods, can be used to gain new insights into reaction rates and help to determine which reactions attain equilibrium and which ones do not. Both individual and combined reactions must be examined.

In summary the soil solution is the resultant of many dynamic equilibria that occur in soils. Some of these reactions attain equilibrium rapidly, while others do not. The soil solution is the key for understanding and interpreting which chemical reactions can occur and which ones do occur. Monitoring the composition of the soil solution with time provides valuable information on how rapidly equilibrium is approached.

6–3 ION ACTIVITY MEASUREMENTS IN SOILS

Ion activity measurements in soils are important because they provide the means of examining equilibrium relationships between ionic species in solution and in solid phases. Ion activities can be used to calculate saturation indices of solids and thus determine whether a given solid is dissolving, is precipitating, or if it is in equilibrium with that soil. Ion activities also are useful for developing solubility diagrams that allow instant examination of many simultaneous equilibria. Total elemental concentrations in soil solution are useful for determining mobility and transport of those elements, but ion activities are necessary to test for equilibrium conditions.

6–3.1 From Elemental Concentrations

One way of obtaining ion activities in soil is to measure total elemental concentrations in solution, then use an appropriate chemical speciation model, to partition total concentrations into specific ionic species. Several speciation models were evaluated and demonstrated in a recent symposium (Loeppert et al.,1995). Most models include a means of determining ionic strength and of calculating ion activity coefficients. This approach is generally acceptable so long as all significant solution species are accounted for, and accurate formation constants are included in the speciation models. A disturbing factor in soils is that many existing natural organic complexes have not been adequately identified, nor have accurate formation constants been determined for them. To the extent that such complexes contribute significantly to total elemental concentrations in soil solution, errors will result in the ion activities calculated therefrom. On the other hand, if the major solution species of a given element are recognized and their formation constants are accurately known and included in the speciation model data base, then accurate ion activities will result. Soil chemists need to be highly discriminating about the equilibrium constant data bases that they use to calculate ion activities in soils. It is not enough to judge a data base by its size alone. There is need to continually examine, test, and, in some cases, remeasure important equilibrium constants. Otherwise, interpretations based on erroneous solubility measurements or constants will continue to be misleading.

6–3.2 From Ion Specific Electrodes

Ion specific electrodes offer a direct approach for measuring ionic activities in soils. Fortunately such electrodes generally sense the activity of a given cation or anion, not its total concentration. Use of the glass electrode to measure H^+ activity is perhaps the most widely used and familiar ion specific electrode. The Pt electrode is frequently used to measure electron activity. Many other ion specific electrodes are now available and have been used in soil research.

Three major problems have been recognized in using ion specific electrodes in soils. First, interferences often occur, so it is necessary to prove that such interferences are minimal in the chemical environments in which the ion specific electrode is used. Secondly, ion specific electrodes are often not sufficiently sensitive to measure the low activity of many ions that are of interest in soils. For example, Cu^{2+}, Pb^{2+}, and Cd^{2+} metal ion electrodes are usually restricted for use under acidic conditions where the activities of these ions are above their detection limits (generally about 10^{-8}). Thirdly, ion specific electrodes are not yet available for many of the ionic species that are important in soils. Future research is needed to examine the problems that are associated with the development, testing, and use of ion specific electrodes in soils. The advent of solid state and microelectrodes has allowed for unique applications to soils and soil solutions. As the sensitivity and robustness of these tools improve, they will become more common place in soils research.

6–3.3 From Metal Chelation

Metal chelation provides an alterative method of measuring metal ion activities in soils that circumvents or reduces some of the problems discussed above. By selecting an appropriate chelating agent (depending on pH and chelate equilibrium constants) a target metal ion and a reference metal ion can be selected and loaded onto the chelating agent to provide a range of initial mole fractions. Changes in the concentration of these initial mole fractions during short periods (5–7 d) permit the selection of an equilibrium ratio of the two chelated metals at which they neither go into nor come out of the chelate. This ratio, along with the known or measured activity of the reference metal ion, allows the activity of the target metal ion to be calculated.

The chelation method can be used to calculate the activities of selected metal ions that are present at extremely low levels. For example, free Fe^{3+} in a neutral soil generally is in the vicinity of 10^{-18} M. When the chelating agent is added in the 10^{-5} to 10^{-4} M range, chelated Fe^{3+} will be present at an easily measurable level. So long as natural organic complexes of the target metal ion are <10% of total chelated metal (approximately 10^{-5} M), these natural complexes can be ignored because their contribution to the total target metal in solution is negligible.

The chelation method of determining Fe^{3+} activity in soils is illustrated in Fig. 6–2 (Norvell & Lindsay, 1982a). The target metal ion in this case is Fe^{3+} and the reference metal ion is Ca^{2+}. The soil pH was near 6.8 and the chelating agent, EDTA (ethylenediaminetetraacetic acid), was selected because both Fe^{3+} and Ca^{2+} are chelated significantly by EDTA at this pH. Eight solutions containing 10^{-4} M EDTA and ranging from 0 to 1 mole fraction of Fe and Ca were prepared. Suspensions containing 10 g of soil and 20 mL of chelate solution were prepared and reacted on a shaker for 7 d. A unique mole fraction of 0.44 Fe on the chelate reflected the equilibrium mole ratio where Fe^{3+} and Ca^{2+} neither went into nor came out of the chelate. At pH 6.8, soluble Fe consists of 95.6% FeL^- and 4.3% $FeOHL^{2-}$.

The equilibrium reactions that apply to this example can be expressed as follows:

$$Ca^{2+} + FeL^- \rightleftharpoons Fe^{3+} + CaL^{2-} \qquad \log K^m_{0.01} = -14.89 \qquad [1]$$

$$\frac{[Fe^{3+}][CaL^{2-}]}{[Ca^{2+}][FeL^-]} = 10^{-14.89} \qquad [2]$$

$$[Fe^{3+}] = 10^{-14.89} [Ca^{2+}] \cdot \frac{[FeL^-]}{[CaL^{2-}]} \qquad [3]$$

where L represents the deprotonated EDTA ligand or chelate species, and $K^m_{0.01}$ is the mixed equilibrium constant for Reaction [1] at an ionic strength of 0.01 (Norvell & Lindsay, 1982a; Lindsay, 1979). Substitution of measured parameters into the right side of Eq. [3] permits calculation of the concentration of Fe^{3+}, and knowing ionic strength permits conversion to Fe^{3+} activity. Since some soils show a preference for the initial adsorption of FeL^- compared with CaL^{2-}, an adjustment can be made to compensate for this differential behavior (Norvell & Lindsay, 1982a).

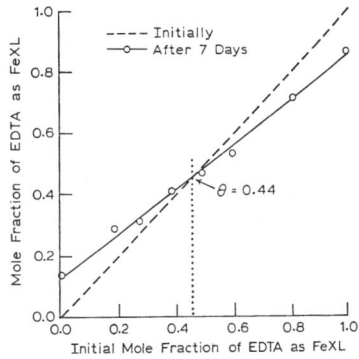

Fig. 6–2. Displacement of Fe^{3+} and Ca^{2+} from EDTA when reacted with a Monticello loam for 7 d. (from Norvell & Lindsay, 1982a).

The chelation method has been used to measure several different metal ion activities in soils (Vlek et al., 1974; Ma & Lindsay, 1990, 1993, 1995; Workman & Lindsay, 1990; El-Falaky et al., 1991; Sachdev et al., 1992; Kalbasi et al., 1995). Although cumbersome to use, the chelation method is perhaps the most sensitive and reliable method presently available for measuring very low metal ion activities in soils. The working range of this method and the considerations involved in selecting metal ions and pH environments have been reviewed (Workman & Lindsay, 1990).

There are many challenges for improving the chelation method of measuring metal ion activities in soils (Oglesby, 1995). There is a need to examine in greater detail the differential adsorption of chelated metals in different types of soils. Another refinement would be to develop quantitative methods of measuring specific metal chelated species in soil solution. These chelated species could then be entered directly into chemical speciation models to calculate chelate equilibrium relationships more precisely. With rapidly developing instrumentation such as capillary electrophoresis, high pressure liquid chromatography (HPLC), and ion chromatography new methods can be developed for quantifying specific chelated metal species in soil solution. Such measurements will be useful for determining simultaneous multimetal ion activities as proposed by Lindsay and Sommers (1997).

6–4 SOLID PHASE CONTROL OF ION ACTIVITIES IN SOIL SOLUTION

6–4.1 Metal Ion Activity Diagram

Solid phases potentially control the composition of soil solution through dissolution, precipitation, and exchange reactions. Figure 6–3 briefly summarizes views of how solid phases in soils control metal ion activities in soil solution. Various methods were used to develop the activity lines shown here, some are based on empirical measurements while others represent the solubilities of minerals taken from the literature. The equilibrium reactions corresponding to these lines are defined and referenced in Table 6–1.

The activity of Ca^{2+} shown at $10^{-2.5}$ at low pH is most likely controlled by cation exchange reactions in soils (Lindsay, 1979). Although Ca^{2+} activity may vary slightly from soil to soil, observed activities are near or slightly below this value. There are no known Ca minerals that are sufficiently insoluble to control Ca^{2+} at this level in acid soils. At higher pH, Ca^{2+} can be controlled by $CaCO_3$ (calcite) and $CO_2(g)$. The calcite line in Fig. 6–3 is drawn for a $CO_2(g)$ partial pressure of 0.3 kPa (0.003 atm.), which is approximately 10 times that of the atmosphere, and corresponds to levels often found in respiring soils where $CO_2(g)$ is given off.

The activity of Mg^{2+} in soil solution generally is slightly lower than that of Ca^{2+}, and exchangeable Mg^{2+} generally is lower than exchangeable Ca^{2+}. The reference activity of Mg^{2+} shown here is 10^{-3} for acid and near-neutral soils, and in calcareous soils, a level similar to that of calcite-dolomite in equilibrium with

CHEMISTRY OF THE SOIL SOLUTION

Table 6–1. Equilibrium reactions used to construct the metal solubility diagram shown in Fig. 6–3.

Equilibrium reaction	Reference	log K^o
Soil-Ca $\rightleftharpoons$ Ca^{2+}	Lindsay, 1979	−2.50
$CaCO_3$(calcite) + $2H^+$ $\rightleftharpoons$ Ca^{2+} + $CO_2(g)$ + H_2O	Lindsay, 1979	9.67
Soil-Mg $\rightleftharpoons$ Mg^{2+}	Lindsay, 1979	−3.00
$MgCa(CO_3)_2$ + $2H^+$ $\rightleftharpoons$ Mg^{2+} + $CO_2(g)$ + H_2O + $CaCO_3$ (calcite-dolomite)	Lindsay, 1979	9.63
$MnCO_3$ (R) + $2H^+$ $\rightleftharpoons$ Mn^{2+} + $CO_2(g)$ + H_2O (rhodochrosite)	Lindsay, 1979	8.08
MnOOH (M) + $3H^+$ + e^- $\rightleftharpoons$ Mn^{2+} + $2H_2O$ (manganite)	Lindsay, 1979	25.27
MnO_2 (P) + $4H^+$ + $2e^-$ $\rightleftharpoons$ Mn^{2+} + $2H_2O$ (pyrolusite)	Lindsay, 1979	41.89
Soil-Cd + $2H^+$ $\rightleftharpoons$ Cd2+	El-Falaky et al., 1991	6.50
$PbCO_3$(cerrusite) + $2H^+$ $\rightleftharpoons$ Pb^{2+} + $CO_2(g)$ + H_2O	Kalbasi et al., 1995	5.32
Soil-Ni + $2H^+$ $\rightleftharpoons$ Ni^{2+}	Aboulroos et al., 1998, unpublished data	6.95
Soil-Zn + $2H^+$ $\rightleftharpoons$ Zn^{2+}	Lindsay, 1979	5.80
Soil-Cu + $2H^+$ $\rightleftharpoons$ Cu2+	Lindsay, 1979	2.80
$Al_2Si_2O_5(OH)_4$ + $6H^+$ $\rightleftharpoons$ $2Al^{3+}$ + $2H_4SiO_4^o$ + H_2O (kaolinite)	Lindsay, 1979	5.45
SiO_2(quartz) + $2H_2O$ $\rightleftharpoons$ $H_4SiO_4^o$	Lindsay, 1979	−4.00
$Al(OH)_3$(gibbsite) + $3H^+$ $\rightleftharpoons$ Al^{3+} + $3H_2O$	Lindsay, 1979	8.04
$Fe(OH)_3$(soil-Fe) + $3H^+$ $\rightleftharpoons$ Fe^{3+} + $3H_2O$	Norvell & Lindsay, 1982a	2.70
Fe^{3+} + e^- $\rightleftharpoons$ Fe^{2+}	Lindsay, 1979	13.04

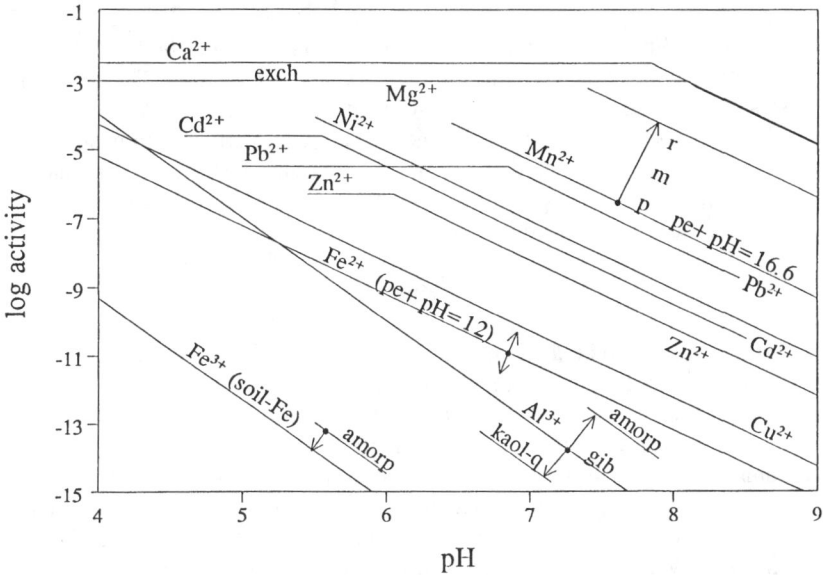

Fig. 6–3. Solid phase controls of metal ions in soils based on equations given in Table 6–1.

$CO_2(g)$ at 0.3 kP (0.003 atm). Mg-montmorillonite in equilibrium with SiO_2(quartz) also can control Mg^{2+} activity near the calcite-dolomite line (Lindsay, 1979). Generally Ca^{2+} and Mg^{2+} dominate the cation exchange of soils except for highly acid soils where exchangeable Al^{3+} becomes significant.

The solubility controls for Mn^{2+} in soils are somewhat more complex than Ca^{2+} and Mg^{2+}. The activities of Mn^{2+} controlled by $MnCO_3$ (rhodochrosite), MnOOH (manganite), and MnO_2 (pyrolusite) depending upon redox (pe + pH) and $CO_2(g)$ are shown in Fig. 6–3. The solubility of Mn also can be depressed by formation of $MnPO_4 \cdot 1.5\ H_2O$ (Boyle & Lindsay, 1986). Thus, the level of Mn^{2+} will, in turn, be influenced by minerals controlling phosphate (Lindsay & Brennan, 1992).

The activities of Pb^{2+}, Ni^{2+}, Cd^{2+}, Zn^{2+}, and Cu^{2+} shown in Fig. 6–3 were determined by the chelation method of El-Falaky et al.(1991), Aboulroos et al. (1998, unpublished data), Norvell and Lindsay (1969), Lindsay (1979), Lindsay and Norvell (1969). Tawfic (1990) observed somewhat similar solubility controls for Cu^{2+} and Pb^{2+}. These metal activities show a strong pH dependence in slightly acid and alkaline soils, reflecting a 2-log-unit decrease in activity for each unit increase in pH. At low pH, the activity of many of these metal ions tends to remain constant as observed by Brennan and Lindsay (1996).

The activity line shown for Al^{3+} in Fig. 6–3 represents equilibrium with $Al(OH)_3$ (gibbsite) with shifts when $Al(OH)_3$ (amorp) or kaolinite-quartz controls Al^{3+} activity (Lindsay, 1979). The solubility line for Fe^{3+} is that in equilibrium with soil-Fe obtained from chelation measurements of Norvell and Lindsay (1982b). The solubility line for Fe^{2+} in Fig. 6–3 represents equilibrium with soil-Fe and the redox couple (Fe^{3+}/Fe^{2+}) at pe + pH 12. This line moves up or down one log unit for each unit change in pe + pH. Again, the activity lines depicted in this diagram can all be expressed mathematically by the equilibrium reactions in Table 6–1.

Figure 6–3 provides a glimpse into our understanding of how solid phases in soils tend to control the activities of metal ions in soil solution. The solids present in soils are often highly variable, some are crystalline minerals, others are amorphous solids, and still others are mixed precipitates whose composition and solubility relationships are ill- defined. Activity measurements in soils are the key parameters that help us decipher the chemical reactions that are occurring and the solubility relationships that are being maintained.

6–4.2 Future Needs

Much remains to be done in characterizing the solid phases that are present in soils and in identifying those phases that are most significant in controlling solubility relationships. The presence of a mineral is no proof that its constituent ions in solution are controlled by that mineral. For example, a soil may contain any number of the following minerals: Fe_2O_3 (hematite), FeOOH (goethite), FeOOH (lipidicrocite), Fe_2O_3 (maghemite), and $Fe(OH)_3$ (amorp), yet only one of these minerals can attain equilibrium with the soil solution at any time. Norvell and Lindsay (1982b) showed that addition of Fe(III) salts to soils initially precipitates $Fe(OH)_3$ (amorp) that slowly reverts to what is referred to as

Fe(OH)$_3$(soil-Fe) having a solubility 0.84 log units lower than Fe(OH)$_3$ (amorp) (Lindsay, 1979). These observations may seem trivial, but they are important in understanding the transformation of Fe oxides in soils and how solid phases control Fe solubility in soils.

There is need to prepare crystalline and amorphous minerals suspected of being present in soils. Measurements of the solubilities of these minerals in the presence and absence of soils will help to confirm solubility measurements and provide supporting evidence of the applicability of chemical equilibrium relationships in soils. Also as inroads are made in spectroscopic confirmation of the nature of reactive surfaces, we will be better equipped to understand surface as well as solid phase chemistry.

6–5 REDOX RELATIONSHIPS IN SOILS

6–5.1 Major Redox Driving Forces

The driving forces that alter the redox status of soils are (i) the release of electrons and protons through biological respiration and (ii) the availability of O_2 as an electron acceptor. The first reaction can be represented as follows:

$$\text{Organic matter} \rightleftharpoons e^- + H^+ + CO_2 + \text{mineral salts} \qquad [4]$$

where electrons, protons, and CO_2 are released to the environment. The electrons combine with numerous receptors or oxidizing agents to carry out biochemical reactions, but O_2 is the ultimate electron acceptor in well-drained soils:

$$e^- + H^+ + 1/4\ O_2 \rightleftharpoons 1/2\ H_2O \qquad [5]$$

Thus, respiration processes release reduced products to the environment, and O_2 combines with the reductants to form water. Photosynthesis recycles CO_2 and H_2O to synthesize glucose and to regenerate O_2.

Reactions 4 and 5 take place continually in soils. If soils are waterlogged, O_2 flow is impeded, and the soil environment becomes reduced. When soils are drained, O_2 flow returns, and soils again become oxidized. These fluctuating redox conditions have a dramatic effect on the chemistry and solubility relationships of redox-sensitive elements in soils. Because of changing redox conditions, soil scientists have often been frustrated in trying to interpret redox measurements in soils.

6–5.2 Fixed pe + pH Poises

An electron titration curve of a soil having a hypothetical mineral composition is shown in Fig. 6–4. This titration curve shows the sequence of redox poises that are encountered when an oxidized soil is slowly reduced as may occur when a soil becomes waterlogged and the supply of $O_2(g)$ diminishes. Key to understanding this development is the redox parameter pe + pH, a term that often intimidates some soil chemists. The flat portions of the curve represent redox poises that result when two minerals containing a common element in different oxidation states are subjected to changing redox conditions.

Let us consider a soil that is initially well-oxidized and contains MnO_2 (pyrolusite), which controls Mn^{2+} solubility. As this soil is slowly reduced, a redox is reached where MnOOH(manganite) can precipitate. The redox at which this can occur can be obtained from the equilibrium reaction corresponding to Reaction 1 in Fig. 6–4 which is:

$$\beta\text{-}MnO_2 + H^+ + e^- \rightleftharpoons \gamma\text{-}MnOOH \qquad \log K^\circ = 16.62 \qquad [6]$$
$$\text{pyrolusite} \qquad\qquad\qquad \text{manganite}$$

This reaction indicates that the two solids can co-exist at equilibrium at pe + pH 16.62. Addition of electrons and protons drives the reaction to the right converting pyrolusite into manganite, however the equilibrium redox potential remains fixed at 16.62 until the transformation is complete. Only after pyrolusite is completely dissolved will pe + pH decrease. The plateau at pe + pH 16.62 identifies the redox poise where this mineral transformation occurs. Above this redox MnO_2(pyrolusite) is stable, and below it, MnOOH(manganite) is stable.

Numerous equilibrium reactions can be examined to identify the pe + pH at which various mineral transformations can take place. Eleven such redox pois-

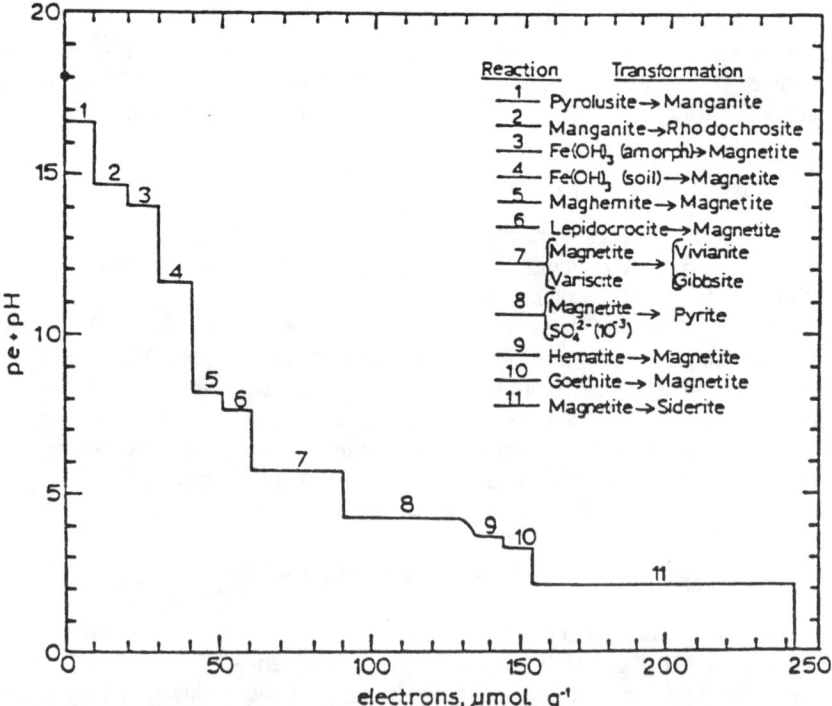

Fig. 6–4. The electron titration curve for a hypothetical soil showing redox poises (Lindsay & Sadiq, 1983).

es are shown in Fig. 6–4. This diagram is very useful for showing the specific redox values at which various minerals can accept electrons and be transformed to minerals of lower oxidation states. When a reduced soil is subjected to oxidizing conditions, the reverse reactions occur, that is, reduced minerals give up electrons at the corresponding pe + pH plateaus. Additional redox transformations can be included in Fig. 6–4 so long as the equilibrium constants for the solid phases that are involved are known.

6–5.3 Experimental Redox Relationships

The results of a first attempt to obtain an electron titration of soils to determine if the predicted pe + pH redox plateaus can be detected and identified are shown in Fig. 6–5. In this study the reductant was $Sn(OH)_2$ and the oxidant was Ag_2O. The redox agents were allowed to react with soil in a 1:2 soil to solution suspension for 7 d in one trial and for 14 d in another. The 14-d curve showed pe + pH plateaus where $Fe(OH)_3$(amorp) is expected to transform to Fe_3O_4(magnetite) at pe + pH 14.0, and an even longer plateau, where $Fe(OH)_3$(soil) is expected to transform to Fe_3O_4 (magnetite) at pe + pH 11.5.

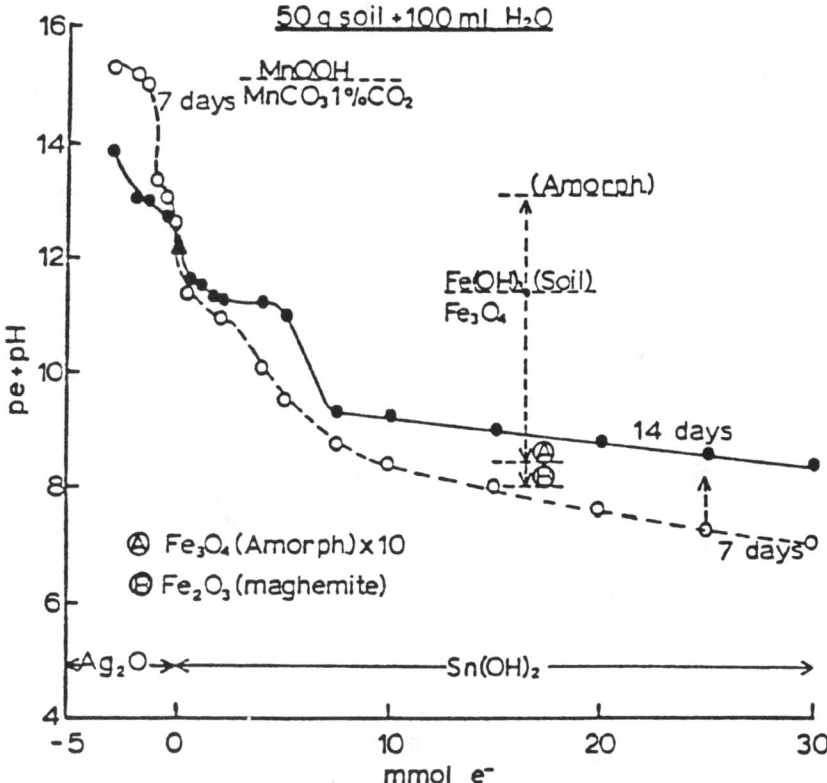

Fig. 6–5. Experimental electron titration curve for soil using Ag_2O as an oxidant and $Sn(OH)_2$ as a reductant showing potential mineral redox poises (Lindsay & Sadiq, 1983).

A longer plateau occurred in the pe + pH range of 8 to 10 and shifted to slightly higher pe + pH values with time. This longer redox poise can be explained by the following equilibrium reactions:

$$3Fe(OH)_3(amorp) + e^- + H^+ \rightleftharpoons Fe_3O_4(amorp) + 5H_2O \qquad \log K^° = 9.09 \qquad [7]$$

$$pe + pH = 9.09 \qquad [8]$$

which shows that when an oxidized soil is reduced, $Fe(OH)_3$(amorp) accepts electrons near pe + pH 9.09 to form Fe_3O_4(amorp). The reverse reaction occurs when a reduced soil is oxidized, that is, Fe_3O_4(amorp) gives up electrons near pe + pH 9.09 to form $Fe(OH)_3$(amorp). The usefulness of Eq. [7] and [8] for describing the large redox poise near pe + pH 9 was demonstrated in a recent study by Brennan and Lindsay (1998). The results from one of their experiments are shown in Fig. 6–6.

In this study a suspension of $Fe(OH)_3$(amorp) was enclosed in a redox cell and first subjected to reduction. This was accomplished by a gas mixture containing 1% H_2, 5% CO_2, and 94% Ar, which was slowly passed through the suspension in a redox cell. The solubility of Fe^{2+} expressed as (log Fe^{2+} + 2pH) moved upward along the $Fe(OH)_3$(amorp) solubility line as reduction occurred and pe + pH was decreased. At low redox the solubility points dropped down to the Fe_3O_4(amorp) solubility line. The suspension was then slowly oxidized by removing H_2(g) from the gas mixture and injecting very small increments of air containing atmospheric O_2 (g). The solubility points then moved down along the

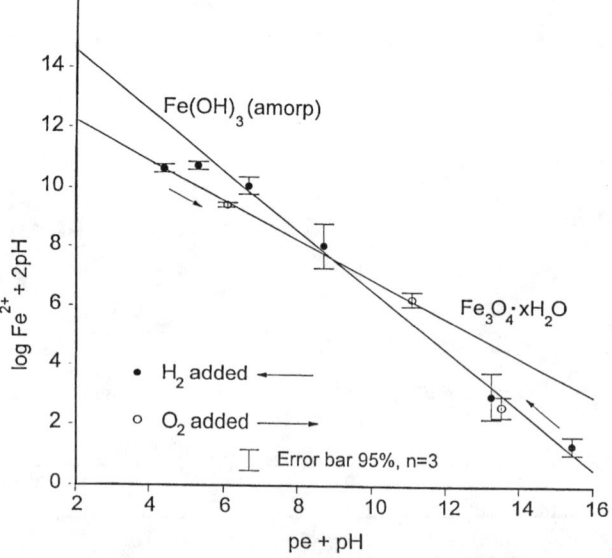

Fig. 6–6. The effect of redox on the solubility of Fe in a Cryaquoll during a reduction cycle with H_2(g) followed by an oxidation cycle with O_2(g) compared with the solubilities of $Fe(OH)_3$(amorp) and Fe_3O_4(amorp; Brennan & Lindsay, 1998).

Fe$_3$O$_4$(amorp) solubility line and eventually dropped back to the more stable Fe(OH)$_3$(amorp) solubility line, thus completing the redox cycle.

Intersection of the two amorphous solubility lines represents the unique pe + pH at 9.09 where the two amorphous solids can co-exist (Eq. [7] and [8]). This study shows that Fe(OH)$_3$(amorp) initially controlled Fe solubility, but upon reduction, the solubility control shifted to the more stable Fe$_3$O$_4$(amorp). When O$_2$ was slowly admitted, Fe solubility followed the Fe$_3$O$_4$(amorp) solubility line but eventually shifted to the more stable Fe(OH)$_3$(amorp). Had the redox changes been made more gradually, Fe solubility probably would have followed very closely along the two amorphous solubility lines. This study helps to confirm the importance of these two amorphous solids in controlling Fe solubility as predicted by Eq. [4] and [5].

The electron titrations and experimental results reported here demonstrate that mineral transformations indeed occur in soils as a result of changing oxidizing and reducing conditions. It is not surprising that many of the redox poises are controlled by amorphous solids that attain equilibrium with soil solution much more rapidly than do many of the crystalline phases. Many of the initial amorphous precipitates can be expected to transform to more stable crystalline minerals, which will eventually control the composition of soil solution. Changing redox conditions often lead to the precipitation of new amorphous phases that delay the attainment of equilibrium among the crystalline phases.

Another important factor affecting redox reactions in soils is the degree to which redox couples attain equilibrium. If a redox reaction requires a high activation energy, that reaction will likely lag theoretical predictions. The electron titration curves predicted in Fig. 6–4 need to be examined to determine the exact pe + pH at which the predicted redox reactions take place. The presence of catalysts and enzymes in soils can be expected to affect selected reactions. We need to know which reactions are affected and how much they are affected. The thermodynamic redox poises as described herein are valuable for setting up experiments and testing the predicted transformations that are expected to occur.

6–6 CHEMICAL SPECIATION MODELS IN SOIL CHEMISTRY

Chemical speciation models and computers have ushered in a new era in soil chemistry. These tools are useful for predicting and interpreting the chemical reactions that occur in soils and for partitioning total elemental concentrations in solution into specific chemical species. Equilibrium calculations of chemical reactions that once took weeks to solve, can now be solved in seconds. A recent SSSA special publication entitled *Chemical Equilibrium Reaction Models* (Loeppert et al., 1995) reflects how recent developments in this area have already impacted soil chemistry. Graduate soil chemistry courses in the future will undoubtedly include the development and use of computer speciation models.

There are different views as to whether or not undergraduate soil chemistry should include extensive work with chemical speciation models. At that stage of learning it is probably more important that students strengthen their background in chemistry and mineralogy to obtain a stronger foundation for dealing with chemical equilibrium modeling at the graduate level. Undergraduate students might best be introduced to equilibrium modeling by setting up a few simple problems on a computer spreadsheet just to get a feel of some of the basic principles of modeling.

We have found that graduate students, on the other hand, need to be challenged with chemical speciation and computer modeling. Examining the dissolution and precipitation of solid phases by using chemical speciation models is an important part of understanding soil chemistry and solubility relationships. In a recent publication Lindsay and Ajwa (1995) demonstrated how MINTEQA2 Version 3.11 equipped with a Revised Data Base dated 11 August 1993, can be used in teaching soil chemistry at the graduate level. The Revised Data Base can be downloaded from the Colorado State University web page at http://www.colostate.edu/Depts/SoilCrop/mnteq.htm.

There are many challenges for soil chemists that relate to the use of chemical speciation models in soil chemistry. There is always the challenge to upgrade software to make it more convenient to use, but just as important is the need to update the solubility and adsorption equilibrium constants that are used in the various speciation models. It is not enough to assume that the constants are correct just because they are included in a large data base. These constants need to be examined, screened, and tested individually to be sure they are suitable for the chemical environments in which they are used. Such upgrading will require years of cooperative effort on the part of many soil chemists, geochemists, and environmentalists.

We have attempted to make selections of these equilibrium constants that are especially useful in soil chemistry (Sadiq & Lindsay, 1979). Since then various supplements have been added: Abdelmonem and Lindsay (1989), Elrashidi et al. (1988), Elrashidi and Lindsay (1984), Karmarker et al. (1992), Michail and Lindsay (1988), Reddy et al. (1989), Reddy et al. (1990), Sadiq et al. (1980). Sadiq and Lindsay (1981a), Sadiq and Lindsay (1981b). Further work is needed to update equilibrium constants for use in soil chemistry for the future.

6–7 CHALLENGES FOR THE FUTURE

In this chapter we have shared a few perspectives on the soil solution, and areas where we think progress can be made in the future. The focus has been on major factors that affect the composition of the soil solution. The examples provided here were taken largely from research investigations from our laboratory during the past few years. The following challenges are offered to future soil chemists:

1. Search for more precise ways to measure ion activities in soils. This search should include the problems of unknown metal-organic complexes, interferences with ion specific electrodes, and improvements in the chelation method of measuring free metal ion activities.
2. Expand our knowledge of the solid phases in soils that are largely responsible for fixing the activities of ionic species in soil solution.

Include among these phases amorphous solids that often dominate solubility relationships in soils. Test these solubility relationships in the presence of soil and in the presence of the hypothesized solid phases. Study the adsorption of solution components onto soil surfaces and express them in quantitative terms that can be used in chemical speciation and adsorption models.
3. Examine redox relationships in soils to dispel the myth that redox measurements in soils are meaningless. Examine further the use of pe + pH as a redox parameter for understanding redox poises and accompanying redox-related mineral transformations. Examine the kinetics of important oxidation and reduction reactions in soils in order to explain differences between observed and predicted equilibrium relationships.
4. Recognize that chemical speciation models and computers are indispensable to soil chemistry and will become even more important in the future. A rigorous course on chemical equilibria and speciation modeling should be required of all graduate students claiming expertise in soil and environmental chemistry.
5. Develop an improved data base of equilibrium constants and adsorption reactions for use in chemical speciation and adsorption models. Test these models to determine their strengths, weaknesses, and application to soil chemical systems.
6. Make use of modern spectroscopic and analytical methods for identifying and characterizing chemical species in soil solution.
7. Examine more critically the importance of organic matter, its surface adsorption and exchange reactions, the stability of organic pollutants, and the identification and characterization of metal–organic complexes that modify the composition of the soil solution and affects the transport of metals in soils and groundwater.

REFERENCES

Abdelmonem, M.A.S., and W.L. Lindsay. 1989. Cobalt supplement to Technical Bulletin 134: Selection of standard free energies of formation for use in soil chemistry. Tech. Bull. LTB89-3. Colorado State Univ. Exp. Stn., Fort Collins.
Brennan, E.W., and W.L. Lindsay. 1996. The role of pyrite in controlling metal ion activities in highly reduced soils. Geochim. Cosmochim. Acta 60:3609–3618.
Brennan, E.W., and W.L. Lindsay. 1998. Reduction and oxidation effect on the solubility and transformation of iron oxides. Soil Sci. Soc. Am. J. 62:930–937.
Boyle, F.W., Jr., and W.L. Lindsay. 1986. Manganese phosphate equilibrium relationships in soils. Soil Sci. Soc. Am. J. 50:588–593.
El-Falaky, A.A., S.A. Aboulroos, and W.L. Lindsay. 1991. Measurement of Cd^{2+} activities in slightly acidic to alkaline soils. Soil Sci. Soc. Am. J. 55:974–979.
Elrashidi, M.A., D.C. Adriano, and W.L. Lindsay. 1988. Selenium supplement to Technical Bulletin 134: Selection of standard free energies for use in soil chemistry. Tech. Bull. LTB88-3. Colorado State Univ. Exp. Stn., Fort Collins.
Elrashidi, M.A., and W.L. Lindsay. 1984. Fluorine supplement to Technical Bulletin 134: Selection of standard free energies of formation for use in soil chemistry. Colorado State Univ. Exp. Stn., Fort Collins.
Kalbasi, M., F.J. Peryea, W.L. Lindsay, and S. Drake. 1995. Measurement of divalent lead activity in lead–arsenate contaminated soils. Soil Sci. Soc. Am. J. 59:1274–1280.

Karmarker, S.V., M.A. Tabatabai, and W.L. Lindsay. 1992. Tungsten supplement to Technical Bulletin 134: Selection of standard free energies of formation for use in soil chemistry. Tech. Bull. LTB92-1. Colorado State Univ. Exp. Stn., Fort Collins.

Lindsay, W.L. 1979. Chemical equilibria in soils, Wiley-Interscience, New York.

Lindsay, W.L., and H.A. Ajwa. 1995. Use of MINTEQA2 in teaching soil chemistry. p. 210–231. *In* R.H. Loeppert et al. (ed.) Chemical equilibrium reaction models. SSSA Spec. Publ. 42. SSSA, Madison, WI.

Lindsay, W.L., and E.W. Brennan. 1992. Effect of redox on phosphorus transformations in soils. p. 7–9. *In* F.J. Sikora (ed.) Future directions for agricultural phosphorus research. TVA Bull. Y-224. Tennessee Valley Authority, Muscle Shoals, AL.

Lindsay, W.L., and W.A. Norvell. 1969. Equilibrium relationships of Zn^{2+}, Fe^{3+}, Ca^{2+}, and H^+ with EDTA and DTPA in Soils. Soil Sci. Soc. Am. Proc. 33:62–68.

Lindsay, W.L., and M. Sadiq. 1983. The use of pe+pH to predict and interpret metal solubility relationships in soils. Sci. Total Environ. 28:169–178.

Lindsay, W.L., and L.E. Sommers. 1997. Complexation of metals by synthetic chelates. *In* D.C. Adriano (ed.) Biogeochemistry of trace metals. Applied Science Publ., Norwood, England.

Loeppert, R.H., S. Goldberg, and A.P. Schwab. 1995. Chemical equilibrium reaction models. SSSA Spec. Publ. 42. SSSA, Madison, WI.

Ma, Q., and W.L. Lindsay. 1990. Divalent zinc activity in arid-zone soils obtained by chelation. Soil Sci. Soc. Am. J. 54:719–722.

Ma, Q., and W.L. Lindsay. 1993. Measurements of free zinc activity in uncontaminated and contaminated soils using chelation. Soil Sci. Soc. Am. J. 57:963–967.

Ma, Q.Y., and W.L. Lindsay. 1995. Estimation of Cd^{2+} and Ni^{2+} activities in contaminated and uncontaminated soils by chelation. Geoderma 68:123–133.

Michail, N.N., and W.L. Lindsay. 1988. Iodine supplement to Tech. Bull. 134: Selection of standard free energies of formation for use in soil chemistry. Colorado State Univ. Exp. Stn., Fort Collins.

Norvell, W.A., and W.L. Lindsay. 1969. Reactions of EDTA complexes of Fe, Zn, Mn, and Cu with soils. Soil Sci. Soc. Am. Proc. 33:86–91.

Norvell, W.A., and W. L. Lindsay. 1982a. Estimation of the concentration of Fe^{3+} and the $(Fe^{3+})(OH^-)^3$ ion product from equilibria of EDTA in soil. Soil Sci. Soc. Am. J. 46:710–715.

Norvell, W.A., and W.L. Lindsay. 1982b. Effect of ferric chloride additions on the solubility of ferric iron in a near-neutral soil. J. Plant Nutr. 5:1285–1295.

Oglesby, J. 1995. Competitive-chelation methods for estimating free metal ion activities. Soil Sci. Soc. Am. J. 59:959.

Reddy, K.J., P.J. Sullivan, M.E. Essington, and W.L. Lindsay. 1989. Strontium supplement to Technical Bulletin 134: Selection of standard free energies of formation for use in soil chemistry. Tech. Bull. LTB89-2. Colorado State Univ. Exp. Stn., Fort Collins.

Reddy, K.J., L. Wang, and W.L. Lindsay. 1990. Molybdenum supplement to Technical Bulletin 134: Selection of standard free energies of formation for use in soil chemistry. Tech. Bull. LTB90-4. Colorado State Univ. Exp. Stn., Fort Collins.

Sachdev, P., W.L. Lindsay, and D.L. Deb. 1992. Activity measurements of zinc in soils of different pH using EDTA. Geoderma 55:247–257.

Sadiq, M., and W.L. Lindsay. 1979. Selection of standard free energies of formation for use in soil chemistry. Tech. Bull. 134. Colorado State Univ. Exp. Stn., Fort Collins.

Sadiq, M., and W.L. Lindsay. 1981a. Arsenic supplement to Technical Bulletin 134: Selection of standard free energies of formation for use in soil chemistry. Colorado State Univ. Exp. Stn., Fort Collins.

Sadiq, M., and W.L. Lindsay. 1981b. Standard free energy of formation of some chemical species: Discrepancies and selections. Arabian J. Sci. Eng. 6:95–104.

Sadiq, M., W.L. Lindsay, and C.G. Enfield. 1980. Thermochemistry of nickel. U.S. Environ. Protection Agency Tech. Serv., Ada, OK.

Tawfic, T.A. 1990. Chemical equilibria of copper and lead in soil environments. Ph.D. diss. Colorado State Univ., Fort Collins.

Vlek, P.L.G., T.J.M. Blom, J. Beek, and W.L. Lindsay. 1974. Determination of the solubility product of various iron hydroxides and jarosite by the chelation method. Soil Sci. Soc. Am. Proc. 38:429–432.

Workman, S.M., and W.L. Lindsay. 1990. Estimating divalent cadmium activities measured in arid-zone soils using competitive chelation. Soil Sci. Soc. Am. J. 54:978–993.

7 Chemistry of Soil Minerals

S. B. Feldman and L. W. Zelazny

Virginia Polytechnic Institute and State University
Blacksburg, Virginia

7–1 INTRODUCTION

Soils are dynamic, open systems that represent the terrestrial interface between the atmosphere and the hydrosphere. About 90% of the volume of most soil solids is comprised of crystalline or paracrystalline, inorganic mineral material (Sposito, 1989) and consist of several trillion square kilometers of mineral surface area that interact with fluids of the Earth's near-surface. Consequently, the study of soils, soil minerals, and weathering processes has been and will continue to be important in the production of food, fiber, and shelter, and is playing an increasingly crucial role in our understanding of complex environmental issues such as surface water quality, reclamation of disturbed lands, and disposal of both hazardous and municipal waste. The reactive surface properties of soil minerals regulate the chemistry of natural waters, for example, by influencing the transport (or immobilization) of excess agricultural nutrients, pesticides, and hazardous leachate in the environment, and by neutralizing many of the potentially detrimental effects of atmospheric (acid) deposition. Soil minerals also provide a medium to store water and support plants in addition to contributing both an available and a labile pool of plant nutrients that often are the only sources of mineral nutrition, besides atmospheric deposition, in unmanaged (e.g., forest) ecosystems. Soil mineral distribution, and regional pedogenic trends documented in residual weathering profiles, transported upland surficial deposits, and relict paleosols also may be among the most permanent records of the geologic history of a soil or landform and are therefore important repositories of information regarding global climate change on the continents. In these respects, the bulk as well as surface chemistry of soil minerals is relevant to virtually all aspects of the applied agricultural and environmental sciences, as well as to the basic sciences of Quaternary geology, geomorphology, geochemistry, pedology, and archaeology.

7–2 ENVIRONMENTAL SIGNIFICANCE OF SOIL MINERALS AND WEATHERING IN NATURAL SYSTEMS

Because of the great importance of clay minerals to industry, agriculture, and the environment, and because of their abundance and the many unique sur-

---

Copyright © 1998. Soil Science Society of America, 677 S. Segoe Rd., Madison, WI 53711, USA.
Future Prospects for Soil Chemistry. SSSA Special Publication no. 55.

face-reactive properties that they impart to soils and sediments, much of the focus of soil mineralogy centers on the formation and properties of phyllosilicates—the clay minerals—and Fe- and Al-oxides, hydroxides, and oxyhydroxides. These minerals govern ion exchange and mobility, provide buffering capacity, and act as adsorbents for organic pollutants, toxic contaminants, and heavy metals in soils and drainage waters. The surface charge properties of clay minerals and oxides also determine the amount and type of water adsorption in soils, and consequently the behavior of soils with respect to their plastic or shrink-swell properties, compressive and shear strength, percolation rate, and permeability to air.

In addition to their influence on soil behavior, clay minerals are used commercially as coatings for paper products, extenders in paints, plastics, and rubbers, as constituents of drilling fluids (for oil–gas exploration) and cracking catalysts in the petroleum industry, as pigments, adhesives, and adsorbents, and as physical barriers to the movements of wastes such as in landfill liners or backfill materials. They also are used extensively in the ceramic industry in the manufacture of glazes, enamels, tiles, and refractories, and in the foundry industry as binding agents in sand molds for metalcast products. High-grade and high brightness specialty clays also are used as pharmaceutical carriers and as the base for some cosmetics.

7–2.1 Soil Development, Weathering Rates, and Application to Geomorphic Problems

Soil minerals are commonly used to indicate weathering rates of surficial materials (White, 1995; Velbel, 1986), determine the age of soils and sediments (Markewich et al., 1987), elucidate mechanisms of soil formation (Pavich, 1989), and reconstruct paleoenvironmental conditions during pedogenesis (Catt, 1991).

Soils form in response to pedogenic processes that are controlled by environmental factors operating through time. As environmental conditions change, soils very often respond to new weathering conditions, recording the magnitude and (or) duration of change. Soil morphology and mineralogy not only respond to environmental change, but soils also can preserve characteristics formed under past conditions thereby providing information about paleoenvironments and previous pedogenic conditions. Deep soils formed under intense weathering conditions, for example, are commonly found in cool-temperate climates where the occurrence of pedogenic kaolinite and Fe-oxides are often attributed to formation under earlier, warmer conditions (Martini & Chesworth, 1992). Preservation of the signal from a formerly dry climate in the soil mineral record is more tenuous, however, as smectites and carbonates, for example, tend not to persist in a change to a more humid climate (Feldman et al., 1999).

Based on comparisons involving such factors as degree of profile development, types of secondary minerals present, the nature of primary mineral surface etching, elemental distribution, and the thickness of rock weathering rinds, soils also have been used to correlate geomorphic surfaces and provide useful estimates of relative ages (Andrews & Miller, 1980; Colman & Pierce, 1981; Harden, 1988). Paleosols formed in loess deposits found throughout central Europe, China, and North America are some of the best known examples of soils used as stratigraphic markers to correlate interglacial periods during the Quaternary (Ellis & Mellor, 1995). Much of the information we have regarding the nature of Quaternary paleoclimates on the continents is derived from loess data.

7–2.2 Surface Water Quality and Watershed Acidification

The chemical composition of surface water is regulated to a large extent by the nature of weathering reactions in soils and by the residence times and various flow paths by which water reaches a stream channel. In other words, because of ion exchange reactions and adsorption–desorption phenomena, water leaving the terrestrial portion of a catchment—or water impacted by runoff from such areas—will have a pH and Al concentration (and concentration of organic species and heavy metals) that reflect the chemistry of the soil horizons through which it flowed (Hendershot et al., 1996).

7–3 SOILS AS UNIQUE WEATHERING PRODUCTS

Clay minerals are not only the most reactive inorganic components of soils, largely governing soil behavior under a wide variety of uses, but they also are among the most complex soil components in terms of their species diversity and variability. In contrast to pure geologic specimens, minerals found in the pedoenvironment are highly altered by weathering, commonly exhibiting an extreme degree of both compositional and structural variability. The variability and complex polymineralic nature of most soils often make identification of mineral suites and quantification of weathering mechanisms and rates exceedingly difficult. In spite of this complexity, however, the need for accurate quantitative information regarding the performance and dynamics of Earth solids has never been greater.

Despite many areas of similarity, soils are different from weathered rock, unweathered sediments, or other geological deposits. Soils usually exhibit distinct vertical zonation (horizons) resulting from relatively aggressive weathering processes that include a complex array of physical, chemical, and biological reactions involving parent materials, usually in a framework of intense hydrological flux. Soils have distinct morphological, chemical, biological, and mineralogical character that is derived from pedogenic, rather than from purely geogenic processes. The pedochemical weathering of soil parent materials results either in the solid-state transformation of inherited (primary) minerals to new phases (Table 7–1), or their dissolution and subsequent formation as new, more stable, authigenic (secondary) species by recrystallization from solution (Table 7–2).

As a consequence of these near-surface weathering processes, minerals found in soils commonly differ in several ways from their counterparts in sediments, rocks, and other geologic environments. Moreover, some minerals found in soils have unique properties and are not found elsewhere. Accordingly, the study of soil mineralogy departs somewhat from traditional petrology or sedimentology and is thus unique in several important respects.

Table 7–1. Common *primary* (residual) minerals present in soil environments in order of increasing stability.

Name	Mineral class	Environment	Ubiquity in soils†	Importance
Pyrite	Sulfides	Tidal marshes (reducing conditions and hard-rock mine tailings (coal and shale beds)	C	Primary mineral (oxidizing conditions) but secondary phase forms in reducing environments; large metal and acidity input to surface waters during weathering
Dolomite	Carbonates	Shallow, young soils formed in limestone	R	Major constituent of limestone parent material; fertilizer source
Pyrophyllite	Phyllosilicates	Low temp. metamorphic and hydrothermal	R	No layer charge—little chemical reactivity; unstable in soils
Talc	Phyllosilicates	Ultramafic rocks	R	No layer charge—little chemical reactivity; unstable in soils
Olivine	Nesosilicates	Basic and acidic igneous rocks	R	Source of Fe, Ca, Mg, and Mn; unstable in highly leached soil
Pyroxenes	Inosilicates	Igneous and contact metamorphic rocks	R	Source of Fe, Ca, Mg, and Mn; unstable in highly leached soil
Amphiboles	Inosilicates	Igneous and metamorphic rocks	R	Source of Fe, Ca, Mg, and Mn; vermiculite precursor
Chlorite (Mg-interlayer)	Phyllosilicates	Metamorphic or igneous rocks	R	Important precursor for 2:1 soil clay minerals
Biotite (Fe^{2+}-bearing micas)	Phyllosilicates	Granitic and high-grade metamorphic rocks	R	Stable in only the youngest or least weathered soils; precursor of other 2:1 soil clay minerals and Fe-oxides; source of K
Feldspars:	Tectosilicates	Wide variety of igneous/metamorphic rocks; persistence in soils and geologic deposits is related to weathering intensity, and the duration of exposure to weathering		Source of Ca, Na, and K; alteration to secondary products is a function of microenvironment; weathering products can be kaolinite, mica, gibbsite, halloysite, smectite, or amorphous materials
Calcic Plagioclase	Tectosilicates		R	
Sodic Plagioclase	Tectosilicates		C	
K-Feldspars	Tectosilicates		C	
Muscovite	Phyllosilicates	Granitic and high-grade metamorphic rocks	C	Vermiculite, smectite, and interstratified 2:1 precursor, K source
Epidote	Sorosilicates	Medium-grade metamorphic and mafic igneous rocks	R	Source of Fe, Ca, Mn; very resistant to weathering
Quartz	Tectosilicates	Nearly all soils and parent materials	U	Concentrated in sand- and silt-fractions; soluble in clay fraction
Garnets	Nesosilicates	High-grade metamorphic/acid igneous rocks	R	Source of Fe, Mg, Ca, Al, and/or Mn/ precursor of Fe-oxides
Tourmaline	Cyclosilicates	Granites and pegmatites; detrital sediments	R	Stable in soils, but may alter to secondary 2:1 clay minerals
Rutile	Oxides	Igneous and metamorphic rocks; detrital sediments; quartz inclusion	C	Very stable in soils, but some mobility of Ti; rarely alters to other oxides
Zircon	Nesosilicates	Acid and basic plutonic, metamorphic rocks	C	Very resistant; used as index minerals in pedologic studies

† U, Ubiquitous; C, Common; R, Rare.

Table 7–2. Common *secondary* silicates and nonsilicates present in soil environments in order of increasing stability.

Name	Mineral class	Environment	Ubiquity in soils[†]	Importance
Halite	Halides	Arid, saline/sodic soils	C	Highly soluble; adverse osmotic effects on plants
Gypsum	Sulfates	Arid soils	C	Moderately-high solubility; used to reclaim sodic soils
Jarosite	Sulfates	Acid sulfate soils; mine overburden, coastal wetlands	C	Product of pyrite oxidation resulting in large production of acidity and toxic metals
Calcite	Carbonates	Arid soils; very limited leaching	C	May act as cementing agent; high P sorption
Pyrite	Sulfides	Tidal marshes (reducing conditions) and hard-rock mine tailings (coal and shale beds)	C	Primary mineral (oxidizing conditions) but secondary phase forms in reducing environments
Allophane/Imogolite	Paracrystalline	Soils derived from volcanic ash deposits	R	Short-range order, highly sorptive
Sepiolite/Palygorskite	Phyllosilicates	Marine sediments, arid soils, high Si and Mg levels	R	Moderately high CEC[‡], surface area, and sorptive properties
Halloysite	Phyllosilicates	Volcanic ash; granitic (feldspathic) saprolite	R	Ephemeral in intensely weathered soils
Vermiculites (dioct.)	Phyllosilicates	Mica alteration in well-drained soils	R	Very high CEC; fixation of K^+; sink for solution Al
Smectites	Phyllosilicates	Mica and/or vermiculite alteration	C	High CEC; high surface area; high shrink-swell capacity
HIV[§]	Phyllosilicates	Acid, highly weathered soil surface horizons	U	Variable CEC (degree of Al interfilling); high anion adsorption
Kaolinite	Phyllosilicates	Desilication of 2:1 clays/feldspathic	U	Low charge but highly pH-dependent; high anion adsorption
Hematite & Goethite	Oxides	Well-drained, near-surface soil	U	Low charge but highly pH-dependent; high anion adsorption
Gibbsite	Oxides	Old, stable soils or feldspar pseudomorphs	C	Low charge but highly pH-dependent; high anion adsorption
Anatase	Oxides	Dissolution of Ti-bearing parent minerals	R	Essentially chemically inert

[†] U, Ubiquitous; C, Common; R, Rare.
[‡] CEC, cation-exchange capacity.
[§] HIV, hydroxy-interlayered vermiculite.

7–3.1 High Clay Content and Dominance of Secondary Phases

Weatherable primary minerals are largely unstable in soils (Loughnan, 1969). Even the most resistant primary minerals in soils (e.g., quartz, zircon, and rutile) are altered to some degree (Carroll, 1953), making identification of the more easily weathered ones considerably more difficult than is the case with fresh, intact geologic specimens. Accurate identification of single grain species or minerals in thin section using a petrographic microscope, for example, is often hampered by an inability to observe crystallographic properties because of secondary coatings, pseudomorphic intergrowths, or intense dissolution etching.

In many soils, particularly those of humid-temperate or tropical regions where weathering intensity is great, nearly all the weatherable primary minerals have either transformed via dissolution and reprecipitation to secondary phases, or they have been significantly altered in the solid state. Secondary minerals in soils are typically so finely divided that analysis and identification requires a combination of physical, chemical, and instrumental techniques.

7–3.2 Compositional and Structural Variability

Soils also are unique in that they contain an impressive array of minerals, dependent primarily on the nature of the parent material, the age and stability of the geomorphic surface, and the intensity of pedochemical weathering (Allen & Hajek, 1989). Soil clay fractions in particular tend to be highly complex polymineralic mixtures, often composed of >10 or 15 different species, each with considerable variability. Many clay minerals exhibit characteristics of a solid-solution series, with continuous variation in properties such as degree of crystallinity, chemical and structural (order-disorder) composition, and magnitude of surface charge, even within a single mineral species. This variability is not surprising when one considers the diverse possibility of chemical conditions controlling dissolution and reprecipitation reactions at the microscopic level within a soil. As mentioned above, mineralogical variability in soils makes the analysis, identification, and partitioning of discrete species, and especially quantification of their abundance, very difficult, even with modern techniques and instrumentation.

Soils undergo weathering processes that are influenced by factors not commonly attendant in geologic environments. These factors include the influence of organic acids in chelating and sequestering metal ions; higher CO_2 levels and fluctuating pH conditions due to plant root and microbial activity (several chapters in Huang & Schnitzer, 1986); the role of intense leaching by geochemically aggressive, low ionic-strength drainage waters; fluctuating redox conditions; and commonly, the existence of preweathered parent materials that have been transported by eolian, colluvial, or fluvial action.

The pseudomorphic (solid-state) replacement of one mineral by another is one example of a relatively poorly understood weathering mechanism responsible for considerable mineralogical variability in soils. Harris et al. (1985a,b), for example, has shown that the dominant alteration product of biotite in mesic (mean annual soil temperature, MAST, 8 to 15°C) and warmer southeastern soils of the USA is kaolinite. This pseudomorph retains the optical properties of biotite when examined under a petrographic microscope (e.g., $2V_\alpha = 5 - 15°$, brown–green pleochroism, and others), but even within single grains, it displays a 7.2Å x-ray diffraction spacing characteristic of kaolinite. Conversely, Feldman et al. (1991) found that vermiculitization through an interstratified intermediary, and not kaolinization of biotite, was the dominant weathering mechanism in the frigid (MAST <8°C) soils at the higher elevations of the southern Blue Ridge. The results of Ghabru et al. (1989) for Canadian Spodosols corroborates this temperature-dependence of biotite weathering, but quantitative limits for this effect and the influence of high levels of soil organic acids are not known. Pseudomorphic replacement of both plagioclase and K-feldspars also has been shown to occur widely in soils. Feldspar weathering products include noncrystalline aluminosilicates, gibbsite, kaolinite, halloysite, smectite, or vermiculite (Cady, 1950; Eswaran & Bin, 1978; Calvert et al., 1980; Douglas, 1989; Weaver, 1989), dependent upon the nature of the pedochemical environment.

Interstratified phases of 2:1 phyllosilicates also commonly occur in both soils and geologic environments; however, in soils these minerals tend to exist only as quasi-stable species, formed as initial weathering products of biotite, muscovite, or chlorite minerals. Interstratified minerals, such as mica-vermiculite, chlorite-vermiculite, or mica-smectite, contain sequences of different kinds of layers stacked along the c-dimension. They form in geologic environments primarily by hydrothermal alteration of micas, but in soils they form from weathering processes involving the partial removal of interlayer K from alternating layers within micas, or by removal of hydroxide interlayers from chlorite (Sawhney, 1989). Accompanying these transformations are structural changes that result in the reduction of overall surface charge. Long-term weathering of interstratified species eventually results in the formation of individual phases of their respective components.

Although vermiculites are common in both geologic and pedologic environments, only the dioctahedral (Al) varieties of this mineral are stable and persist in soils (Kittrick, 1977). Trioctahedral (Mg) varieties of vermiculite are primary minerals commonly found in sedimentary rocks, but these minerals are found only in the least weathered environments. Dioctahedral (Al) vermiculites form as alteration products of muscovite, biotite, or chlorite and are very stable even in severe weathering environments (Gilkes & Suddhprakarn, 1979; Douglas, 1989; Fordham, 1990; Pozzuoli et al., 1994; Feldman et al. 1999). Under acid weathering conditions, soil vermiculites, unlike their geologic–trioctahedral counterparts, act as sinks for Al released to solution by weathering. They almost invariably have a hydroxy-Al (and/or Fe) interlayer that imparts a great deal of stability to the structure (Rich & Obenshain, 1954; Rich, 1958). In fact, hydroxy-interlayered dioctahedral vermiculite (HIV) is often recognized as the most abundant clay mineral in surface horizons of soils containing mica in the parent material (Douglas, 1989). The degree of interlayer filling (by hydroxy-Al polymers) in this mineral reflects the intensity of the weathering environment, and the age and acidity of the soil. There is a continuum between pedogenic chlorite (fully interlayered) and the non-hydroxy-interlayed vermiculitic end-members. In addition to increasing stability, the Al-interlayers also create a 'wedge site' that has an adsorption specificity for ions of small hydrated size (i.e., K^+, , H_3O^+, Ag^+).

Authigenic formation of secondary aluminosilicates by recrystallization from solution is a common weathering mechanism; however, in soils this process often involves coprecipitation or incorporation of a trace element within a host mineral if the size and valency of the substituting element are comparable to those of the element replaced (Sposito, 1989). Coprecipitation of Al with Fe-oxides such as goethite and hematite is a nearly ubiquitous phenomenon in soils. A low degree of crystallinity also has been shown to facilitate incorporation of foreign elements into existing Fe-oxide crystal lattices (Kühnel et al., 1975). In goethite, Al substitution ranges from 0 to 33 mole percent, reaching a maximum in highly weathered soils of subtropical and tropical areas (Fitzpatrick & Schwertmann, 1982; Schwertmann, 1985). Higher levels of Al within Fe-oxide structures have been shown to result in changes in crystal growth and morphology (Fontes & Weed, 1991) and a corresponding increase in thermodynamic and kinetic stability than their unsubstituted counterparts (Fitzpatrick & Schwertmann, 1982; Bryant & Macedo, 1990). These factors have a direct bearing on the sorptive characteristics of Fe-oxide minerals and therefore their interaction with hazardous materials and pollutants in the environment. Whereas Al substitution in soil Fe-oxides is quite common, it is interesting to note that there are no reports of similar Fe for Al substitution in gibbsite and other Al-oxides in soils.

Many coprecipitation or surface adsorption reactions that are mineral specific and ion specific have been reported. Small trivalent metals like Cr^{3+} and Mn^{3+} coprecipitate with both Fe and Al oxides and hydroxides, while Co^{2+} is often associated with Mn oxides in soils (McBride, 1994). The trace metals Cu^{2+}, Zn^{2+}, Ni^{2+}, and Co^{2+} substitute readily into magnetite during its crystallization but are excluded from hematite (McBride, 1994). The small divalent metals Zn^{2+}, Cu^{2+}, and Mg^{2+} form solid solutions with Al hydroxides with structures (hydrotalcites) containing high anion exchange capacities (McBride, 1994). Surface coatings on mineral grains, including precipitates of Al-, Fe-, and Mn- oxyhydroxides (Coston et al., 1995; Zachara et al., 1995) or secondary minerals including carbonates (Zachara et al., 1988) also can play an important role in the uptake of trace metals in natural systems. Apatite mineral additions to Pb, Zn, and Cd contaminated soils have been used to induce geochemically stable trace-metal precipitates resulting in acceptable Pb bioavailability levels (Ma et al., 1993).

In addition to coprecipitation of trace elements within secondary phases, solid-state substitutions of ions commonly occur within the lattice positions of many soil phyllosilicates. This solid-state substitution of lower valence for higher valence ions, such as Fe^{3+} or Al^{3+} for Si^{4+} in tetrahedral sheets, and Mg^{2+}, Fe^{2+}, or Li^+ for Al^{3+} or Fe^{3+} in octahedral sheets, results in considerable compositional variability and inconsistent degree of surface charge even within a single mineral species. Although this process of surface charge development has often been referred to as isomorphous substitution, it is now recognized that micro-surface morphology has dramatically been altered. These alterations include an increase in base surface corrogation, changes in the ditrigonal silanol cavity size and coordination, interlayer overshifts, octahedra distortions, hydroxyl orientation shifts and sheet thinning (Bailey, 1984).

Solid-state substitution accounts for the majority of permanent surface charge of the phyllosilicates, and consequently, much of the cation-exchange capacity of mineral soils except Oxisols. Pyrophyllite and talc, the unsubstituted or zero-permanent-charge end-members of a solid-solution series of the 2:1 phyllosilicates, are found in ultramafic rocks and hydrothermal deposits, but they are rarely found in near-surface environments (White & Zelazny, 1986). Exceptions have been reported by Harris et al. (1984) and others who noted that lithic fragments of talc, inherited from ultramafic parent material, were being preserved in a highly-weathered soil by the armoring effects of secondary Fe-oxide coatings formed during weathering.

Finally, additional soil components that are not commonly found elsewhere include the amorphous and paracrystalline phases, or those aluminosilicates whose repeating structures persist only over very short distances (Wada, 1989). These materials form in volcanic ash deposits or in young soils with freshly precipitated aluminosilicates. They range continuously from noncrystalline to paracrystalline in nature, with little enhancement of observed x-ray diffraction peaks due to preferred orientation effects. With the exception of infrared spectroscopy and transmission electron microscopy (TEM), many of the conventional instrumental techniques used in soil mineralogical analysis, and consequently, the atomic structures and morphology of amorphous materials in soils have traditionally been poorly understood. Quantitative, or semiquantitative measurement of these materials is usually approached by using selective dissolution techniques, which vary in their efficacy of selectivity (Jackson, et al. 1986).

7–4 APPROACHES TO THE STUDY OF SOIL MINERALOGY

Mineralogical characterization has traditionally been used to assess the macroscopic physical and chemical behavior of soils in order to predict fertility status for production agriculture or the engineering properties of soils. More commonly during the last two decades, studies involving the surface charge, ion sorption, mobility and bioavailability, and weathering characteristics of clay minerals have taken on a new significance as the critical role of soils is being recognized in global environments ranging from geochemical cycling of elements to the reconstruction of Quaternary paleoclimates. Many of the recent advances in the fundamental and theoretical understanding of geochemical reactions occurring at the solid–aqueous interface have been driven by the development and application of surface-sensitive instrumentation, analytical techniques, and modeling.

7–4.1 Traditional Methods of Analysis

Detailed characterization of their constituent minerals is essential for accurately understanding pedological and geochemical processes occurring in soils; however, because of their great complexity in weathering environments, one or more complementary analytical techniques is usually required to accurately identify clay mineral phases and noncrystalline aluminosilicates in soils. Quantifying the amounts of individual clay species in complex polymineralic mixtures, however, poses quite another problem. Suitable reference standards with which to compare instrumental peak areas, or other analytical parameters, are often simply unavailable. The mined, geologic materials that are used as reference standards for quantitative analysis are often chemically or crystallographically dissimilar to the highly variable mineral phases under investigation in soils. Extraction of an identical mineral reference standard directly from a subsample of the soil being analyzed is ideally the best way to obtain accurate quantitative results; however, it is either impossible to obtain an accurate separation for certain minerals, or, in cases where exact separation is plausible, the procedure is prohibitively time-consuming for routine analysis. Despite many technological advances in analytical instrument electronics and data handling capabilities, quantitative clay mineral analysis remains a formidable obstacle in the soil and environmental sciences.

Chemical and/or mechanical pretreatments are initially required to prepare soil samples for mineralogical analysis. Organic matter, Fe (Al)-oxide or carbonate cementing agents, and excess soluble salts commonly prevent effective dispersion and hinder size fractionation. They also typically coat individual mineral particles that obscures microscopic examination of optical properties and causes anomolous instrumental effects. These interferences are normally removed by wet oxidation, reductant-soluble chelation, and washing with low ionic-strength salt solutions, respectively. Additionally, because the properties of clay minerals can vary appreciably as a function of analytical conditions, the practice of saturating clays with known index cations and/or organic solvating agents, and of conducting analyses at specific temperatures is essential to the correct identification of discrete species. Many of these techniques are not routinely used in the geological–sedimentalogical laboratory where analyses tend to be conducted on materials that are less weathered, more well crystalline, and lower in clay content than soils.

In addition to chemical extraction, traditional methods of analysis available to the soil mineralogist include x-ray powder diffraction (XRD) for identification and quantification of crystalline species present, differential thermal analysis (DTA) or differential scanning calorimetry (DSC), which measures energy changes associated with heating or cooling of specimen minerals, fourier transform infrared analysis (FT-IR) for studying chemical bonding on the surface as well as within mineral structures, specific surface area analysis (SSA) for determination of total reactive surface area and pore size distribution, scanning electron microscopy (SEM) and transmission electron microscopy (TEM) for visual observation of submicroscopic structures, x-ray fluoroescence (XRF) for analysis of the elemental composition of solids, and petrographic microscopy for measurement of the optical properties of sand- and silt-size (e.g., >10µm) minerals.

7–4.2 Recent Advances in Instrumentation

Rapid advances in the development of modern spectroscopic, microscopic, and synchotron-based x-ray techniques within the last ≈15 yr have resulted in a new level of sophistication with which chemical reactions occurring at the mineral–water interface can be observed. This has created an unprecedented opportunity for scientists to explore important Earth surface geochemical processes such as mineral dissolution and precipitation, adsorption–desorption phenomena, and ion-exchange reactions in greater detail than has ever been possible before.

It is beyond the scope of this chapter to contribute a detailed review of the manifold advanced analytical and spectroscopic techniques available in soil mineralogical research. Here, we briefly discuss some of the tools and techniques for studying surface composition, microtopography, and atomic structure. More detailed discussion of these and additional techniques and their applications in soil mineralogy and environmental chemistry can be found in Nagy and Blum (1994), Vaughan and Pattrick (1995), Bertsch and Hunter (1998, this publication) and other reviews.

High-vacuum spectroscopic techniques, such as x-ray photoelectron spectroscopy (XPS) and auger electron spectroscopy (AES) have been used to document leached (cation depleted) layers on minerals ranging from as little as the top few Angstroms (1 or 2 monolayers) (Hochella, 1990) to 10–30Å thickness in the case of artificially weathered albite (Hellman et al., 1990). XPS also has been widely used to determine mechanisms of adsorption–desorption phenomena at the mineral–water interface, and to provide structural analysis of sorbed complexes on mineral surfaces (Brown, 1990). These techniques also have been used to determine surface compositional modification of O_2 species on redox sensitive minerals (Stucki et al., 1976). While the level of sensitivity reported in these studies was previously not possible with most other surface analytical elemental depth profiling techniques, high-vacuum, or invasive, spectroscopic techniques may enhance changes in the coordination chemistry or oxidation state of the samples in question (Bertsch & Hunter, 1998, this publication). In-situ or noninvasive spectroscopic methods, such as fluorescence, vibrational, x-ray absorbance (XAS), and Mössbauer spectroscopies eliminate many of these shortcomings.

Imaging of soil mineral surfaces by a class of techniques collectively known as scanning probe microspcopy (SPM) has revolutionized conception of atomic structure and mineral surface microtopography (Nagy & Blum, 1994). Techniques such as scanning tunneling microscopy (STM) and variations of scanning force microscopy (SFM; Bertsch & Hunter, 1998, this publication) have shown that mineral growth and/or dissolution surfaces are comprised of flat areas, or terraces, interspersed with atomic- and molecular-scale steps, corners, kinks, and atomic protrusions, even on perfect cleavage surfaces (Hochella, 1990). STM imaging has shown these kink sites, edges, and crystal defects to be the most highly reactive sites where dissolution preferentially occurs.

Atomic resolution of STM has been used both for imaging surface structures, including atomic vacancies (Eggleston & Hochella, 1990), and for spectroscopic analysis of individual atoms on surfaces (Avouris, 1990). STM has permitted direct analysis of the electronic properties of atomic-scale point defects, and the extent of their reactive influence across surfaces (Hochella, 1995).

The availability of synchotron-based x-ray sources within the last several years has resulted in the development of many powerful techniques, often involving x-ray absorption techniques (Bertsch & Hunter, 1998, this publication). X-ray absorption near edge structure spectroscopy (XANES) has been used to characterize oxidation state and configuration of cations and neighboring coordinating anions in mineral polyhedral groups. Another technique, extended x-ray absorption fine structure spectroscopy (EXAFS), has been used to provide specific information on the identity and distance of surrounding atoms (Schultze & Bertsch, 1995).

7–5 CONCLUSIONS

Soil mineralogy has played a major role in our understanding of soil acidity and the influence of soils on plant nutrition; however, it has evolved considerably since the time when traditional production agriculture held sway. Recent advances in soil mineralogy, driven largely by the development and use of modern spectroscopic techniques, have resulted in substantial contributions to our understanding of soil surface chemistry, weathering and pedogenesis, nutrient and pollutant dynamics, and environmental remediation strategies. In addition to helping to identify best management practices for nonrenewable soil resources, future challenges for soil mineralogists include being able to more accurately describe and quantify the fundamental chemical bonding mechanisms involved with crystal growth and dissolution reactions, and both ion exchange and sorption–release mechanisms that govern the movement of potentially hazardous materials in the environment. New analytical tools continue to be developed and existing techniques are continually refined in response to new demands, such as the need for nondestructive sample analysis and in situ measurement of solid and liquid flux in soils.

As our understanding of the physicochemical processes occurring at the colloidal, molecular, and even atomic level in soils continues to grow as well as our ability to integrate and model these complex heterogeneous systems, so will our ability to assess the long-term effects of global change on terrestrial ecosystems, and to predict and mitigate the effects of radionuclides, pesticides, fertilizers, metals, industrial chemicals, sludges, and manures in soils and hydrologic systems.

REFERENCES

Allen, B.L., and B.F. Hajek. 1989. Mineral occurrence in soil environments. p. 199–278. *In* J.B. Dixon and S.B. Weed (ed.) Minerals in soil environments. 2nd ed. SSSA Book Ser. 1. SSSA, Madison, WI.

Andrews, J.T., and G.H. Miller. 1980. Dating Quaternary deposits more than 10,000 years old. p. 263–287. *In* R.A. Cullingford and J. Lewin (ed.) Timescales in geomorphology. John Wiley & Sons, New York.

Avouris, P. 1990. Atom-resolved surface chemistry using the scanning tunneling microscope. J. Phys. Chem. 94:2246–2256.

Bailey, S.W. 1984. Crystal chemistry of the true micas. p. 13–60. *In* S.W. Bailey (ed.) Micas. Reviews in Mineralogy no. 13. Mineralogy Soc. of Am., Washington, DC.

Bertsch, P.M., and D.B. Hunter. 1998. Elucidating fundamental mechanisms in soil and environmental chemistry: The role of advanced analytical, spectroscopic, and microscopic methods. p. 103–122. *In* P.M. Huang et al. (ed.) The future prospects for soil chemistry. SSSA Spec. Publ. 55. SSSA, Madison, WI (this publication).

Brown, G.E., Jr. 1990. Spectroscopic studies of chemisorption reaction mechanisms at oxide-water interfaces. p. 309–363. *In* M.A. Hochella, Jr. and A.F. White (ed.) Mineral–water interface geochemistry. Reviews in Mineralogy no. 23. Mineralogical Soc. of Am., Washington, DC.

Bryant, R.B., and J. Macedo. 1990. Differential chemoreductive dissolution of iron oxides in a Brazilian Oxisol. Soil Sci. Soc. Am. J. 54:819–821.

Cady, J.G. 1950. Rock weathering and soil formation in the North Carolina Piedmont region. Soil Sci. Soc. Am. Proc. 15:337–342.

Calvert, C.S., S.W. Buol, and S.B. Weed. 1980. Mineralogical characteristics and transformations in a vertical rock-saprolite-soil sequence in the North Carolina Piedmont: II. Feldspar alteration products. Their transformations through the profile. Soil Sci. Soc. Am. J. 44:1104–1112.

Carroll, D. 1953. Weatherability of zircon. J. Sed. Petrol. 23:106–116.

Catt, J.A. 1991. Soils as indicators of Quaternary climatic change in mid-latitude regions. Geoderma 51:167–187.

Colman, S.M., and K.L. Pierce. 1981. Weathering rinds on andesitic and basaltic stones as a Quaternary age indicator, western United States. U.S. Geol. Surv. Prof. Pap. 1210. U.S. Gov. Print. Office, Washington, DC.

Coston, J.A., C.C. Fuller, and J.A. Davis. 1995. Pb^{2+} and Zn^{2+} adsorption by a natural aluminum- and iron-bearing surface coating on an aquifer sand. Geochim. Cosmochim. Acta 59:3535–3547.

Douglas, L.A. 1989. Vermiculities. p. 635–674. *In* J.B. Dixon and S. B. Weed (ed.) Minerals in soil environments. 2nd ed. SSSA Book Ser. 1. SSSA, Madison, WI.

Eggleston, C.M., and M.F. Hochella, Jr. 1990. Scanning tunneling microscopy of sulfide surfaces. Geochim. Cosmochim. Acta. 54:1511–1517.

Ellis, S., and A. Mellor. 1995. Soils and environment. Routledge, London.

Eswaran, H., and W.C. Bin. 1978. A study of a deep weathering profile on granite in Peninsular Malaysia: III. Alteration of feldspars. Soil Sci. Soc. Am. J. 42:154–158.

Feldman, S. B., L.W. Zelazny, and J.C. Baker. 1991. High-elevation forest soils of the southern Appalachians: II. Geomorphology, pedogenesis, and clay mineralogy. Soil Sci. Soc. Am. J. 55:1782–1791.

Feldman, S.B., L.W. Zelazny, M.J. Pavich, and H.T. Millard, Jr. 1999. Late-Pleistocene eolian activity and post-depositional alteration in the Piedmont of northern Virginia. *In* G.M. Clark et al. (ed.) Regoliths in the Appalachian highlands beyond the Wisconsinan Glacial Borders Symp. Geol. Soc. Am. Spec. Pap. Mineralogical Soc. Am., Washington, DC (in press).

Fitzpatrick, R.W., and U. Schwertmann. 1982. Al-substituted goethite: An indicator of pedogenic and other weathering environments in South Africa. Geoderma 27:335–347.

Fontes, M.P.F., and S.B. Weed. 1991. Iron oxides in selected Brazilian Oxisols: I. Mineralogy. Soil Sci. Soc. Am. J. 55:1143–1149.

Fordham, A.W. 1990. Weathering of biotite into dioctahedral clay minerals. Clay Minerals 25:51–63.

Ghabru, S.K., A.R. Mermut, and R.J. St. Arnaud. 1989. Layer-charge and cation exchange characteristics of vermiculite (weathered biotite) isolated from a Gray Luvisol in northwestern Saskatchewan. Clays Clay Miner. 37:164–172.

Gilkes, R.J., and A. Suddhiprakarn. 1979. Biotite alteration in deeply weathered granite: I. Morphological, mineralogical, and chemical properties. Clays Clay Miner. 27:349–360.

Harden, J.W. 1988. Genetic interpretations of elemental and chemical differences in a soil chronosequence, California. Geoderma 43:179–193.

Harris, W.G., L.W. Zelazny, and J.C. Baker. 1984. Depth and particle size distributions of talc in a Virginia Piedmont Ultisol. Clays Clay Miner. 32:227–230.

Harris, W.G., L.W. Zelazny, J.C. Baker, and D.C. Martens. 1985a. Biotite kaolinization in Virginia Piedmont soils: I. Extent, profile trends, and grain morphological effects. Soil Sci. Soc. Am. J. 49:1290–1297.

Harris, W.G., L.W. Zelazny, J.C. Baker, and D.C. Martens. 1985b. Biotite kaolinization in Virginia Piedmont soils: II. Zonation in single grains. Soil Sci. Soc. Am. J. 49:1297–1302.

Hendershot, W.H., F. Courchesne, and D.S. Jeffries. 1996. Aluminum geochemistry at the catchment scale in watersheds influenced by acidic precipitation. p. 419–449. *In* G. Sposito (ed.) The environmental chemistry of aluminum. 2nd ed. Lewis Publ., Boca Raton, FL.

Hellman, R., C.M. Eggleston, M.F. Hochella, Jr., and D.A. Crerar. 1990. Formation of leached layers on albite during dissolution under hydrothermal conditions. Geochim. Cosmochim. Acta. 54:1267–1281.

Hochella, M.A., Jr. 1990. Atomic structure, microtopography, composition and reactivity of mineral surfaces. p. 87–132. *In* M.A. Hochella, Jr. and A.F. White (ed.) Mineral–water interface geochemistry. Reviews in Mineralogy no. 23. Mineralogical Soc. Am., Washington, DC.

Hochella, M.F., Jr. 1995. Mineral surfaces: Their characterization and their chemical, physical and reactive nature. p. 17– 86. *In* D.J. Vaughan and R.A.D. Pattrick (ed.) Mineral surfaces. Chapman & Hall, London.

Huang, P.M., and M. Schnitzer. 1986. Interactions of soil minerals with natural organics and microbes. SSSA Spec. Publ. 17. SSSA, Madison, WI.

Jackson, M.L., C.H. Lim, and L.W. Zelazny. 1986. Oxides, hydroxides and aluminosilicates. p. 101–150. *In* A. Klute (ed.) Methods of soil analysis. Part 1. 2nd ed. Agron. Monogr. 9. ASA and SSSA, Madison, WI.

Kittrick, J.A. 1977. Mineral equilibria and the soil system. p. 1–25. *In* J.B. Dixon and S.B. Weed (ed.) Minerals in soil environments. SSSA, Madison, WI.

Kühnel, R.A., H.J. Roorda, and J.J. Steensma. 1975. The crystallinity of minerals: A new variable in pedogenetic processes: A study of goethite and associated silicates in laterite. Clays Clay Miner. 23:349–354.

Loughnan, F.C. 1969. Chemical weathering of silicate minerals. Elsevier Scientific Publ. Co., New York.

Ma, Q.Y., S.J. Traina, and T.J. Logan. 1993. In-situ lead immobilization by apatite. Environ. Sci. Technol. 27:1803–1810.

Markewich, H.W., M.J. Pavich, M.J. Mausbach, R.L. Hall, R.G. Johnson, and P.P. Hearn. 1987. Age relations between soils and geology in the Coastal Plain of Maryland and Virginia. U.S. Geol. Surv. Open-File Rep. 1589-A. U.S. Gov. Print. Office, Washington, DC.

Martini, I.P., and W. Chesworth. 1992. Weathering, soils and paleosols. Developments in earth surface processses. 2. Elsevier, Amsterdam.

McBride, M.B. 1994. Environmental chemistry of soils. Oxford Univ. Press, New York.

Nagy, K.L., and A.E. Blum. 1994. Scanning probe microscopy of clay minerals. CMS Workshop Lectures Vol. 7. Clay Mineral Soc., Boulder, CO.

Pavich, M. J. 1989. Investigations of the characteristics, origin, and residence time of the upland residual mantle of the Piedmont of Fairfax County, Virginia. U.S. Geol. Surv. Prof. Pap. 1352. U.S. Gov. Print. Office, Washington, DC.

Pozzuoli, A., E. Viela, E. Franco, A. Ruiz-Amil, and C. De LaCalle. 1994. Weathering of biotite to vermiculite in quaternary lahars from Monti Ernici, central Italy. Clay Minerals 27:175–184.

Rich, C.I. 1958. Muscovite weathering in a soil developed in the Virginia Piedmont. Clays Clay Miner. 5:203–212.

Rich, C.I., and S.S. Obenshain. 1954. Chemical and clay mineral properties of a Red-Yellow Podzolic soil derived from muscovite schist. Soil Sci. Soc. Am. Proc. 190:334–339.

Sawhney, B.L. 1989. Interstratification in layer silicates. p. 789–828. *In* J.B. Dixon and S.B. Weed (ed.) Minerals in soil environments. SSSA Book Ser. 1. SSSA, Madison, WI.

Schultze, D.G., and P.M. Bertsch. 1995. Synchotron x-ray techniques in soil, plant, and environmental research. Adv. Agron. 55:1–66.

Schwertmann, U. 1985. The effect of pedogenic environments on iron oxide minerals. Adv. Soil Sci.1:172–200.

Sposito, G. 1989. The chemistry of soils. Oxford Univ. Press, New York.

Stucki, J.W., C.B. Roth, and W.E. Baitinger. 1976. Analysis of iron-bearing minerals by electron spectroscopy for chemical analysis (ESCA). Clays Clay Miner. 24:289–292.

Vaughan, D.J., and R.A.D. Pattrick. 1995. Mineral surfaces. Chapman & Hall, London.

Velbel, M.A. 1986. Influence of surface area, surface characteristics, and solution composition on feldspar weathering rates. p. 615–634. *In* J.A. Davis and K.F. Hayes (ed.) Geochemical processes at mineral surfaces. ACS Symp. Ser. 323. Am. Chem. Soc., Washington, DC.

Wada, K. 1989. Allophane and imogolite. p. 1051–1088. *In* J.B. Dixon and S.B. Weed (ed.) Minerals in soil environments. 2nd ed. SSSA Book Ser. 1. SSSA, Madison, WI.

Weaver, C.E. 1989. Clays, muds, and shales. Developments in sedimentology no. 44. Elsevier, New York.

White, A.F. 1995. Chemical weathering rates of silicate minerals in soils. Rev. Mineral.31:407–462.

White, G.N., and L.W. Zelazny. 1986. Charge properties of soil colloids. p. 39–81. *In* D.L. Sparks (ed.) The physical chemistry of soils. CRC Press, Boca Raton, FL.

Zachara, J.M., P.L. Gassman, S.C. Smith, and D. Taylor. 1995. Oxidation and adsorption of Co(II)EDTA^{2-} complexes in subsurface materials with iron and manganese oxide grain coatings. Geochim. Cosmochim. Acta 59:4449–4463.

Zachara, J.M., J.A. Kittrick, and J.B. Harsh. 1988. The mechanism of Zn^{2+} adsorption on calcite. Geochim. Cosmochim. Acta 52:2281–2291.

8 New Ideas on the Chemical Make-Up of Soil Humic and Fulvic Acids

Morris Schnitzer
Agriculture and Agri-Food Canada
Ottawa, Ontario, Canada

Hans-Rolf Schulten
Institute for Soil Science
University of Rostok
Rostock, Germany

8–1 SOIL ORGANIC MATTER

The objective of this chapter is to present an account of our current knowledge of the chemical make-up of soil humic (HA) and fulvic (FA) acids. Special attention will be paid to humic substances, the major soil organic matter components, and to HA, the most studied humic compound in soils. After a brief review of the main theories on the synthesis of HAs, we will describe the development of a novel structural concept for HA based on long-term chemical, ^{13}C nuclear magnetic resonance spectroscopic (NMR), pyrolysis-field ionization mass spectrometric (Py-FIMS), and, finally, Curie-point pyrolysis-gas chromatographic–mass spectrometric (Py-GC–MS) investigations of extracted HAs and whole soils. With the aid of HyperChem software (Hypercube, Gainesville, FL), the two-dimensional HA structure has recently been converted to a three-dimensional one. Interactions of this structure with biological molecules and biocides will be discussed. This chapter is not an exhaustive review of the literature but rather a description of recent developments of a structural concept of HA and soil organic matter (SOM). For more comprehensive reviews of the chemistry of HA, see Hayes et al., (1989); Schnitzer, (1991); Stevenson, (1994); Senesi and Miano, (1994).

8–1.1 Gross Chemical Composition of Soil Organic Matter

The following are estimates of the chemical composition of soil organic matter (Schnitzer, 1991): Carbohydrates, 10%; N-components (including proteins, peptides, amino acids, purines, pyrimidines, and heterocyclics), 10%; lipids (alkanes, alkenes, fatty acids, esters), 10%; and humic substances, 70%. These

---

Copyright © 1998. Soil Science Society of America, 677 S. Segoe Rd., Madison, WI 53711, USA. *Future Prospects for Soil Chemistry.* SSSA Special Publication no. 55.

percentages are averages for agricultural soils; deviations from these values may occur depending on environmental conditions. While in studies on carbohydrates, N-components, and lipids, soil chemists have benefitted from advances made by organic chemists and biochemists specializing in these fields, advancing the chemistry of humic substances became the domain of soil chemists, and to a much smaller extent, of coal chemists.

8–2 HUMIC SUBSTANCES

8–2.1 Extraction and Fractionation

More than 200 years ago, Archard (1786) proposed the use of dilute aqueous NaOH solution for the extraction of humic substances. In the 1830s, Berzelius (1839) attempted to separate humic substances into three fractions. These were: (i) HA, which was soluble in alkaline solution; (ii) humin, which was insoluble in base and supposedly inert; and (iii) crenic and apocrenic acids (currently referred to as fulvic acid), which had the ability of forming salts and complexes with di- and tri-valent metal ions. The extraction and separation scheme currently accepted by most specialists on humic substances is based on their solubility in alkali, neutral salt, and acid. Thus, (I) HA, is that humic fraction that is soluble in dilute alkali or neutral salt solution but is coagulated when the alkaline or neutral salt extract is acidified; (II) FA, which is that humic fraction that remains in solution when the alkaline or neutral salt solution is acidified, that is, it is soluble under alkaline, neutral, and acidic conditions, and (III) humin, which is that humic fraction that cannot be extracted from soils by dilute base, neutral salt solution, or acid.

8–2.2 Hypotheses for Humic Acid Synthesis

The synthesis of HAs is hypothesized to occur by one of the following two pathways: (i) the degradation of plant biopolymers to form the central HA core, and (ii) condensation–polymerization reactions in which plant biopolymers are first degraded to small molecules that then repolymerize (Hatcher & Spiker, 1988). According to Fischer and Schrader (1921), lignin is most likely the mother substance of humic materials. It soon became apparent that microorganisms played an essential role in HA synthesis (Waksman, 1938). The main problem with the lignin theory is that it cannot account for the high N content of humic substances. Waksman (1938) amended the lignin theory by proposing that microbially-produced protein is chemically linked to microbially-modified lignin to form the core of HA. Another approach was advocated by Flaig (1964) who proposed that lignin is oxidatively degraded to simple phenolic monomers, which then undergo oxidative polymerization to produce HA. Along similar lines, oxidative polymerization of simple phenolic compounds to form HA can be catalyzed by enzymes (Flaig et al., 1975; Haider et al., 1975; Sulflita & Bollag, 1981) and also by abiotic catalysis (Wang et al., 1986; Huang, 1990).

Table 8–1. Analytical characteristics of a Mollisol humic acid (HA) and a Haplaquod fulvic acid (FA) (Schnitzer, 1995).

	HA	FA
Element, g kg^{-1}		
C	564	509
H	55	33
N	41	7
S	11	3
O	329	448
Functional groups, cmol kg^{-1}		
Total acidity	660	1240
COOH	450	910
Phenolic OH	210	330
Alcoholic OH	280	360
Quinonoid C=O	250	60
Ketonic C=O	190	250
OCH$_3$	30	10
E_4/E_6	4.3	7.1

Maillard (1913) suggested that HA results from interactions of reducing sugars with amino acids or amines. The brown to dark polymers so produced are rich in N. According to Maillard (1913), HA is formed by purely chemical reactions in which microorganisms do not play a direct role except to produce sugars from carbohydrates and amino acids from proteins. During the years, a number of chemical structures have been proposed for HA (Dragunov et al., 1948; Flaig, 1964; Stevenson, 1994; Kleinhempel, 1970; Felbeck, 1965). None of the structures proposed so far has found wide acceptance so that the search for more representative model structures continues.

8–2.3 Analytical Characteristics

From the mid-1950s to the mid-1980s, the senior author and his associates (Schnitzer & Khan, 1972; Schnitzer, 1978) initiated what may be termed the analytical approach to studies on the chemical structure and characteristics of HAs, FAs, and humins. The first investigations dealt with the elemental and functional group analyses of HAs and FAs extracted from widely differing soils on the earth's surface. In the case of functional groups, it was necessary to establish methods for their analyses.

The main differences between HAs and FAs (Table 8–1) are the following: (i) HAs contain more C but less O than FAs; (ii) HAs contain more H, N, and S than FAs; (iii) the total acidity and the CO$_2$H content of FAs are appreciably higher than those of HAs; (iv) FAs are richer in phenolic OH and ketonic C=O than HAs but the latter contain, per unit weight, more quinonoid C=O groups; (v) more of the O in FAs than in HAs can be accounted for in functional groups; and (vi) HAs have higher molecular weights than FAs.

From analytical data published on humins, it appears that soil humins are similar to HAs (Schnitzer, 1991; Preston et al., 1989), except that soil humins are strongly complexed with metals and clays and so become difficult to extract.

Other workers (Hatcher & Spiker, 1988) consider humins large molecules made up of microbiologically refractory substances such as lignin, cutin, suberin, paraffins, melanins, and other complex components. They believe that in the soil humins are oxidatively degraded to HAs, which in turn, are oxidatively degraded to FAs.

8–2.4 Oxidative Degradation of Humic Acids, Fulvic Acids, and Humin

During the 1960s and 1970s the senior author and his coworkers (Schnitzer & Khan, 1972; Schnitzer, 1978) undertook extensive investigations on the oxidative degradation of humic materials of widely differing origins. Methylated and unmethylated HAs, FAs, and humins were oxidized with alkaline $KMnO_4$ solution. The somewhat milder degradation with alkaline CuO, as well as sequential oxidations with CuO-NaOH + $KMnO_4$, and with CuO-NaOH + $KMnO_4$ + H_2O_2 also were employed. In addition, humic substances were degraded under acidic conditions with HNO_3 and peracetic acid. Other oxidants included alkaline nitrobenzene, sodium hypochlorite, and H_2O_2 solutions. Major oxidation products under alkaline as well as under acidic conditions were benzenecarboxylic acids (especially the tri-, tetra-, and penta-forms; see Fig. 8–1), phenolic acids (with between 1 and 3 OH groups and between 1 and 5 CO_2H groups per aromatic ring; see Fig. 8–2), and aliphatic mono-, di-, tri-, and tetra-carboxylic acids.

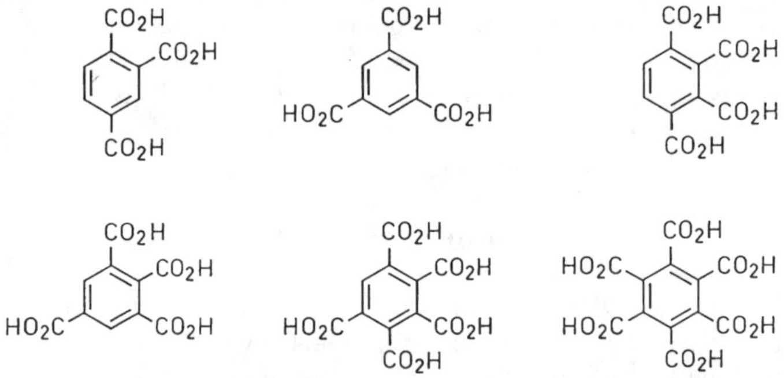

Fig. 8–1. Major benzenecarboxylic oxidation products of humic substances (modified from Schnitzer, 1978).

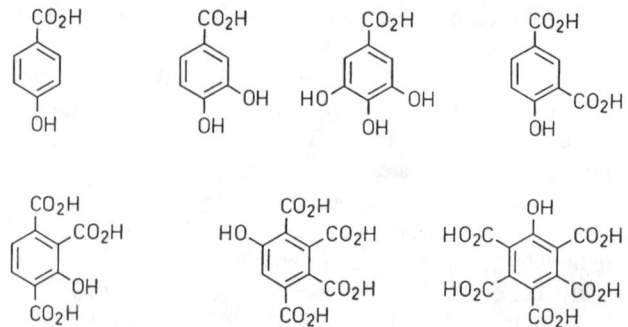

Fig. 8–2. Major phenolic oxidation products of humic substances (modified from Schnitzer, 1978).

8–2.5 Chemical Structure for Humic Substances Based on Oxidation Products

From the oxidation products identified by Schnitzer and coworkers (Schnitzer & Khan, 1972; Schnitzer, 1978), it appeared that the aromatic rings were cross-linked by paraffinic chains (Fig. 8–3). On oxidation, the aliphatic C closest to the ring became the C of CO_2H groups and remained attached to the ring, while the other C in the aliphatic chain were oxidized to aliphatic acids. The purpose of methylation prior to oxidation was to protect phenolic OH groups against attacks by electrophilic oxidants. This made it possible to isolate and identify phenolic in addition to benzenecarboxylic acids.

Two conclusions can be drawn from the oxidative degradation experiments : (i) isolated aromatic rings are important structural units of all humic substances; and (ii) aliphatic structures are linking aromatic rings to form an alkyl-aromatic network.

8–2.6 Reductive Degradation of Humic Acids and Fulvic Acids

Of the reductive degradation methods employed by Schnitzer et al. (Schnitzer & Khan, 1972; Schnitzer, 1978), the most interesting information was obtained from Zn-dust distillations. The major products were methyl-substituted naphthalene, anthracene, phenanthrene, pyrene, and perylene. The methyl groups on the polycyclic rings are probably the remains of longer alkyl chains linking the polycyclics in HA and FA structures.

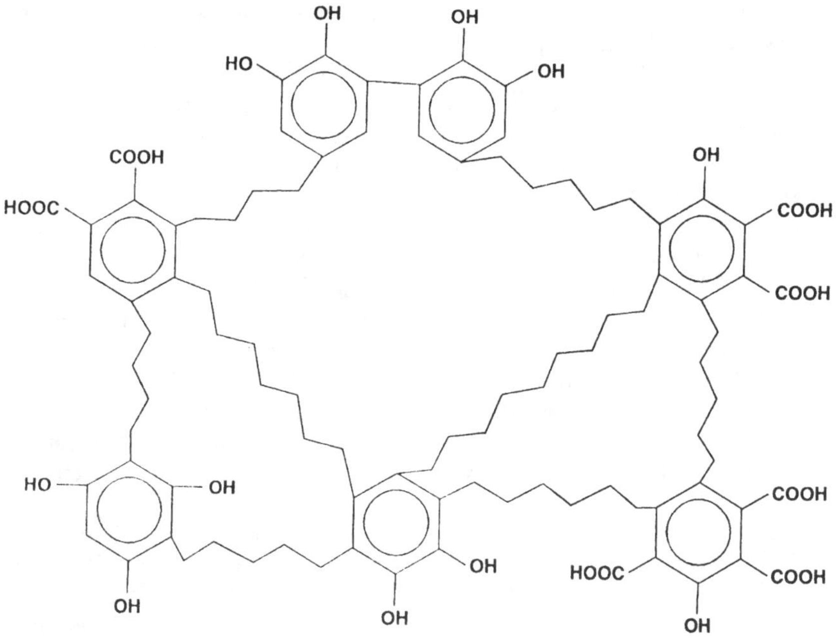

Fig. 8–3. Chemical structure of humic acids indicated by oxidation products.

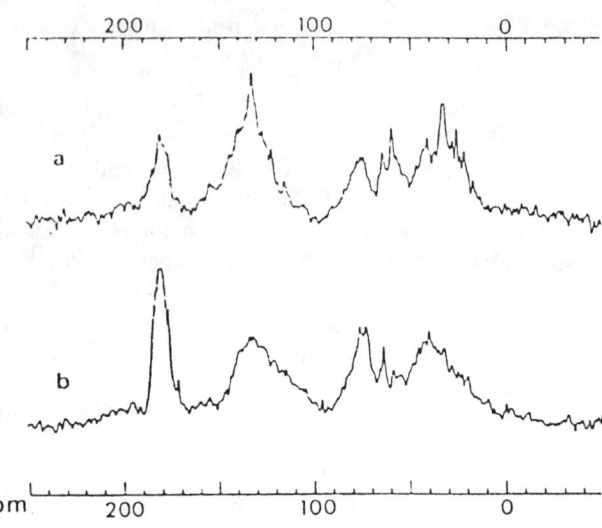

Fig. 8–4. ^{13}C NMR spectra of (a) Mollisol humic acid and (b) Haplaquod Bh fulvic acid (modified from Schnitzer, 1991).

8–2.7 ^{13}C NMR Spectroscopy of Humic Acids and Fulvic Acids

Until about 15 yr ago, when the use of liquid- and solid-state ^{13}C NMR for the analysis of HAs and FAs became more widespread, most soil chemists thought that the chemical structure of humic materials was predominantly aromatic. It was ^{13}C NMR that demonstrated that aliphatic structures in humic substances were often as important or, at times, more important than aromatic structures (Norwood, 1988; Schnitzer et al., 1991). Aromaticities of HAs extracted from soils from widely differing pedological origins range from 30 to 60%, with many in the 47 to 60% range. A substantial portion of aliphatic C in HAs consists of paraffinic C. Of considerable interest are the prominent resonances in both liquid-state and solid-state ^{13}C NMR spectra of humic substances near 130 to 132 ppm, which can be assigned to C in aromatic rings that are not substituted by strong electron donors such as O and N but by C. Alkylbenzenes are typical structures that yield such resonances (Breitmaier & Voelter, 1978).

One of the advantages of ^{13}C NMR is that it indicates the presence in humic substances of a wide variety of structures whose determinations by other methods would either be laborious and time-consuming or not possible at all. In this sense, ^{13}C NMR offers unique possibilities. Even more valuable information on the chemical structure of humic materials can be obtained in an integrated approach by combining ^{13}C NMR with chemical and mass spectrometric methods.

Of considerable interest is a comparison of solid-state ^{13}C NMR spectra of HA and FA. The ^{13}C NMR spectra of a HA extracted from a Mollisol Ah horizon and a FA extracted from a Haplaquod Bh horizon are shown in Fig. 8–4. Spectrum (a) exhibits several distinct peaks in the aliphatic (0–105 ppm), aromatic (106–150 ppm), phenolic (151–170 ppm), and carboxyl (171–190 ppm)

Table 8–2. Distribution of C (%) in a Mollisol humic acid (HA) and a Haplaquod fulvic acid (FA) as determined by 13C NMR (Schnitzer, 1995).

Chemical shift range (ppm)	% of C	
	HA	FA
0–40	24.0	15.6
41–60	12.5	12.8
61–105	13.5	19.3
106–150	35.0	30.3
151–170	4.5	3.7
171–190	10.5	18.3
Aliphatic C (0–105 ppm)	50.0	47.7
Aromatic C (106–150 ppm)	39.5	34.0
Phenolic C (151–170 ppm)	4.5	3.7
Aromaticity†	44.1	41.6

† [(aromatic C + phenolic C)/(aromatic C + phenolic C + aliphatic C)] × 100

regions. The signals at 16.5, 21.1, 25.0, 27.1, and 31.3 ppm are probably due to alkyl C in chains. The peak at 16.5 ppm is characteristic of terminal CH_3 groups, and that at 31.3 ppm of $(CH_2)_n$ in long paraffinic chains. The resonance at 40.2 ppm could include contributions from both alkyl C and amino acid C. In the 50 to 105 region, aliphatic C substituted by O and N are usually observed. The broad signal at 52.6 ppm and the sharper one at 58.8 ppm may be due to C in OCH_3. Amino acid-C also may contribute in this region (Breitmaier & Voelter, 1978). Carbohydrates in HA would be expected to produce signals in the 60–65, 70–80, and 90–105 regions, although other types of aliphatic C bonded to O also could do so. The aromatic region contains a relatively sharp maximum near 130 ppm, due to alkylaromatics. The peak at 155 ppm in spectrum (a) indicates the presence of O- and N-substituted aromatic C (phenolic OH and/or aromatic NH_2 groups). The broad signal near 180 ppm is attributed to C in CO_2H groups. Amides and esters also could contribute to this peak.

The ^{13}C NMR spectrum of the FA (b) consists of a number of aliphatic resonances in the 20 to 50 ppm region, followed by signals from OCH_3 groups, amino acids, amino sugars, and carbohydrates between 50 and 85 ppm. The presence of alkylaromatics is indicated by a broad signal between 130 and 133 ppm. The preponderance of CO_2H groups is shown by signals between 171 and 182 ppm. In general, fewer sharp signals are observed in the ^{13}C NMR spectrum of the FA than in that of the HA, possibly because of more H-bonding in the FA.

The ^{13}C NMR spectra are summarized in Table 8–2. The spectra are divided into the following regions: 0 to 40 ppm (C in straight-chain, branched, and cyclic aliphatics); 41 to 60 ppm (C in branched aliphatics, amino acids, and OCH_3 groups); 61 to 105 ppm (C in carbohydrates, and in aliphatics containing C bonded to OH, ether oxygens, or occurring in five- or six-membered rings bonded to O); 105 to 150 ppm (aromatic C); 151 to 170 ppm (phenolic C), and 171 to 190 ppm (C in CO_2H groups).

An inspection of the data in Table 8–2 shows a similar C distribution in the two humic fractions. The HA is slightly more aromatic than the FA, but the FA is considerably richer in CO_2H groups, which appears to be the main difference between the two substances. Other differences are that the HA is richer in paraffinic C but poorer in carbohydrate-C than the FA. But on the whole, the main structural features such as aromaticity and aliphaticity, are similar.

8–2.8 Summary of the Structural Information Generated by Chemical, Degradative, and ^{13}C NMR Methods

From the information presented so far, the following conclusions can be drawn: (i) isolated aromatic rings are important structural components of humic substances; (ii) aliphatic structures are more numerous than had been assumed previously; (iii) aromatic rings in humic substances are cross-linked by aliphatic chains; (iv) on oxidation, the aliphatic C closest to the aromatic ring becomes the C of a CO_2H group; the remaining C is oxidized to aliphatic acids;
(v) aromaticities of soil HAs range from 30 to 60%, with many in the 47 to 53% range; (vi) a substantial portion of the aliphatic C in humic substances consist of paraffinic C; and (vii) the prominent resonances in ^{13}C NMR spectra of humic substances at 130 to 132 ppm are due to C in aromatic rings not substituted by electron-donating O and N but by C as in alkylaromatics. The latter appear to be important structures in all humic substances.

8–3 PYROLYSIS-FIELD IONIZATION MASS SPECTROMETRY OF HUMIC SUBSTANCES AND SOIL ORGANIC MATTER

8–3.1 Mass Spectra of Humic Acids, Fulvic Acids, Humin, and Whole Soil

The pyrolysis-field ionization (Py-FI) mass spectrum of the Armadale HA in Fig. 8–5a (Schnitzer & Schulten, 1992) shows the presence in this material of the following four major components: carbohydrates, phenols, lignins, and n-fatty acids. Especially noteworthy is the prominence of the n-C_{24} (m/z 368), n-C_{26} (m/z 396), n-C_{27} (m/z 410), n-C_{28} (m/z 424), and n-C_{30} (m/z 452) fatty acids. Other components present in smaller amounts are monomeric lignins, n- C_{10} to n-C_{20} diesters, and the n-C_{44} to n-C_{50} alkyl monoesters, of which the n-C_{45} monoester (m/z 662) is the most abundant. Relatively weak signals characteristic of N-components are m/z 59 (acetamide), 79 (pyridine), 81 (methylpyrrole), 93 (methylpyridine), 117 (indole), 131 (methylindole), and 167 (most likely chitin-derived 3-acetamido-6-methylpyron).

The Py-FI mass spectrum of the Armadale FA in Fig. 8–5b (Schnitzer & Schulten, 1992) is dominated by carbohydrates and phenols, followed by lignins. The most intense peaks are m/z 58 (acetone) and m/z 60 (acetic acid). Both compounds are thermally eliminated from methyl-ethyl ketones and carbohydrates and fatty acids above 300°C. Also, smaller amounts of n- fatty acids (m/z 256, 284, 312, and 382), sterols (m/z 386, cholesterol; m/z 414, β-sitosterol), n- alkyl diesters, and monomeric and dimeric lignins appear to be present in the FA. No intense peaks due to N-compounds are present.

The spectrum of the Armadale humin (Fig. 8–6a; Schnitzer & Schulten, 1992) shows the significant presence in this material of carbohydrates, phenols, monomeric and dimeric lignins, alkylbenzenes, and alkyl esters. In addition, a homologous series of fatty acids ranging from C_{16} to C_{27} can be detected. Of special interest is the series of *n*-alkylbenzenes with signals at m/z 316, 330, 344, 358, 372, 386, 400, 414, and 428, which appear to indicate the presence of $C_6H_5 \cdot C_{17}H_{35}$ to $C_6H_5 \cdot C_{25}H_{51}$ alkylbenzenes, respectively. Molecular ions m/z 206 and 220 could be due to di- and tri-methylphenanthrene. Intense signals probably due to *n*-C_{10} to *n*-C_{20} alkyl diesters are observed from m/z 202 to 342. Except for weak signals at m/z 67 (pyrrole) and m/z 81 (methylpyrrole), no signals due to N-compounds were detected in this spectrum.

The Py-FI mass spectrum of the Armadale soil (Fig. 8–6b; Schnitzer & Schulten, 1992) is dominated by carbohydrates, phenols, monomeric and dimeric lignins, and *n*-alkyl esters. Molecular ions m/z 394 and 408 appear to indicate the presence of small amounts of *n*-C_{28} and *n*-C_{29} alkanes, whereas the weak signals at m/z 442, 456, and 470 appear to be due to $C_6H5 \cdot C_{26}H_{53}$ to $C_6H_5 \cdot C_{28}H_{57}$ *n*-alkylbenzenes, respectively. This soil appears to contain suberin-derived aromatic esters at m/z 446, 474, 502, and 530 (Hempfling et al., 1988). The signals at m/z 170 and 184 arise most likely from tri- and tetra-methylnaphthalene, respectively, while m/z 178, 192, 206, 220, and 234 are due to phenanthrene, methyl-, dimethyl-, trimethyl-, and tetramethyl-phenanthrene, respectively. Similar to the Armadale humin, *n*-C_{10} to *n*-C_{18} alkyl diesters are present in the Py-FI mass spectrum of the Armadale soil. The occurrence of N-compounds is indicated by m/z 59, 67, 81, 93, 103, 117, 131, and 167.

A summary of the compounds identified in the three humic fractions and the initial (whole) soil is shown in Table 8–3. An inspection of these data shows that in all materials carbohydrates, phenols, lignin monomers and dimers, and, to

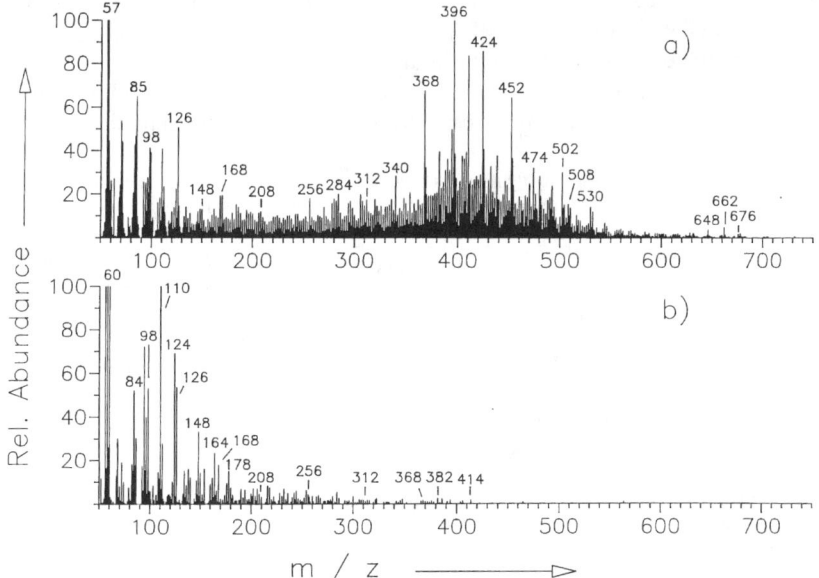

Fig. 8–5. Py-FI mass spectrum of (a) the Armadale humic acid; (b) the Armadale fulvic acid. (Schnitzer & Schulten, 1992).

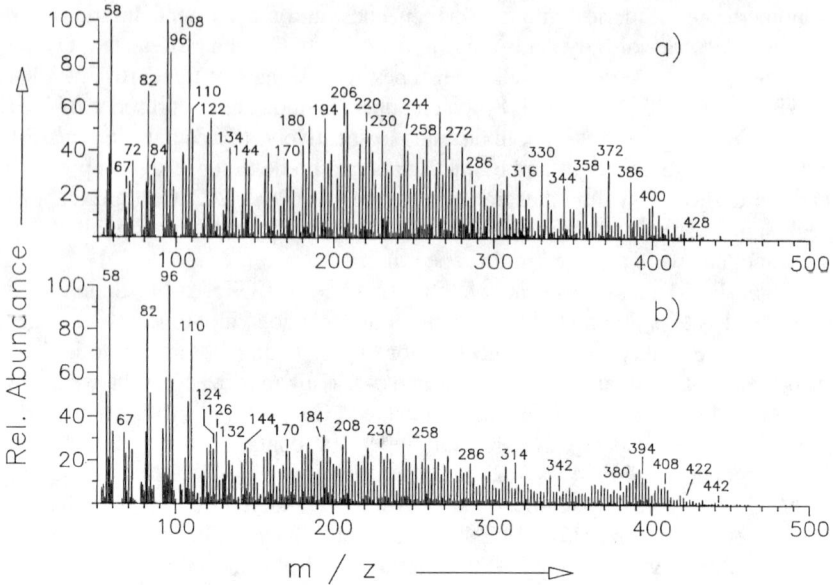

Fig. 8–6. Py-FI mass spectrum of (a) the Armadale humin; (b) the initial Armadale soil (Schnitzer & Schulten, 1992).

a lesser extent, *n*-fatty acids are the major components. Minor components include *n*-alkyl mono- and di-esters, *n*-alkylbenzenes, *n*- alkenes, methylnaphthalenes, methylphenanthrenes, and N-compounds. Similar compounds are produced from all materials except that HA tends to be enriched in fatty acids and humin in alkylbenzenes.

From the data presented herein it appears that from the analytical point of view, the most suitable material to be analyzed is the whole soil. The Py-FI mass spectrum of the whole soil produces more identifiable compounds than that of

Table 8–3. Compounds identified by Py-FI mass spectrometry in the initial Armadale soil and in the humic acid, fulvic acid, and humin fractions isolated from the same soil (Schnitzer & Schulten, 1995).

Compound identified	Soil	HA	FA	Humin
Carbohydrates	++†	++	++	++
Phenols	++	++	++	++
Lingin monomers	++	++	++	++
Lignin dimers	++	++	++	++
n-Fatty acids	+	+++	+	++
n-Alkyl monoesters				
n-Alkyl diesters	++	+	+	++
n-Alkyl benzenes	+		+	++
Methylnaphthalenes	+			+
Methylphenanthrenes	+			+
N compounds	+	+		+
n-Alkanes	+			

† +, weak (relative intensity <20%); ++, intense (relative intensity 20–60%); +++, very intense (relative intensity >60%).

any of the fractions. This obviates the need for laborious and possible damaging extraction, separation, and purification procedures. Py-FI mass spectrometry is possibly the first and only procedure currently available that allows soil chemists to do comprehensive organic matter analyses at the molecular level on undisturbed, air-dry soils without any pretreatment. This represents a significant advance in soil organic matter analysis and characterization.

8–4 CURIE-POINT PYROLYSIS–GAS CHROMATOGRAPHY–MASS SPECTROMETRY OF HUMIC ACIDS

8–4.1 The Development of a Novel Two-Dimensional Structural Concept for Humic Acids

Curie-point Pyrolysis-Gas Chromatography–Mass Spectrometry (Py-GC–MS) is a valuable method for structural studies on humic materials (Schulten & Schnitzer, 1992; Schnitzer & Schulten, 1995) because the transfer of thermal energy to the sample is fast, with temperature rises on the order of milliseconds. The resulting thermal shock produces small, stable organic molecules as pyrolysis products. Their identification is based on two independent data sets: (i) gas chromatographic retention times and (ii) computer-assisted library searches of mass spectra in standard libraries.

Curie-point Py-GC–MS analyses were performed by Schulten and Schnitzer (1992) on two HAs that had previously been examined intensively by pyrolysis-soft ionization mass spectrometry. As shown in Table 8–4, the major compounds produced from the two HAs are benzenes and n-alkylbenzenes. Of special interest is the series of C_1 to C_{13} n-alkylbenzenes. In addition, ethylmethylbenzene, methylpropylbenzene, methylheptylbenzene, methyloctylbenzene, essentially members of the same series alkylbenzenes, also were produced. It is likely that alkylbenzenes play important functions in the structural network of humic substances. Other compounds are trimethyl- and tetramethylbenzenes, alkylnaphthalenes, and alkylphenanthrenes. The alkyl substitution on naphthalene ranges from one to five methyls, while on phenanthrene it extends from one to four methyls.

The alkylaromatic compounds identified in Table 8–4 consist of aromatic rings that are covalently linked to aliphatic chains. Schulten et al. (1991) propose that these building blocks are released during pyrolysis from an alkylaromatic structural network that is made up of the constituents listed in Table 8–4. The resulting structure contains voids of various dimensions that can trap and bind other organic and inorganic components. Note that the Armadale HA is richer in most components listed in Table 8–4 than the Bainsville HA. This may be due to differences in the origins of the two HAs. The Armadale HA was extracted from the Bh horizon of a Haplaquod, about 25 cm below he surface, while the Bainsville HA was extracted from the surface of the Bainsville soil, a Haplaquoll. One of the striking features of the Bh horizon is its low microbial activity, which may have led to a better preservation of the compounds listed in Table 8–4.

Table 8–4. Building blocks of Bainsville and Armadale humic acids (HA) identified by Curie-point pyrolysis GC–MS (Schulten et al., 1991).

Intensity†		
Armadale	Bainesville	Compounds
+++++	+++	Benzene
+++++	+++++	Toluene
++++	++++	Ethylbenzene, xylenes
+	+	Ethylmethylbenzene
++	++	Propylbenzene
++	++	Trimethylbenzenes
+		Butylbenzene
+		Methylpropylbenzene
+	+	Tetramethylbenzene
++	+	Pentylbenzene
+	+	Hexylbenzene
++	+	Heptylbenzene
++	+	Octylbenzene
+		Methylheptylbenzene
+	+	Nonylbenzene
+		Methyloctylbenzene
++	+	Decylbenzene
+		Methylmonylbenzene
+	+	Undecylbenzene
+		Methyldecylbenzene
++		Dodecylbenzene
+		Methylundecylbenzene
+		Tridecylbenzene
+		Tetradecylbenzene
+		Pentadecylbenzene
+		Hexadecylbenzene
+		Heptadecylbenzene
+		Octadecylbenzene
+		Nonadecylbenzene
+		Eicosylbenzene
+		Hemicosylbenzene
+		Docosylbenzene
+++	+++	Styrene
+	+	Methylstyrene
++	+	Indene
+++	++	Indane
+		Fluorene
++	+	Naphthalene
++	+	Methylnaphthalenes
++	+	Dimethylnaphthalenes
++	+	Trimethylnaphthalenes
+	+	Tetramethylnaphthalenes
+		Pentamethylnaphthalene
+		Phenanthrene
+		Mehtylphenanthrene
+		Dimethylphenanthrene
+		Trimethylphenanthrene
+		Tetramethylphenanthrene

† Intensity of peak height: +++++, 80–100%; ++++, 60–80%; +++, 40–60%; ++, 20–40%; +, observed.

A chemical structure for the HA skeleton, based on alkylbenzenes, alkylnaphthalenes, and alkylphenanthrenes, is shown in Fig. 8–7.

The following, more complete version of the HA structure is displayed in Fig. 8–8, in which O, H, and N atoms have been inserted in conformity with analytical data obtained on many naturally-occurring soil HAs. Oxygen is present in carboxyls, phenolic and alcoholic hydroxyls, ester and ether groups, and N is present in heterocyclic structures and as nitriles (Schulten et al., 1997; Schulten & Schnitzer, 1998). The elemental composition of the HA structure in Fig. 8–8 is $C_{308}H_{328}O_{90}N_5$, with a molecular weight of 5539.9720 g mol^{-1} and an elemental analysis of 66.78% C, 5.97% H, 25.99% O, and 1.26% N.

There are different views in the literature on SOM as to whether carbohydrates and proteinaceous materials are adsorbed or loosely retained by HA or whether they are bonded covalently to HA. Regardless of which mechanisms are considered, carbohydrates and proteinaceous materials are HA components for analytical purposes because their presence affects the elemental analysis and functional group content of HAs. Carbohydrates have been reported to constitute about 10% of the HA weight (Lowe, 1978); a similar value has been suggested for proteinaceous materials in HA (Khan & Sowden, 1971). Thus, Schulten and Schnitzer (1993) assume that a molecular weight of HA interacts with 10% carbohydrates and 10% proteinaceous materials. The resulting HA has an elemental composition of $C_{342}H_{388}O_{124}N_{12}$, with a molecular weight of 6650.8492 g mol^{-1} and an elemental analysis of 61.76% C, 5.88% H, 29.83% O, and 2.53% N. As more carbohydrates and proteinaceous materials are added to the HA, the C content decreases but the O content increases.

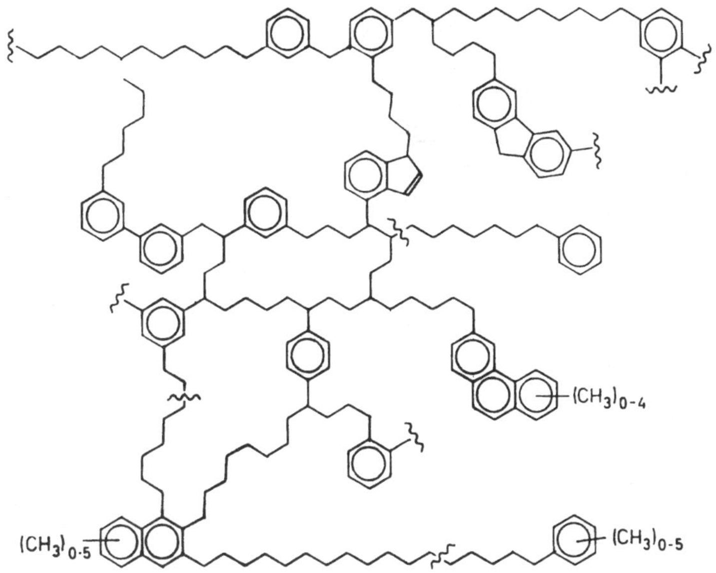

Fig. 8–7. Chemical structure of humic acids based on alkylaromatic building blocks (Schulten et al., 1991).

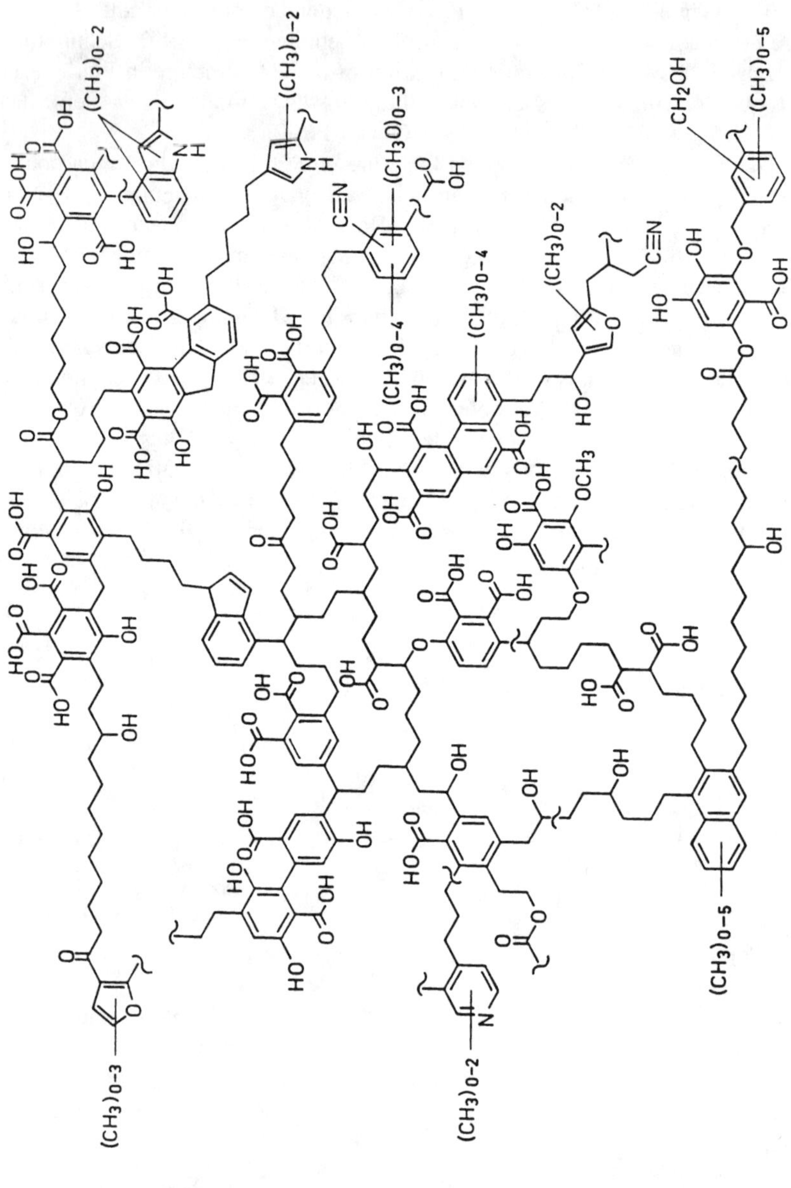

Fig. 8–8. Two-dimensional chemical structure for humic acids (Schulten & Schnitzer, 1993).

The HA structure in Fig. 8–8 is in agreement with chemical (Schnitzer & Khan, 1972; Schnitzer, 1978), oxidative and reductive degradation (Schnitzer & Khan, 1972; Schnitzer, 1978), colloid-chemical (Gosh & Schnitzer, 1980), electron-microcopic (Stevenson & Schnitzer, 1982), and ^{13}C NMR and x-ray (Schnitzer et al., 1991) investigations done on HAs over many years, and with exhaustive consultations of the voluminous literature on the subject. For the development of the structure, Schnitzer and Schulten (1995) assume that carbohydrates and proteinaceous materials are adsorbed on external HA surfaces and in internal voids and that H bonds play an important role in their immobilization.

8–4.2 The Three-Dimensional Structure of Humic Acids

The two-dimensional HA structure (Fig. 8–8) was converted to a three-dimensional structural model and geometrically optimized with the aid of HyperChem (Hypercube, Inc.) software. Details of how this conversion was achieved have been published elsewhere (Schulten, 1995a,b; Schnitzer & Schulten, 1995).

The resulting structure is shown in Plate 8–1. Of special interest are the spaces in the HA structure that can trap and retain both organic and inorganic molecules. To test this hypothesis Schulten and Schnitzer (1995) modeled geometrically optimized three-dimensional structures of two biological molecules. One was a trisaccharide (66 atoms) that could be regarded as a cellulose subunit that may have been present in soil organic matter, whereas the second structure was a hexapeptide (92 atoms), which is made up of amino acids characteristic of soil organic matter. Following this, the three-dimensional HA structure was merged with the structures of the two biological molecules (Schulten and Schnitzer, 1995), showing that the latter were trapped by the HA. With decreasing total energy, the HA structure appeared to strangle the biological molecules in the narrowing voids and channels making chemical treatments including extractions, increasingly difficult if not impossible as entrances to the voids were covered by inorganics and/or organo-mineral complexes. This experiment showed that biological substances could be trapped and bound in voids of the three-dimensional HA structure. Thus, the earlier hypothesis of a sponge-like HA structure that could occlude biological molecules (Schulten et al., 1991; Schulten & Schnitzer, 1993; Schnitzer, 1994; Schulten, 1994) was supported by independent computational chemical analysis.

The proposed structural HA model should be considered as the first approximation in the development of a structural concept of soil HA and soil organic matter based on a combination of soil chemistry and computational chemistry. Further refinements in soil organic matter research, in particular in view of the rapid increase of personal computer (PC) capacity and powerful commercial software packages for computational chemistry, are under way to narrow the gap between theory and practice. Our main aims are to explain soil processes and to predict properties of soil organic matter and whole soils. Thus, some guidelines for future prospects for soil chemistry are outlined on the basis of PC-based molecular modeling. In view of the need for further advancing the frontieres of knowledge of the chemistry of soil organic matter, and to contribute to solving agricultural, environmental, and ecological problems, a few preliminary examples in these directions are illustrated in this review.

8–5 MOLECULAR MODELING OF HUMIC ACIDS AND SOILS

8–5.1 Quantitative Structure–Activity Relationships of Humic Colloids In Vacuo

Humic substances with sizes between 1 nm and 1 μm can be regarded as humic colloids. These macromolecular HA structures offer a vast number of different structural variations with high capacities for trapping and binding inorganic (minerals, water, gases), biological (carbohydrates, peptides, lignins, and others) and anthropogenic substances. As typical example for the three-dimensional structure of a humic macromolecule, the humic colloid constructed from nine HA subunits is illustrated in the color plot in Plate 8–2a. Geometrical optimization (and thus energy minimization) was performed using the HyperChem software (release 5.1; technical terms are given in brackets and *italics*). Following the molecular mechanics calculations (*All Atoms*), using the algorithm of Polak-Ribiere and an all atom force field (*MM+*), the humic colloid had a total energy of 46240.01 kJ mol^{-1} and reached the convergence limit of 41.9 kJ mol^{-1} nm^{-1} for the termination of the calculations. Beside the total energy of the molecular system, the calculations allow us to evaluate six characteristic energy contributions (in kJ mol^{-1}) such as: bond lengths and stretching = 11179.55; angles bending = 2611.91; dihedral angles, torsions 2599.37; van der Waals interactions = 21104.02; bond stretch-angle bending = 750.94; and electrostatic interactions (bond dipoles) = 81.19. The HA colloid obtained by PC-based molecular modeling in Plate 8–2a has the elemental composition of $C_{2837}H_{3133}O_{806}N_{45}$, the elemental analysis of 67.13% C, 6.22% H, 25.41% O, and 1.24% N, and the corresponding molecular mass of 50758.9027 g mol^{-1} (Standard Atomic Weights, IUPAC, 1994) The smallest rectangular box enclosing this macromolecule has the spatial dimensions of $x = 9.8293$ nm, $y = 7.5196$ nm, and $z = 16.0575$ nm ($\Sigma = 1186.41$ nm^{+3}).

As shown in Plate 8–2b (*Sticks*) a wide range of voids and channels in the colloid is observed that is formed by association of two tetramer molecules I and II and the monomer molecule III. Recently the formation and structure of HA oligomers has been described (Schulten, 1996a) and the importance of inter- and intramolecular H bonds has been stressed as a characteristic structual feature of HAs. Hydrogen bonds are formed if the H-donor distance is <0.32 nm and the angle made by covalent bonds to the donor and acceptor atoms is <120 degrees. In the case of the described aggregate of nine HA subunits, two intermolecular H bridges are generated during geometry optimization and approach the local energy minimum of the structure shown in Plate 8–2. These important links which hold the whole colloid together between molecules I and III and between molecules II and III are indicated by arrows. In Fig. 8–9 the intermolecular hydrogen bond between molecules II and III is shown at a very high magnification. Although only an area of 1.9 × 1.14 nm is covered, the positions of the atoms

NEW IDEAS ON CHEMICAL MAKE-UP

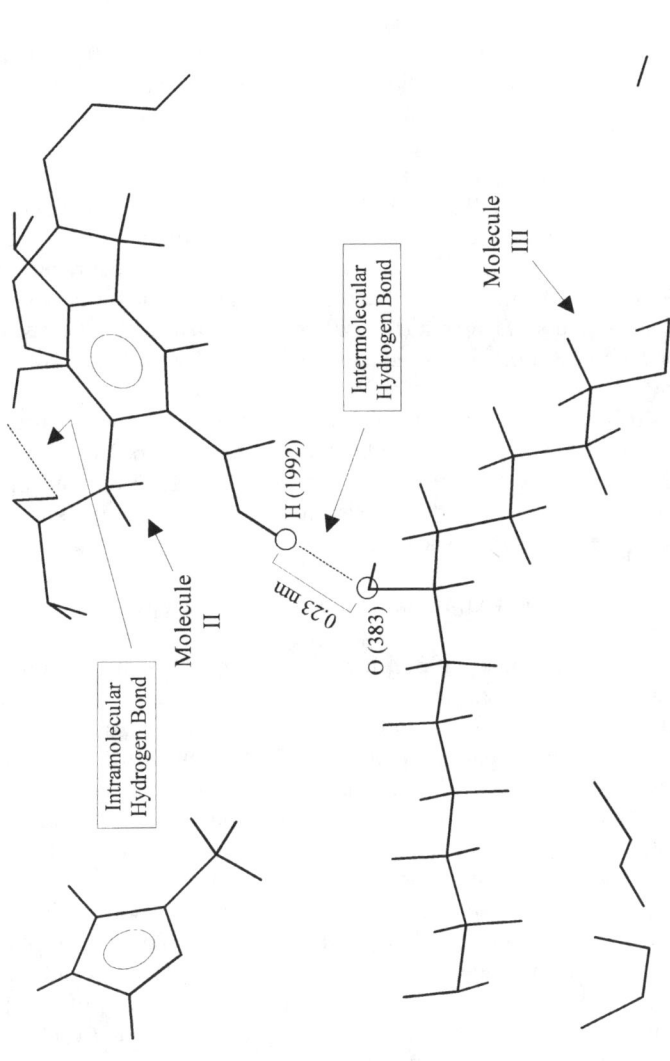

Fig. 8–9. Magnified section of humic acids association with details of the intermolecular H bond between a monomer and tetramer HA structure (Schulten, 1997, unpublished data).

forming the H bond, i.e., O number 383 from the long-chain aliphatic hydroxyl group and the H number 1992 from an aromatic carboxyl function are clearly discerned. The distance between both atoms is calculated as 0.234357 nm and the H bond energy as 1450.874 kJ mol^{-1}. In addition eight intramolecular H bonds within molecules I, II, and III are identified. There is no problem to determine additional distances, bonds, partial charges, angles and energies at the nanochemistry level.

A wide variety of quantitative structure activity relationship (QSAR) properties can be calculated using the additional ChemPlus software program (release 1.5). QSAR investigations allow us to correlate molecular structures with humic and soil properties such as surface area, volume, partial charges (electronegativity), polarizability, refractivity, hydrophobicity, hydration energy, and mass (Schulten & Schnitzer, 1997, 1998). Two alternatives are offered to calculate the molecular surface and volume of the colloid. Firstly, the solvent-accessible data can be obtained by fast empirical calculations. In this example, the radius of the solvent was chosen as 0.15 nm that is the default parameter for water as solvent. The solvent accessible surface area (*ChemPlus*; exact *Grid* method) was determined to be 6981.08 nm^{+2} and the surface-bounded molecular volume = 678.8081 nm^{+3}. Secondly, the van der Waals surface area (*Grid*) of 1842.867 nm^{+2} and a surface-bound molecular volume of 95.47 nm^{+3} were calculated. Additional calculated data were obtained for refractivity = 12.601 nm^{+3}; polarizability = 5.274 nm^{+3}; log P (partitioning coefficient) = 1373.85; and mass = 50760.11 amu (ChemPlus). The determination of hydration energy is available but is limited at present to amino acids and peptides.

8–5.2 Geometry Optimization of Whole Soil Particles

According to Greenland (1965), most of the organic matter in predominantly inorganic soils is present in the form of metal and clay complexes. Thus, the formation and characteristics of soil organo-mineral complexes are of fundamental importance in soil science (Schnitzer, 1995). Preliminary studies to model these high complexity three-dimensional structures have been reported (Leinweber & Schulten, 1994; Schulten, 1995b; Schulten & Schnitzer, 1995; Schulten et al., 1998). In modeling soil particles, it is necessary to consider the metal bridges between humic substances and the mineral surface. In particular Al, Fe, Ca, and Mg cations play an important role. In Plate 8–3 a soil particle (3196 atoms) is simulated consisting of a HA plus two trapped biological molecules, a trisaccharide, and a hexapeptide. Trapping of the sugar ($C_{18}H_{32}O_{16}$) and the peptide ($C_{26}H_{46}O_{10}N_{10}$) in the voids of the model HA and results of molecular mechanics calculations of this SOM complex have been reported (Schulten & Schnitzer, 1995). The model silica matrix is constructed using eight silica sheets with $285 \times 8 = 2280$ atoms (Schulten, 1995b). The smallest rectangular box of this soil particle has the dimensions of $x = 7.44$ nm, $y = 4.27$ nm, and $z = 9.34$ nm ($\Sigma = 296.72$ nm^{+3}) and a molecular mass of 46924.4308 g mol^{-1}. The elemental composition of $C_{352}H_{698}N_{15}O_{1506}Si_{616}Fe_5Al_4$ and elemental analysis of 9.01% C, 1.50% H, 0.45% N, 51.35% O, 36.87% Si, 0.60% Fe, and 0.23% Al were determined. For the molecular mechanics calculations and geometry optimization (*All*

Atoms) of the complex of 11 molecules, the capacity of presently available PCs plus attachments is sufficient. The obtained total energy is 16089.31 kJ mol^{-1} at a convergence gradient of 26.89 kJ mol^{-1} nm^{-1}. In order to work with geomacromolecules such as humic and soil particles, faster processors and much more random access memory is needed. For future PC-based modeling, in particular the sophisticated inorganic structures will pose problems although initial software (e.g, *crystal builder*, ChemPlus) is already offered. Evaluation of molecules of this and even larger size and complexity are possible. Plate 8–3 displays the soil particle with the HA and the trapped biological molecules in the center. The surrounding mineral matrix is open at the upper left side and allows invasion and trapping of large molecules (biological, anthropogenic), microorganisms and inorganic and/or organic particles (e.g., plant material). Four bonds of trivalent Fe cations to the silica sheets are shown and are essentially -Si-O-Fe(OH)-OC=O- bonds. The fifth Fe still carries the free hydroxyl group HO-Fe(OH)-OC=O-. Similarly the four Al cations are linked to the silica sheets. Obviously the distances between the mineral surface and the functional groups of the HA are mainly between 0.2 and 2 nm. In total 15 H bonds are observed and these underline the importance of H bonding in humic substances and soil organic matter.

8–5.3 Binding of Anthropogenic Substances to Three-Dimensional Humic Models in Water

Recently the HA models described above were further developed in the direction of SOM and whole soils (Schulten & Schnitzer, 1997). Increase in O content mainly due to carboxylic and phenolic functions, and the use of averaged lengths of aliphatic chains between the aromatic units were considered. Furthermore, results of particle-size fractions and whole soils by Py-GC–MS in combination with N-selective detectors required modifications in terms of heterocyclic components (Schulten et al., 1997; Schulten & Schnitzer, 1998). Thus, a novel model of Total Humic Substance (THS, 755 atoms) with an elemental composition of $C_{305}H_{299}O_{134}N_{16}S_1$, elemental analysis: 57.56% C, 4.74% H, 3.52% N, 33.68% O, 0.5% S, and molecular weight of 6364.8225 g mol^{-1} was put forward.

Since all investigations described above were performed by calculations in vacuo, it is of interest to evaluate to what extent the gas phase experiments can be transferred to the condensed phase such as water and whether direct chemical modeling of humic substances and their interactions with biocides can be performed in water. Initial simulations of molecular model structures of humic acids for trapping and binding of water molecules and humic-contaminant complexes with pentachlorophenol dissolved in water have been illustrated (Schulten, 1996b). Molecular mechanics calculations are reported (i) to visualize interactions between the HA model (Plate 8–1) and water as well as HA and contaminant; (ii) to describe the nanochemistry of HAs and their complexes in vacuum and in aqueous solution.

For work in the condensed phase, a complex of THS with pentachlorophenol (PCP; C_6HOCl_5; molecular weight 266.3368 g mol^{-1}; 13 atoms) was constructed and optimized geometrically in vacuo (Schulten, 1996b; Schulten et al.,

1998). As shown in Plate 8–4a this molecular complex (768 atoms) is placed in a periodic box containing water molecules (*Set Up*; *Periodic Box*) using HyperChem software. The water box has the dimensions $x, y, z = 5.61041$ nm and can contain a maximum number of 5833 water molecules. The minimum distance between solvent and solute atoms is set to 0.23 nm. When the THS + PCP complex is submerged into the water box, 5479 water molecules are added to the box (*cut off* from 2.4 to 2.8 nm) which holds now in total 17 205 atoms. The elemental composition of $C_{311}H_{11258}N_{16}O_{5614}S_1Cl_5$, elemental analysis of 3.55% C, 10.77% H, 0.21% N, 85.27% O, 0.03% S, 0.17% Cl, and summed molecular weight of 105336.8785 g mol^{-1} were determined. Although the arrow in Plate 8–4a points to the center of the water box and the position of the dissolved THS + PCP complex, it is practically not easy to identify the complex in the haystack of 17 205 atoms (*Sticks*). Therefore a sphere with a radius of 1.3 nm was cut out of the water box and the color plot in Plate 8–4b shows the selected atoms in the space-filling disk mode. A weak but clearly identifiable green spot allows at least the location of one of the chlorine atoms of PCP. As illustrated by the pentachlorophenol arrow in Fig. 8–10a, the almost horizontal positioning of the benzene ring of PCP can be found. More important is the fact that computational chemistry allows to perform energy calculations and geometry optimization on (i) total humic substance dissolved in water; (ii) PCP dissolved in water; (iii) the dissolved THS+PCP complex; and (iv) portions of THS in vacuo or the phase transition area.

In order to give a very simplified view of the reactions in the condensed phase, in Fig. 8–10b the H bond between the phenolic O of PCP (donor) and the H (acceptor) of one water molecule is shown. During molecular mechanics calculations the hydrophilic and hydrophobic moieties of the THS + PCP complex are clearly displayed. The presently ongoing evaluations of the results of computational chemistry strongly support future detailed investigations of the reactions of biocides with humic substances, SOM and whole soils.

Considerable help will come from software readily available on the Internet (e.g., U.S. Environmental Protection Agency home page). Several hundreds of state-of-the-art three- dimensional structures of compounds relevant for aquatic toxicologists have been processed by high-performance computing resources (Renner, 1997). Thus, using QSAR for predicting the toxicity, but also agricultural, ecological, and others, properties of chemicals in soils, might be one of the forthcoming trends in PC-based molecular modeling of soil properties.

The following general observations can be made:

1. With increasing geometry optimization, and thus decreasing total energy of the complex during molecular-mechanics calculations, the space available to the host molecule decreases. Following an initial period of fixation by H bonds, the occluded molecules are finally completely surrounded by the organic-mineral complex and trapped.
2. The voids in soil particles are usually of larger dimensions between the SOM and the mineral matrix compared to voids inside the humic particles. Soil properties at nanochemistry level can be studied and allow the simulation of reactions on soil surfaces. Hydrophilic and hydrophobic

NEW IDEAS ON CHEMICAL MAKE-UP

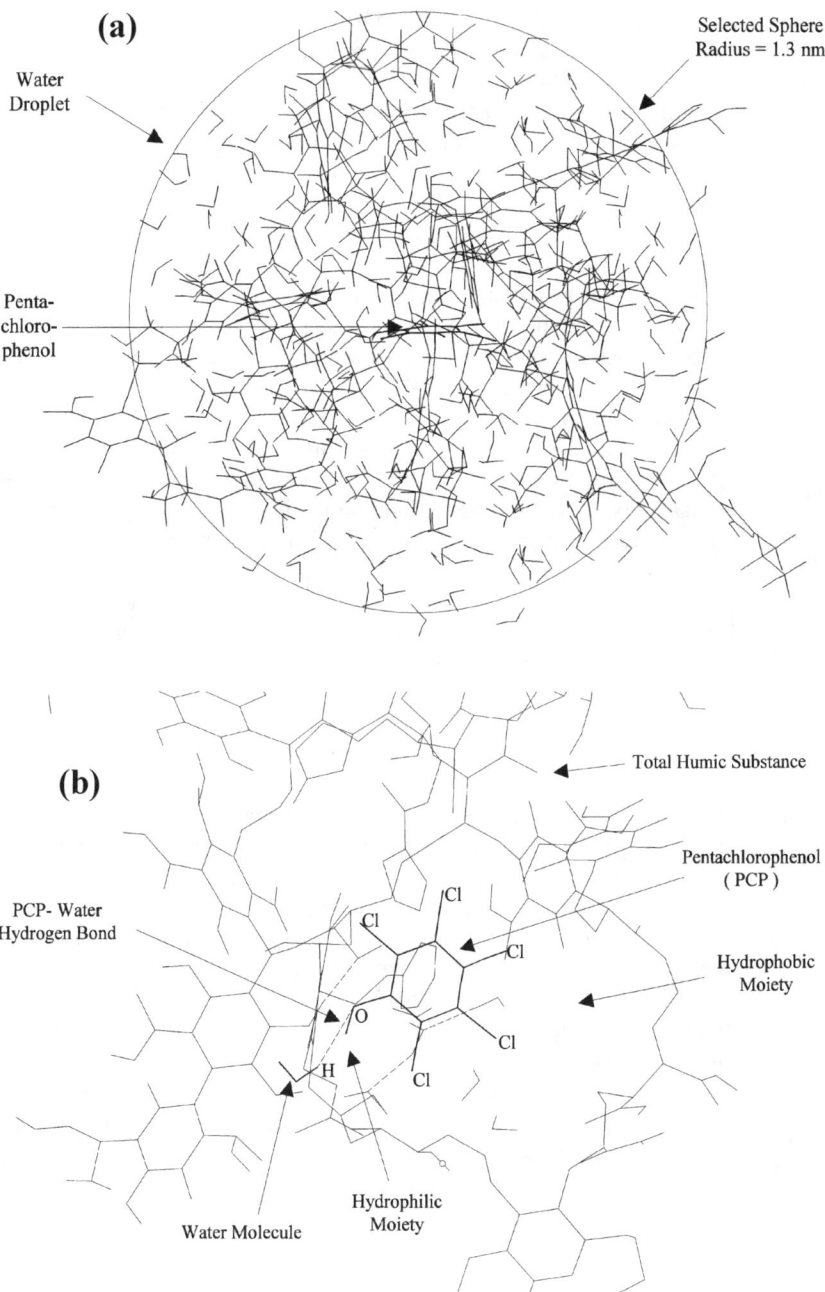

Fig. 8–10. (a) Spheric droplet of the humic pentachlorophenol complex described in Plate 8–4b following further selection and geometry optimization (2048 atoms); and (b) selecting the plane of the aromatic ring of the pentachlorophenol in the water box (see Plate 8–4a), the chemical interaction of the biocide with water and total humic substance can be studied (606 atoms) (Schulten, 1997, unpublished data).

reaction sites are observed on the inner and outer surfaces which adapt to biological and anthropogenic substances during geometry optimization. For immobilization H bonding plays a major role.
3. Quantitative structure–activity relationship (QSAR) properties can be determined and the approach to correlate properties of molecular structures with a specific kind of chemical or biochemical activity is widely used. For investigations of key physical and chemical processes, applications of QSAR in soil science and agricultural chemistry appear promising.
4. At this stage of preliminary reconstruction of the inorganic matrix, only a very simplified model of a band silicate has been used in order to illustrate the organic-mineral bonds, but future work with software designed for inorganic lattices and complex matrices will be possible (e.g., crystal builder, ChemPlus).
5. Studies on interactions and processes in vacuo and in the condensed phase are now available. The principal solvent is water, but other solvents can be used; however, at present the number of atoms and size of data sets using water boxes for humic substances and SOM are still difficult for PC-based computational chemistry. Future application could include investigation of soil processes such as the persistence and decomposition of biological and/or anthropogenic substances in soils and the binding capacities for heavy metals.

8–6 CONCLUDING COMMENTS

1. By Py-FI mass spectrometry the molecular composition of soil organic matter can be determined directly on whole soils, without extracting the organic matter, separating the latter into HA and FA, and considering the organic matter in the residual soil as humin. Whether or not it is advisable at this time to abandon the fractionation of soil organic matter into HA, FA, and humin is another matter. It is clear, however, that the development of Py-FI mass spectrometry and its application to the analysis of whole soils without any pretreatments constitutes a significant advance in SOM chemistry.
2. The hypothetical HA structures proposed in this chapter contain voids of sufficient dimensions and chemical properties that can trap and bind biological molecules such as carbohydrates and peptides as well as biocides and possible also inorganics.

8–7 FUTURE PROSPECTS FOR HUMIC SUBSTANCES AND SOIL ORGANIC MATTER RESEARCH

Studies on the chemistry of HAs, FAs, humin, and SOM go back more than 200 yr. Yet in spite of a voluminous literature, there exists at this time no widely accepted concept of the chemical structures of these materials. One can ask the following question: why is chemical structure important? During the years, studies in organic chemistry have shown that there is a close relationship between structure and reactivity. Let us take an example from soil science: among the most fundamental reactions in soils are those between HAs, FAs, humins, and SOM, with metals, hydrous oxides of Fe and Al, and clay minerals. Usually these reactions are thought to occur via O-containing functional groups on HAs and FAs, mainly CO_2H and phenolic OH groups, with metals and minerals. But before these functional groups can interact, the H at the end of CO_2H and OH groups needs to dissociate. The latter process is governed by the solvent (water in the case of soils), the pH, and the chemical structure to which the functional group is attached. Thus, while there are numerous reports in the literature on metal-HA or metal-FA complexing, the chemical structures of the ligands remain unknown. To better understand these reactions, it would certainly be advantageous to know more about the chemical structures of HAs, FAs, and humin. Note that metal- and clay-organic interactions are important in soil genesis, soil structure formation, availability of macro- and micro-nutrients, and their mobilization, transport, and immobilization in soil profiles (Schnitzer, 1995).

The HA model (Fig. 8–8) which we propose in this chapter is based on long-term chemical, ^{13}C NMR, colloid-chemical (surface tension, surface pressure, viscosity), electron microscopic, x-ray, and mass spectrometric studies. One of the most striking features of this structure is that it contains functional groups on the outside as well as inside the internal voids. The overall impression is that of a relatively open and flexible structure.

With the aid of computational chemistry we evaluated and characterized the structural model from the viewpoints of geometry, energy, and structure-property correlations. (Plate 8–1). The internal voids in the structure are clearly visible. We measured the molecular dimensions of the voids as well as those of a triose, a hexapeptide, and of atrazine, a herbicide. The latter three compounds fitted into the voids of the HA structural model. This means that relatively small molecules can penetrate internal voids in the HA structure and remain there for a long period of time. Our study explains the persistence of strongly bound pesticide residues, especially in organic soils, by their retention in HA voids. Our observations confirm suggestions by Schnitzer (1978) that HAs could accumulate in internal spaces significant amounts of toxic chemicals.

From the examples presented in this paragraph it becomes clear that the combination of classical and computational chemistry can contribute significantly to solve theoretical and practical problems in soil organic chemistry. Our model HA structure constitutes an early attempt to use computational chemistry. There is little doubt that with the passage of time, advances in computational chemistry, and improvements in the designs of ^{13}C NMR spectrometers, mass spectrometers, and other instruments, our HA structural model will be refined, modified, and hopefully improved. As soil organic chemists we have to be aware of and take advantage of the rapid advances in computational chemistry so that our research can be productive and effective in solving problems related to agriculture and the environment. These problems may become more urgent as we enter the 21st century. We need to be ready to provide solutions.

ACKNOWLEDGMENTS

This work was financially supported by the German Research Association (Deutsche Forschungsgemeinschaft, DFG) and the Ministry of Science and Technology, Bonn-Bad Godesberg, Germany.

REFERENCES

Achard, F.K. 1786. Chemische untersuchungen des torfs. Crell's Chem. Ann. 2:391–403.
Berzelius, J.J. 1839. Lehrbuch der chemie. Arnoldische Buchhandlung, Dresden.
Breitmaier, E., and W. Voelter. 1978. ^{13}C NMR spectroscopy. Verlag Chemie Int., Deerfield Beach, FL.
Dragunov, C.C., H.H. Zhelokhovtseva, and E.J. Strelkova. 1948. A comparative study of soil and peat humic acids. Pochvovclenie 7:409–420.
Felbeck, G.T. 1965. Structural chemistry of soil humic substances. Adv. Agron. 17:327–368.
Fischer, F., and H. Schrader. 1921. The origin and chemical structure of coal. Brennstoff Chem. 2:37–45.
Flaig, W. 1964. Chemische untersuchungen an huminstoffen. Zeitschr. Chem. 4:253–265.
Flaig, W., H. Beutelspacher, and E. Rietz. 1975. Chemical composition and physical properties of humic substances. p. 1–211. In J.E. Gieseking (ed.) Soil components. Vol. 1. Organic components. Springer Verlag, New York.
Ghosh, K., and M. Schnitzer. 1980. Macromolecular structures of humic substances. Soil Sci. 129:266–276.
Greenland, D.J. 1965. Interactions between clays and organic compounds: I. Mechanisms of interactions between clays and defined organic compounds. Soils Fert. 28:415–425.
Haider, K., J.P. Martin, and Z. Filip. 1975. Humus biochemistry. Vol. 4. Marcel Dekker, New York.
Hatcher, P.G., and E.C. Spiker. 1988. Selective degradation of plant biomolecules. p. 59–74. In F.H. Frimmel and R.F. Christman (ed.). Humic substances and their role in the environment. John Wiley & Sons, Chichester.
Hayes, M.H.B., P. MacCarthy, R.I. Malcolm, and R.S. Swift (ed.) 1989. Humic substances: II. In search of structure. John Wiley & Sons, Chichester.
Hempfling, R., W. Zech, and H.-R. Schulten 1988. Chemical composition of the soil organic matter in forest soils: 2. Model profile. Soil Sci. 146:262–276.
Huang, P.M. 1990. Role of soil minerals in transformation of natural organics and xenobiotics in soils. p. 29–115. In J.-M. Bollag and G. Stotzky (ed.) Soil biochemistry. Vol. 6. Marcel Dekker, New York.
IUPAC. 1994. Atomic weights of the elements. Commission on Atomic Weights and Isotopic Abundances. Pure Appl. Chem. 66:2423–2425.
Khan, S.U., and F.J. Sowden. 1971. Distribution of nitrogen in the black solonetzic and black chernozemic soils of Alberta. Can. J. Soil Sci. 51:185–193.
Kleinhempel, D. 1970. Ein beitrag zur theorie des huminstoffzustandes. Albrecht Thaer Archive 14:3–14.
Leinweber, P., and H.-R. Schulten. 1994. Zur bedeutung pedogener Oxide für die Bindung organischer Substanzen. Mitt. Dtsch. Bodenkundl. Gesellsch. 74:383–386.
Lowe, L.E. 1978. Carbohydrates in soils. p. 65–94. In M. Schnitzer and S.U. Khan (ed.). Soil organic matter. Elsevier, Amsterdam.
Maillard, L.C. 1913. Formation de matières humiques par action de polypeptides sur sucres. C. R. Acad. Sci. 156:148–149.
Norwood, D.L. 1988. Critical comparison of structural implications from degradative and nondegradative approaches. p. 133–148. In F.H. Frimmel and R.F. Christman (ed.). Humic substances and their role in the environment. John Wiley & Sons, Chichester.
Preston, C.M., M. Schnitzer, and J.A. Ripmeester. 1989. A spectroscopic and chemical investigation on the deashing of a humin. Soil Sci. Soc. Am. J. 53:1442–1447.
Renner, R. 1997. Analyzing chemical toxicity in 3-D on the Web. Environ. Sci. Technol. 31:32A–33A.
Schnitzer, M. 1978. Humic substances: chemistry and reactions. p. 1–64. In M. Schnitzer and S.U. Khan (ed.). Soil organic matter. Elsevier, Amsterdam.

Schnitzer, M. 1991. Soil organic matter: The next 75 years. Soil Sci. 151: 41–58.
Schnitzer, M. 1994. A chemical structure for humic acid. Chemical, ^{13}C NMR, colloid chemical, and electron microscopic evidence. p. 57–69. In H. Senesi and T.M. Miano (ed.) Humic substances in the global environment and implications to human health. Elsevier, Amsterdam.
Schnitzer, M. 1995. Organic-inorganic interactions in soils and their effects on soil quality. p. 3–19. In P.M. Huang et al. (ed.) Environmental impact of soil component interactions. Vol. 1. Natural and anthropogenic organics. CRC Press/Lewis Publ., Boca Raton, FL.
Schnitzer, M., and S.U. Khan. 1972. Humic substances in the environment. Marcel Dekker, New York.
Schnitzer, M., H. Kodama, and J.A. Ripmeester. 1991. Determination of the aromaticity of humic substances by x-ray diffraction analysis. Soil Sci. Soc. Am. J. 55:745–750.
Schnitzer, M., and H.-R. Schulten. 1992. The analysis of soil organic matter by pyrolysis-field ionization mass spectrometry. Soil Sci. Soc. Am. J. 56:1811–1817.
Schnitzer, M., and H.-R. Schulten. 1995. Analysis of organic matter in soil extracts and whole soils by pyrolysis-mass spectrometry. Adv. Agron. 55:167–217.
Schulten, H.-R. 1994. A chemical structure for humic acid. Pyrolysis gas chromatography/mass spectrometry and pyrolysis-soft ionization mass spectrometry evidence. p. 43–56. In H. Senesi and T.M. Miano (ed.) Humic substances in the global environment and implications to human health. Elsevier, Amsterdam.
Schulten, H.-R. 1995a. The three-dimensional structure of humic substances and soil organic matter studied by computational analytical chemistry. Fresenius J. Anal. Chem. 351:62–73.
Schulten, H.-R. 1995b. The three-dimensional structure of soil organo-mineral complexes studied by analytical pyrolysis. J. Anal. Appl. Pyrolysis 32:111–126.
Schulten, H.-R. 1996a. A new approach to the structural analysis of humic substances in water and soils: Humic acid oligomers. p. 42–56. In J.S. Gaffney et al. (ed.) Humic and fulvic acids: Isolation, structure and environmental role. Symp. Ser. 651. Am. Chem. Soc., Washington.
Schulten, H.-R. 1996b. Three-dimensional, molecular structures of humic acids and their interactions with water and dissolved contaminants. Int. J. Environ. Anal. Chem. 64:147–162.
Schulten, H.-R., B. Plage, and M. Schnitzer. 1991. A chemical structure for humic substances. Naturwissenschaften 78:311–312.
Schulten, H.-R., P. Leinweber, and M. Schnitzer. 1998. Analytical pyrolysis and computer modeling of humic and soil particles. p. 281–324. In P.M. Huang et al. (ed.) Structure and surface reactions of soil particles. Vol. 4 of IUPAC series of Analytical and Physical Chemistry of Environment Systems. John Wiley & Sons, Chichester, London.
Schulten, H.-R., and M. Schnitzer. 1992. Structural studies on soil humic acids by Curie-point pyrolysis-gas chromatography/mass spectrometry. Soil Sci. 153:205–224.
Schulten, H.-R., and M. Schnitzer. 1993. A state of the art structural concept for humic substances. Naturwissenschaften 80:29–30.
Schulten, H.-R., and M. Schnitzer. 1995. Three-dimensional models for humic acids and soil organic matter. Naturwissenschaften 82:487–498.
Schulten, H.-R., and M. Schnitzer. 1997. Chemical model structures for soil organic matter and soils. Soil Sci. 162:115–130.
Schulten, H.-R., and M. Schnitzer. 1998. The chemistry of soil organic nitrogen: A review. Biol. Fertil. Soils 26:1–15.
Schulten, H.-R., C. Sorge-Lewin, and M. Schnitzer. 1997. The structure of "unknown" soil nitrogen investigated by analytical pyrolysis. Biol. Fertil. Soils 24:249–254.
Senesi, N., and T.M. Miano (ed.). 1994. Humic substances in the global environment and implications on human health. Elsevier, Amsterdam.
Stevenson, F.J. 1994. Humus chemistry. John Wiley & Sons, New York.
Stevenson, I.L., and M. Schnitzer. 1982. Transmission electron microscopy of extracted fulvic and humic acid. Soil Sci. 133:179–185.
Sulflita, J.M., and J.-M. Bollag. 1981. Polymerization of phenolic compounds by a soil-enzyme complex. Soil Sci. Soc. Am. J. 45:297–302.
Waksman, S.A. 1938. Humus, origin, chemical compositions and importance in nature. Williams & Wilkins, Baltimore.
Wang, T.S.C., P.M. Huang, C.-H. Chou, and J.-H. Chen. 1986. The role of soil minerals in the abiotic polymerization of phenolic compounds and formation of humic substances. p. 251–281. In P.M. Huang and M. Schnitzer (ed.) Interactions of soil minerals with natural organics and microbes. SSSA Spec. Publ. 17. SSSA, Madison, WI.

9 Chemistry of Soil–Nutrient Interactions and Future Agricultural Sustainability[1]

Stanley A. Barber
Purdue University
West Lafayette, Indiana

9–1 INTRODUCTION

This chapter involves the portion of soil chemistry that relates to the supply of nutrients from the soil to the plant root surface. This is a highly significant phase of soil chemistry since it is involved in the growth of all plant species growing in all soils throughout the world. While we are usually concerned with those species used directly or indirectly for food, this phase of soil chemistry also is concerned with all plant species that grow in soil, for example, the growth of trees for their many uses from newsprint and lumber to their use for scenic beauty and air purification. It also is important for golf courses, athletic fields, flowers, and the many other uses we make of plants. In addition to the chemistry of nutrient supply to the root it is involved with supply of toxic elements that may interfere with plant growth as well as be harmful to humans, animals and birds, and others, which eat plants or their parts. Hence, the chemistry of the soil–plant root interface is important for food, shelter, sports, health, and others, world wide.

Involvement in soil chemistry during my career has primarily been with its influence on nutrient supply to the root surface. Hence, this paper in the symposium "Whither Soil Chemistry" will be based on impressions gained during my 45 yr research career. During this time I have been asked to give talks at International and National meetings that were primarily plant oriented as well as at meetings that were primarily soil oriented, since this area involves both plant chemistry, soil chemistry, and their interaction. Hence, I will briefly discuss some of the soil chemistry that has helped me understand plant nutrition problems, then I will give my opinions on soil chemistry research that may be promising areas for future research.

9–2 DEVELOPING A NUTRIENT UPTAKE MODEL

Developing a concept of soil nutrient supply to plant roots and publishing it was important in the development of my research, so I recommend it. It focused

---

[1] Contribution from the Department of Agronomy, Purdue University, West Lafayette, IN.

Copyright © 1998. Soil Science Society of America, 677 S. Segoe Rd., Madison, WI 53711, USA. *Future Prospects for Soil Chemistry.* SSSA Special Publication no. 55.

my research and got others interested who both tested my concepts and helped in the development of the research area. Some of the highlights of this research are given here. In 1958, a graduate student of mine, John Walker, successfully obtained one of the first autoradiographs of ion depletion in the soil perpendicular to a maize (*Zea mays* L.) root using ^{86}Rb labeled Rb mixed with the soil (Walker, & Barber, 1962). The ^{86}Rb pattern on the x-ray film clearly demonstrated that ^{86}Rb was reaching the root by diffusion. Years later, Claassen et al. (1986) made microtome sections of soil perpendicular to the root that clearly showed that unlabeled K gave the same pattern. Stimulated by this early result and by information found in the library while on sabbatical, I wrote an article entitled "A Mass Flow and Diffusion Concept of Soil Nutrient Availability" (Barber, 1962) where I discussed the significance of these methods of moving nutrients to the root surface. This became a classic article in Citation Index and, surprisingly, is occasionally being referred to in articles being published at this writing. This article stimulated research in several countries (I was recently told by those doing research in this area). Their research contributed to the progress of the research in this area.

In 1969, a theoretical study of the distribution of nutrients around the root was published by Nye and Marriott (1969). Their article gave a method for describing mathematically the distribution of nutrients in the soil perpendicular to the root with time that result from uptake by mass flow and diffusion. In 1972, with the help of a computer scientist, we developed (Claassen & Barber, 1976) a mechanistic computer program to calculate the amount of nutrient uptake with time by a plant growing in soil. The model used both soil and plant parameters as follows:

Soil parameters
  C_{li}, initial concentration of nutrient in the soil solution
  b, buffer power, which is C_{si}, concentration of labile nutrient in the soil volume, which includes C_{li}, divided by C_{li} (this differs from buffer capacity)
  D_e, effective diffusion coefficient. It can be closely approximated from the expression $D_l \theta f 1/b$, where D_l is the diffusion coefficient in water, θ is the volumetric fraction of water in the soil, f is the tortuosity factor, calculated in silt loam or lighter texture soils using $f = 1.60\theta - 0.172$ (Barraclough & Tinker, 1981); for heavier textured soils, soil bulk density enters into the calculation. Hence the principal soil measurements needed are: C_{li}, C_{si}, and θ, and possibly soil bulk density.

Plant factors
  v_o, mean water influx into the root
  r_o, mean root radius
  r_1, mean one-half distance between roots
  L_o, initial root length
  k, root length growth rate
  I_{max}, maximal nutrient influx rate
  C_{min}, minimal nutrient concentration where In, influx = 0
  K_m, solution concentration – C_l where $In = 1/2\ I_{max}$

With some plant species or low P soils, P uptake is affected by root hairs, mycorrhiza, or rhyzosphere soil pH changes, while K uptake may not be affected by any of these in the same experiment. This is primarily due to the much lower D_e for P or the effect of pH on P solubility. Measuring both P and K in the same growth experiment can give additional information about the processes occurring when the model is used to predict uptake of both in the same experiment.

The model uses the following equation in calculating ion distribution perpendicular to the root.

$$\frac{\partial C_1}{\partial t} = \frac{1}{r} \frac{\partial}{\partial r} \left[rD_e \frac{\partial C_1}{\partial r} + \frac{r_o v_o C_1}{b} \right] \qquad [1]$$

where r is radial distance. The inner boundary condition at the root surface $r = r_o$ can be developed assuming that uptake follows Michaelis-Menten kinetics. This is the kinetics commonly used by plant physiologists in describing nutrient uptake. This gives the following equation for the inner boundary condition at the root surface,

$$D_e b \frac{\partial C_1}{\partial t} + v_o C_1 = I_{max} \frac{C_1 - C_{min}}{K_m - C_1 - C_{min}}, \qquad r = r_o, t > 0. \qquad [2]$$

If the roots do not compete with adjacent roots for the nutrient studied, then the outer boundary at r_1, will remain constant. If the concentration gradients extending from adjacent roots overlap then $J = 0$, at $r = r_1$, at $t > 0$ where J is ion flux toward the root. Cushman (Barber & Cushman, 1981) rewrote the uptake model including the competition between roots so that nutrient flow to the root would only come from the soil volume about each root. The technique for solving the model is given in Barber and Cushman (1981) and also in Barber (1995). Accumulated uptake with time as the plant grows can be calculated with additional expressions given in the last two references. The assumptions used in the calculation are given in detail in Barber (1995) and also in Barber and Cushman (1981). The assumptions do not detract from the use of the model, but indicate where the model can be used. The mathematical solution of the model requires iteration, however, this is done rapidly using newer PC computers and software so that calculations for each problem are finished in seconds rather than in many minutes or even hours as occurred with earlier computers and software.

The accuracy of the model for determining nutrient uptake can be evaluated by comparing uptake predicted by the model with actual uptake measured from plant analysis as shown in Fig. 9–1. Many comparisons have shown close agreement (Barber, 1995). Comparisons also have been made using plant uptake experiments measuring both P and K uptake in the same experiment. Examples are given in Table 9–1. Comparisons also have been made where time of growth was constant while soil or plant species or both were varied (Barber, 1995). In addition to experiments conducted in pots grown in controlled climate facilities, experiments were conducted in the field. In pot experiments the time of growth

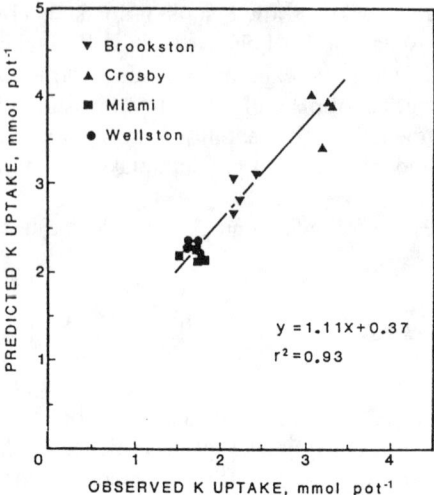

Fig. 9–1. The relation between K uptake by maize grown in four replicates of 3 L pots of four different soils, varying in cation-exchange capacity, in a controlled climate facility and predicted uptake from these pots of soil calculated with the Barber-Cushman uptake model. All pots had approximately the same amount of exchangeable K. The soils were Brookston silt loam (fine-loamy mixed mesic Typic Argiaquall), Crosby silt loam (fine mixed mesic Aeric Ochraqualf), Miami silt loam (fine loamy mixed Typic Haplaqualf), and Wellston silt loam (fine silty mixed mesic Ultic Hapludalf; Shaw et al., 1983).

was kept short enough so that root density in the soil was similar to what occurs in the field. This is important because with extended time periods the root length density in the soil in the pot becomes large and the distance between roots very small so that uptake may be influenced more by C_{si} than C_{li}. For annual plant species root length density in the soil in the field is usually 1 to 3 cm of root cm^{-3} of soil. Hence, time of growth of four plants in 6 L pots should only be about 10 to 25 d depending on species. Results for predicted and observed potassium uptake by maize growing in 3 L pots of four soils varying in cation-exchange capacity (Fig. 9–1) indicate the accuracy of the model where parameters were measured carefully. Although predicted uptake agreed closely with observed uptake in many experiments, there were some where agreement was not obtained. In this case there was usually an additional factor affecting uptake that needed to be investigated further (Barber, 1995).

The model assumes that the roots absorb nutrients from solution, so the concentration in solution at the root surface, C_{lo}, determines uptake rate. The value of C_{lo} depends on the flux of ions to the root by mass flow and diffusion relative to the rate of ion absorption by the root. The flux to the root surface is buffered by the equilibrium between ions in solution and the total diffusible ions in the soil. This is why the mathematical computations include b, the buffer power. The value of b also affects the size of D_e.

Hence size of b is important in determining flux to the root. Looking at this from a different perspective the smaller the size of C_{li} at the same size of C_{si} the larger b will be This is why measurements of both C_{li} and C_{si} are important. It also

Table 9–1. Examples of the accuracy of the mechanistic model for predicting nutrient uptake.

1. Maize grown in pots in the growth chamber for 13 d using four Indiana soils and an andosol from Columbia. Phosphorus measured y (predicted uptake) = 1.02 x (observed uptake) – 4.03 (r^2 = 0.98; Cox, 1991).
2. Sorghum grown in pots in the growth chamber for 10, 14, and 17 d using six Malaysian oxisols and ultisols and an Indiana molisol; y = 1. 16 x +27 (r^2 = 0.94) when root hairs were measured (Bidin & Barber, 1985).
3. Maize grown in pots in the growth chamber for 13 d in a mollisol at five pH levels. When effect of solution pH on phosphate uptake as affected by charge of the phosphate ion accounted for, y = 1.7 + 0.97 x (r^2 = 0.97; Chen & Barber, 1990).
4. Same as 3 except measurements on K. y = 3 + 0.94 x (r^2 = 0.99; Chen & Barber, 1990).

indicates the significance of measuring their true values rather than a relative value; however, where C_{li} is relatively high, mass flow may supply most of the nutrients to the root and accumulation even occurs at the root. This may occur for Ca.

The following briefly describe methods of measurement of C_{li} and C_{si} for K and P in the soil. The total amount of available solution and exchangeable K in the soil, C_{si}, that buffers that in solution, C_{li}, may be measured by extracting K from the soil with a solution of 1 mol ammonium acetate, pH 7.0, and then calculating the result as concentration per unit volume of soil. Solution ions are measured by displacing the solution from the soil by a method that removes the solution as is, so as to give the concentration of the ions in the solution as it exists in the soil and not diluting by using water or salt extraction. The total concentration of diffusible P in the soil may be obtained by shaking the soil with anion exchange resin, separating the resin from the soil by sieving, and displacing the P from the resin by shaking with warm 1 mol NaCl for 24 h. These are examples. More detailed information on measuring C_{li} and C_{si} is given by Barber (1995). Calculations for C_{li} and C_{si} are given on a volume basis because ions move by mass flow and diffusion through a soil volume toward the area of root surface.

9–3 SENSITIVITY ANALYSIS

An example of the use of the model is to make a sensitivity diagram constructed for P as shown in Fig. 9–2 (Silberbush & Barber, 1983) in order to determine which model parameter is having the greatest effect on predicted uptake. Looking at the effect of changing each of the soil factors individually, the effects on predicted uptake were C_{li} > b > D. Using the data for calculated P uptake from Raub silt loam by soybeans [*Glycine max* (L.) Merr.], the data for each parameter are separately changed by 0.5, 1.0, 1.5, and 2.0 times the original while the values for the remaining parameters are left at their original value, then predicted uptake determined with the uptake model for each and plotted vs. the degree of change in the value. Similar calculations are made for each of the 10 parameters and plotted on the same figure. The results shown in Fig. 9–2 indicate the effect of change of parameter size on predicted phosphorus uptake. It also shows whether the effect is positive or negative. In addition to that shown in Fig. 9–2, values for two or

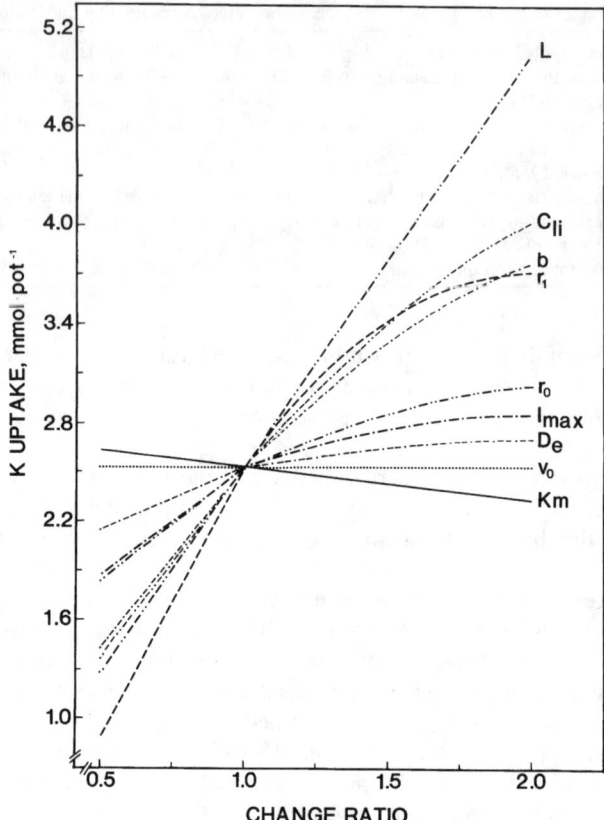

Fig. 9–2. A sensitivity diagram showing the effect on predicted phosphorus uptake of changing each parameter separately to 0.5, 1.0, 1.5, and 2.0 of the original value while keeping the remaining parameters of the model at their original values. Experimental data from experiment for P uptake by soybeans from Raub soil (fine-silty mixed, mesic Aquic Argiudall; Silberbush, & Barber, 1983).

more parameters can be changed at the same time, before determining predicted uptake. A study of this type is important where effect of parameters on uptake are confounded (see Barber, 1995 for examples).

The original model, written for a main frame computer was transferred to rotary pascal and written for use on the PC computer by Oates (Oates & Barber 1987). This form made the model more readily available than the main frame version.

9–3.1 Applications of the Mechanistic Nutrient Uptake Model

Since the size of the water content of the soil plays a large part in determining D_e, Cox and Barber (1992) studied the effect between soil water content and P and K uptake using four soils varying in soil water fraction, from 0.13 to 0.44 at −33kPa soil water potential. For these four soils, each with four rates of added P, predicted uptake obtained with the uptake model agreed with observed uptake obtained from analysis of maize plants grown on these soil treatments ($y = 1.024 - 4.0$, $x^2 = 0.98$). The level of available soil P of each soil needed to give the same predicted uptake for the four soils was determined from the soil data obtained from the four rates of added P to get the relation between the increase of C_{li} and C_{si} and P added. The soil with the lower soil water fractions required greater C_{li} and C_{si} levels to give predicted uptake equal to that of the soils with a higher water fraction. This example shows that a mechanistic model rather than a regression model where a coefficient is calculated from the data, gives a model that can be used to explain differences in nutrient uptake among soils that vary in water content. There are many similar applications that show where the mechanistic uptake model can be used if the mechanisms of the system are sufficiently understood.

9–4 APPLICATIONS

While many additional applications related to plant growth have been studied, only those relating to soil chemistry are listed here.

1. Influence of soil pH on P uptake (Chen & Barber 1990). At soil pH levels between 4.0 and 6.0, predicted uptake by maize agreed with that obtained experimentally. At higher soil pH levels observed P was much less than that calculated with the model. This discrepancy could be corrected by accounting for the effect of pH of the soil solution on the ionic form of P present. As solution pH increases the P changes from diphosphate to monophosphate and rate of monophosphate uptake is only one tenth of the uptake rate for diphosphate. When measurements were made with K, no correction was needed as pH increased since no charge change occurs.
2. The effect of fraction of the soil fertilized with P on P uptake. A given amount of P applied per pot of soil can be mixed with all the soil in the pot or with a fraction of the soil volume. As the fraction of the fertilized soil is reduced, rate of addition to the soil fertilized increases. When fertilizing 0.02 of the soil, the rate of fertilization is increased 50 times and C_{li} and D_e of the fertilized soil increased. The degree of change depends on soil chemistry, hence varies with soil and needs to be measured or predicted. Root growth is stimulated in the fertilized volume according to the ratio of the C_{si} levels in the fertilized soil and the unfertilized soil and root growth stimulation is predictable (Zhang & Barber, 1992). Root absorption of P increases at a curvilinear rate with increase in P level of C_l at the root until it reaches I_{max}. This is predictable after initial measurement (Barber, 1995). It varies with species and plant age. As the proportion of soil fertilized is increased (Fig. 9–3), uptake increases because more root surface is fertilized until P uptake reaches a maximum and then declines due to the rate of decline in P availability in the fertilized soil being greater than the increase in root surface area as soil volume fertilized increases. Hence there is an optimum degree of place-

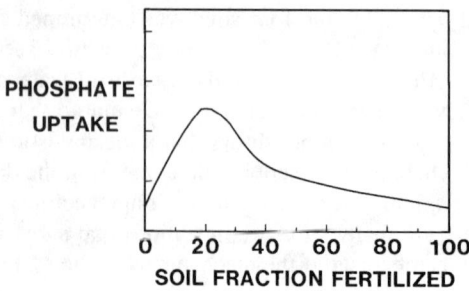

Fig. 9–3. Effect of changing the fraction of the soil fertilized with the same amount of P on the uptake of P by maize from soil as calculated using the uptake model (Kovar & Barber, 1987).

ment for each situation where P uptake is maximized. This can be predicted with the mechanistic uptake model and has been verified (Kovar & Barber, 1987).

3. Influence of root exudate into the rhizosphere. Common root exudates are protons that decrease soil pH. This frequently increases phosphorus in C_{li} so that measured uptake is greater than uptake calculated with the uptake model. A frequent reason for proton exudate is a greater uptake of cations than of anions by the root. This is more common with legumes. Measuring the increase in P uptake can be used to measure the decrease in pH needed to give observed P uptake (Li & Barber, 1991).
4. Effect of other nutrients such as N or K on P availability sometimes occurs. The increase in measured uptake over predicted uptake indicates a positive effect has occurred. The increase often occurs an increase of cation uptake over anion uptake causes a proton release and a decrease in rhizosphere soil pH. When studying effect of adding N on P uptake, measuring both P and K allows K to act as an internal standard for P uptake since N effects seldom occur with K.
5. Studies with widely varying soil types may illustrate which soil differences influence nutrient flux. In our studies we found that predicted uptake usually agreed with observed uptake. This enables us to determine the reason for observed differences among soils when differences occur.

The studies listed above are some of those I have used where soil chemistry was involved. There are many I used where changes in root physiology were the main factor involved in changing nutrient uptake. The examples given show how a mechanistic model can be more useful than a regression model in obtaining basic information about nutrient flux to plants.

9–5 FUTURE NEEDS

The following discusses where soil chemistry can be involved in new approaches to developing our understanding of nutrient availability.

1. Kinetics of the flux of ions moving from firmly held positions in the solid phase to the exchangeable to the solution phase and vice versa. It would be very useful to predict reaction rates without having to measure them for each situation. The section on kinetics by D.L. Sparks in this volume is an excellent discussion of this area.

    An example of an important kinetic effect is the flux of K to roots of perennial crops during the winter interval and in the soil between growth of annual crops. During long periods when roots are not absorbing appreciable amounts of a nutrient the level in the soil may recover toward the original level of C_{li} in some soils. This varies with soil and depends on soil chemistry supplying nutrients to the plant or where excesses may be detrimental to plant growth. There are many more ions to investigate.
2. Values for C_{li} and C_{si} have been obtained for only a few of the nutrients or ions that may be important in pollution or in damaging plant growth because of excesses. These have been reviewed by Barber (1995). Also, values for heavy metals and any ions that may occur at harmful levels in the food chain need investigation. Obtaining these values may involve the development of special analytical techniques since values for C_{li} may be very low in the soil solution and difficult to measure by present techniques.
3. There is room for the development of new mechanistic models that have greater flexibility. A few have been developed; however there is room for more. The one I present was developed many years ago.
4. Development of rapid simple analytical techniques to obtain the values for C_{li} and C_{si} and θ. This would encourage scientists to use the mechanistic approach.
5. Develop models that would allow people working with producers and who are able to obtain the necessary inputs to be able to use the model to answer the producers' questions, hence the mechanistic approach could be extended beyond its use in research. Computers are widespread so more people may become interested in using the modeling approach.

9–6 AGRICULTURAL SUSTAINABILITY

Soil chemistry and the mechanistic model are related to agricultural sustainability in a number of ways. The amount of growth of crops on a soil affects the level of soil organic matter that will be maintained. Experiments in Indiana and in Iowa have shown that when soil is left bare of vegetation the soil organic matter in the soil was lost at a rate of 1.9 to 2.0% per year (Barber, 1979). When maize was grown, about 11% of the C in residues after grain harvest was synthesized into C in new soil organic matter. In a rate of N fertilization experiment on continuous maize about 8% of the C in the organic matter in the residues left after grain removal was synthesized into C in soil organic matter. In some soils, maintaining level of organic matter is important for maintaining crop production and reducing soil losses due to soil erosion.

Where soils vary in their chemical makeup, the mechanistic model is important in determining the fertilization needed to have economical production without conducting a series of field experiments for each particular soil or soil group since a soil test can frequently be incorrect where not calibrated for the particular soils. These remarks are particularly important where new areas are being developed for future agricultural production. The ability to investigate for a series of plant nutrients by laboratory examination is much more economical than conducting a series of field experiments.

Agricultural sustainability is affected by a number of factors. This chapter concerns the use of soil chemistry to develop ways of supplying nutrients to the plant root so that plant growth is not limited and in removing ions that adversely affect plant growth. When plants are well supplied with nutrients and water the crop provides sufficient growth that soil erosion is reduced during growth and residues left after harvest minimize soil erosion between cropping periods. In addition, adequate crop growth and residues improve or maintain soil structure. These remarks may apply more in developing countries where nutrient applications are affected by economics. Keeping available nutrients at levels that produce the most economic yields without adding additional amounts will reduce the danger of pollution from water runoff or leaching through the soil.

9–7 RESEARCH GOALS

It is important to have the ability to obtain all needed basic measurements, including kinetic, so that answers to nutrient uptake problems can be obtained economically by using mechanistic computer models. Also to do basic research that will increase our understanding, so that lack of nutrient supply soil chemistry information for a widely diverse range of soils does not limit our progress. Then we need to be able to use this information in a mechanistic computer program where the solution gives the nutrient application rate and method of application for any crop in any soil and environment that gives the producer the greatest economic return and has a negligible deleterious effect, in fact improves soil sustainability.

9–8 SUMMARY AND CONCLUSIONS

This chapter is a discussion of soil chemistry involving the flux of plant nutrients through the soil to the root surface where they can be adsorbed. The discussion uses a mechanistic uptake model involving the parameters of this flux. The soil parameters involved in nutrient ion flux to the root surface are: soil solution nutrient concentration, C_{li}, soil volume concentration of diffusible nutrients, C_{si}, soil volumetric water fraction, θ, and soil bulk density. These parameters are used to calculate buffer power of nutrients, b, and effective diffusion coefficient, D_e used along with C_{li} in the mechanistic uptake model. The future of soil chemistry in this area may involve extension of the model to nutrients and other ions not presently being studied in this manner. Additional kinetic research is needed on the rate of flow of ions among soil fractions to account for those ions moving more slowly between soil phases. Procedures for more rapid measurement of the above mentioned parameters would allow the model to be used more widely and move it from a strictly research model to enable it to be used in applied areas. More precise nutrient applications to the soil will increase crop growth, reduce pollution of soil drainage water and reduce soil erosion which will help improve soil sustainability.

REFERENCES

Barber, S.A. 1962. A diffusion and mass flow concept of soil nutrient availability. Soil Sci. 93:39–49.

Barber, S.A. 1979. Corn residue management and soil organic matter. Agron. J. 71:625–627.

Barber, S.A. 1995. Soil nutrient bioavailability: A mechanistic approach. 2nd ed. John Wiley & Sons, New York.

Barber, S.A., and J.H. Cushman. 1981. Nitrogen uptake model for agronomic crops. p. 382–409. *In* I.K. Iskander (ed.) Modeling waste water renovation: Land treatment. Wiley Interscience, New York.

Barraclough, P.B., and P.B. Tinker. 1981. The determination of ionic diffusion coefficients in field soils in relation to water content and bulk density. J. Soil Sci. 32:225–236.

Bidin, A.A., and S.A. Barber. 1985. Phosphate in Malaysian utisols and oxisols as evaluated by a mechanistic model. Soil Sci. 106:500–504.

Chen, J.H., and S.A. Barber. 1990. Soil pH and phosphorus and potassium uptake by maize as evaluated with an uptake model. Soil Sci. Soc. Am. J. 54:1032–1036.

Claassen, N., and S.A. Barber. 1976. Simultaneous model for nutrient uptake from soil by a growing plant root system. Agron. J. 68:961–964.

Claassen, N., K.M. Syring, and A. Jungk. 1986. Verification of a mathematical model by simulating potassium uptake from soil. Plant Soil 95:209–220.

Cox, M.S. 1991. Predicting soil phosphorus levels needed for equal uptake in soils with differing water levels. M.S. thesis. Purdue Univ., West Lafayette, IN.

Cox, M.S., and S.A. Barber. 1992. Soil phosphorus levels needed for equal P uptake from four soils with different water contents at the same soil water potential. Plant Soil 143:93–98.

Kovar, J.L., and S.A. Barber. 1987. Placing phosphorus and potassium for greatest recovery. J. Fert. Issues 4:1–6.

Li, Y., and S.A. Barber. 1991. Calculating changes of legume rhizosphere soil pH and soil solution phosphorus from phosphorus uptake. Commun. Soil Sci. Plant Anal. 22:955–973.

Nye, P.H., and F.H.C. Marriott. 1969. A theoretical study of the distribution of substances around roots resulting from simultaneous diffusion and mass flow. Plant Soil: 30:451–472.

Oates, K., and S.A. Barber. 1987. Nutrient uptake: A microcomputer program to predict nutrient adsorption from soil by plant roots. J. Agron. Ed. 16:65–68.

Shaw, J.K., R.K. Stivers, and S.A. Barber. 1983. Evaluation of differences in potassium availability in soils of the same exchangeable potassium level. Commun. Soil Sci. Plant Anal. 14:1035–1049.

Silberbush, M., and S.A. Barber. 1983. Sensitivity of simulated phosphorus uptake to parameters used by mechanistic-mathematical model. Plant Soil 74:93–100.

Walker, J.M., and S.A. Barber. 1962. Uptake of rubidium and potassium from soil by corn roots. Plant Soil 17:243–259.

Zhang, J., and S.A. Barber. 1992. Maize root distribution between phosphorus-fertilized and unfertilized soil. Soil Sci. Soc. Am. J. 56:819–822.

10 Impact of Soil Chemical Reactions on Food Chain Contamination and Environmental Quality

Terry J. Logan

School of Natural Resources
Ohio State University
Columbus, Ohio

10–1 INTRODUCTION

The theme of this SSSA Special Publication is "Future Prospects For Soil Chemistry", an examination of what soil chemists are likely to be doing in the next few years and on into the next millennium. The advent of this rare historical marker has prompted many in society to examine what our immediate future holds for us. At the same time, there are enormous forces at work that will dramatically change the way in which professionals like soil chemists function, and we must analyze those forces carefully in light of our own situations. We in academia have a particular responsibility to anticipate future professional developments because we are charged with the responsibility of training the next generation of soil chemists who will have to deal with these changes. My particular topic in this publication is the future role of soil chemists in studying and managing chemical contamination of soils and its implications for food chain contamination and environmental degradation.

I was asked to speak to this topic—and I feel qualified to do so—because of my own professional development as a soil chemist. I earned my Ph.D. in soil chemistry in 1971 as part of the first generation of soil chemists to study environmental issues, and I was one of the first soil chemists to take an academic position that had environmental quality as its research and teaching missions. My generation of soil chemists, and soil scientists in other subdisciplines, carved out a niche for environmental studies in our professional societies, ASA and SSSA, that were previously overwhelmingly devoted to the study and management of soils for agricultural production. We founded Division A-5 in ASA, Divisions S-10 and S-11 in SSSA, and *Journal of Environmental Quality*. From the 1970s to the present, we produced a number of SSSA Special Publications dealing with various aspects of environmental soil science. Today, environmental topics dominate our annual meetings and the *Soil Science Society of America Journal*.

Copyright © 1998. Soil Science Society of America, 677 S. Segoe Rd., Madison, WI 53711, USA. *Future Prospects for Soil Chemistry*. SSSA Special Publication no. 55.

Where do we stand in 1996, and what can we say about the immediate future? In this chapter, I will examine the technical questions asked of soil chemists with regard to food chain contamination and environmental quality, but I also will consider some of the larger societal forces that will have great impact on the role of soil chemists in this area.

10–2 SOIL CHEMICAL CONTAMINATION

Soil is the interface of man's activities and the land. We live on the land, we extract minerals from the earth and deposit the residues on the land surface, we manipulate soil to produce our food and fiber, and we dispose of our wastes on land. We chemically contaminate and pollute soil. Miller (1991), as cited by Pierzynski et al. (1994), defines pollution as "any undesirable change in the characteristics of the air, water, soil, and food that can adversely affect the health, survival, or activities of humans or other living organisms . . ." Miller (1994) gives a more recent discussion on this topic. Pierzynski et al. (1994) give a working definition of a pollutant as "a chemical or material out of place or present at higher than normal concentrations that has adverse affects (sic) on any organism." The authors point out that the distinction between contaminant and pollutant is in the last part of their definition above ". . . that has adverse affects on any organism." This distinction is vague in that humans deliberately add pesticides to soil to suppress or destroy pest organisms. So the distinction between contaminant and pollutant is a human one and is based on human judgement and the human value system. We declare some organisms to be pests and contaminate the soil in order to control them. We do not consider this to be pollution. From an ecological standpoint, however, we should be concerned with any level of chemical in soil that has adverse effects on soil biota, and on plant growth and composition, that are not easily reversible; and on levels of the chemical in soil that increase transfer of the chemical to air and water in amounts to cause adverse environmental impact.

One of the great environmental challenges today, and one in which soil chemists have a role to play, is determining if soil contamination will lead to adverse effects on beneficial organisms (those that humans value). One of the great strengths of modern environmental chemistry is our ability to detect the presence of chemicals in environmental matrices at vanishingly low concentrations. This also is a curse because the mere presence of a chemical in air, water, soil, biota, or food can raise public fears and a call for regulation of that chemical or its removal from the environment in which it was found. Environmental monitoring may be able to show that the chemical is a contaminant because concentrations are above some established background (even organics like DDT (dichlorodiphenyltrichloroethane), PCB (polychlorinated biphenyl), and dioxin have background levels in soils and biological tissues as a result of their persistence and global dispersion). What is not known is if these contamination levels constitute pollution as defined above. In order to determine if the measured contamination level can have an adverse effect on an organism, we must use the tools

Table 10–1. Pathways for human exposure to soil-borne contaminants. The most exposed individual (MEI) of the population is also identified. Based on U.S. Environmental Protection Agency exposure assessment for land applied contaminants in sewage sludge (U.S. Environmental Protection Agency, 1995).

Pathway	MEI
Soil → Plant → Human	Human lifetime plant ingestion; general population
Soil → Plant → Human	Human lifetime plant ingestion; home gardener
Soil → Human	Child
Soil → Plant → Animal → Human	Human lifetime ingestion of animal products; animals raised on forage
Soil → Animal → Human	Human lifetime ingestion of animal products; animals ingest soil
Soil → Dust → Human	Human lifetime dust inhalation
Soil → Surface water → Human	Human lifetime ingestion of surface water and fish
Soil → Ground water → Human	Human lifetime ingestion of groundwater
Soil → Air → Human	Human lifetime inhalation of volatilized contaminants

of risk assessment. Risk assessment combines *effects* assessment (how does the chemical harm the organism? how much exposure to the chemical is required to cause harm?) with *exposure* assessment (how is the organism exposed to the chemical and how much exposure is there?). Soil chemists have an important role to play in assessing exposure to soil-borne chemicals.

10–3 SOIL CHEMISTRY AND ENVIRONMENTAL EXPOSURE ASSESSMENT

Assessing exposure to soil contaminants includes determining the pathways to exposure (inhalation, drinking water, food or direct soil ingestion), and the extent of contaminant transfer from soil to the various pathway components. Possible soil pathways of human contaminant exposure are given in Table 10–1. A transfer coefficient (K) must be determined for each component of the pathway, i.e., the fraction of the contaminant that is transferred from one component to another (e.g., from soil to crop or from soil to groundwater). Such an approach is used by Mackay in his general fugacity model (Mackay & Paterson, 1991). The transfer coefficient can be viewed as an availability index, although other terms can be used as well (extractability, solubility, accessibility, volatility), depending on the mechanism of transfer. In the case of soil-borne contaminants, the K values for water can be described for inorganics as adsorption–desorption coefficients or as ion activity products, depending on the known or perceived mechanism of soil-water exchange, and as partition coefficients (K_{oc}) for hydrophobic organics. Soil-air exchange can be described as a Henry's Law constant (K_H), and plant uptake as an uptake slope (U.S. Environmental Protection Agency, 1995) or a bioconcentration factor (BCF; the ratio of plant contaminant concentration to soil concentration). Uptake slope is a measure of the incremental increase in contaminant concentration with incremental increase in contaminant soil concentration or loading, and is determined by regression analysis for multiple loadings (or a delta ratio if there is only one loading). BCF is the ratio of plant to soil concentration and does not reflect incremental plant response to soil contaminants. Similar coefficients are not available for transfer of contaminants from soil to higher organisms (like humans) by direct digestion. In this case, a bioavailability or bioaccessibility index can be calculated that compares transfer of the contaminant from soil to the organism to a reference form of the contaminant. For example, transfer of Pb from soil to rat is compared with uptake from Pb acetate when both materials are added to rat feed at constant dietary Pb concentrations (Logan et al., 1995).

What role does soil chemistry play in soil environmental exposure assessment? Soil chemists have developed many of the techniques that are used to determine *macroscopic* assessments of soil-contaminant exchange. These may have been developed to provide information on phytoavailability of nutrients, but in the last three decades have been adapted to the study of soil contamination. An example is the DTPA (diethylenetriaminepentaacetic acid) extraction test for phytoavailable trace metals (Lindsay & Norvell, 1978). Originally developed to assess phytoavailability of Zn, Mn, Fe, and Cu, it has been used with varying success to assess phytoavailability of toxic trace metals like Cd and Pb (O'Connor, 1988). These macroscopic empirical techniques are used by civil and chemical engineers to provide site-specific quantitative data for exposure assessment, often without a fuller knowledge of the chemical characteristics of the soils themselves. What is not often clearly stated is that these empirical chemical soil extractants are only indices of phytoavailability; each has utility across a limited range of soils and plants, and most importantly, the absolute value of the index is only proportional to plant uptake and is dynamic, i.e., the pool of chemical represented by the index can be replenished by shifting equilibrium from less soluble forms.

Soil chemists bring a perspective to the development and use of macroscopic soil chemical techniques that can provide a more mechanistic and predictive approach to soil contaminant exposure assessment. We are able to place the importance of particular soil chemical characteristics in the context of soil genesis, classification and land use. For example, partitioning of xenobiotics in soil is characterized by the K_{oc} of the compounds. A site-specific K_D is determined by multiplying K_{oc} by percentage of soil organic carbon (OC). The soil chemist can provide a more intuitive and predictive assessment of site K_D by considering known pedogenic effects on OC (e.g., landscape position, soil genesis) as well as land use impacts (e.g., erosion rate).

Macroscopic approaches to determining soil-contaminant interactions will continue to evolve, and it is hoped that soil chemists continue to play a leadership role in their development. Those of us who strive for a more mechanistic understanding of soil chemical processes should realize the utility and power of simple macroscopic tests, even where the technical basis for the test is limited. Nowhere is this more apparent than in the Toxicity Characteristic Leaching Procedure (TCLP) for identifying hazardous wastes and soils (Code of Federal Regulations, 1986). Based on a very simple view that many soil contaminants are primarily acid labile, the test uses a highly standardized extraction protocol. It is the standardization that gives it utility, and its power derives from the fact that it is used as one of the legal ways of identifying hazardous materials.

Our knowledge of soil-borne contaminants will only evolve, however, when we can determine the chemical forms (speciation) of contaminants in the soil matrix. We can only answer the questions of long-term fate of chemicals in soil if we know the form of that chemical. For this we must use *microscopic* and *macroscopic* measurements. We also must use models and molecular probes (where appropriate) to infer the chemical environment of the chemical. If we are considering the fate of toxic trace elements in soil, we must know if the element is primarily complexed with soil humic substances or coprecipitated with inorganic solids. If the element is complexed with organic matter, will it be released into the environment as the soil organic matter decomposes (McBride, 1995)? If the trace element is associated with the mineral fraction of soil, is it in its most thermodynamically stable form? If we can identify the mineral phase in which the trace element is dominantly found, we can use geochemistry and geological records to make confident predictions as to long-term fate of the trace element.

A knowledge of chemical speciation can be used with kinetic measurements to determine the dynamic nature of the chemical in soil. A reasonable environmental soil paradigm is that biologically and geochemically stable (refractory) chemicals have less immediate environmental impact than more labile chemicals, although long-term stability (residence time) in soil carries the burden of long-term management (e.g., long-lived, geochemically stable radionuclides). Reversible and irreversible chemical kinetics, when used with chemical speciation, can be used to assess the lability of the chemical and thus its impact on biological uptake, water solubility, volatilization, weathering, and other transformation-transport processes.

10–4 SOIL CHEMICAL SPECIATION IN ENVIRONMENTAL EXPOSURE ASSESSMENT

Chemicals in soil can exist in solid, liquid, and gaseous phases, and in inorganic and organic forms (Fig. 10–1). The solid phase includes primary and secondary minerals, undecomposed and stable organic matter and biomass, and sorbed species. The liquid phase (predominantly water but also organic fluids like solvents and petroleum) includes dissolved ions, neutral inorganic species, dissolved organic matter (humic and fulvic substances and discrete organic compounds like organic acids), dissolved xenobiotics (e.g., pesticides), small polymers (e.g., hydroxy-Al, detergents), and microparticulates (e.g., colloidal Fe oxide). The soil atmosphere contains normal soil gases (N_2, O_2, CO_2, NO_x), NH_3, volatile organics (pesticides, volatile amines, gasoline), and, under reducing conditions, H_2S, H_2, and CH_4. Components of the solid, liquid, and gaseous phases interact dynamically by chemical, physical, and biological processes.

Current models can not fully predict chemical speciation because of the enormous complexity of these dynamic processes, but a combination of chemical speciation modeling and advanced analytical techniques have given soil chemists the tools to predict soil chemical speciation more accurately than ever before.

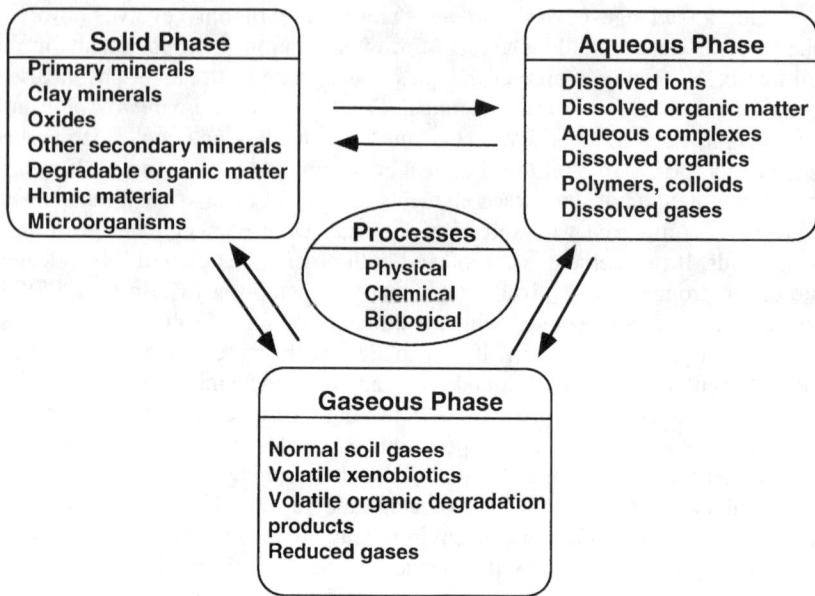

Fig. 10–1. Chemical composition of soil solid, aqueous and gaseous phases.

10–4.1 Aqueous Phase Speciation

A general axiom in soil chemistry, particularly with respect to the trace elements and other inorganic elements like Al, is that the *free* (uncomplexed) form of the element is most bioavailable (e.g., McBride, 1994; Sposito, 1989; Pierzynski et al., 1994). While this primarily refers to biological uptake from soil solution, knowledge of chemical activity also is important in predicting specific chemical reactions such as precipitation and complexation. Estimation of chemical activity can be made for some species with potentiometric methods (ion selective electrodes), anodic stripping voltammetry, fluorescence spectrometry, colorimetry, electrophoresis, or membrane dialysis. Another approach is to use total solution composition data with thermodynamic equilibrium association models like GEOCHEM (Sposito & Mattigod, 1980; Parker et al., 1987; Sposito & Coves, 1988) or MINTEQ (Felmy et al., 1983; Allison et al., 1990) to calculate the distribution of chemical species. Combined with empirical equations to calculate activity coefficients (e.g., Davies equation), free ion activities can be estimated.

An important consideration in the speciation of soil solutions is to have a complete chemical analysis. Not only should all major metals and ligands be analyzed but also important minor constituents. Concentration of dissolved organic C (DOC) is rarely measured, but DOC is often the major ligand complexing polyvalent metals in soil solutions. Preanalytical simulations of soil solution speciation using GEOCHEM or MINTEQ can show if a minor metal or ligand is likely to affect the speciation, but highly reactive metals and ligands (e.g., Cu^{2+} or

F⁻) should always be considered. Preanalytical chemical speciation can be effectively used to determine which analytes should be measured.

10–4.2 Solid Phase Speciation

Solid phase soil chemical constituents exist as sorbed species, as discrete mineral phases, as humic substances, as coprecipitated mineral phases, or as metal-organic complexes (Fig. 10–1). These constituents do not exist as separate entities in soil, but rather as complex mixtures. Attempts to treat soil as simple mixtures of mineral and organic components have been unsuccessful because they fail to account for these interactions. Likewise, it is virtually impossible to physically or chemically separate soil solid phase components for further characterization. Cleaning up the soil by selective chemical treatment (e.g., removal of organic matter with H_2O_2) results in artifacts that make extrapolation of analytical results to real systems highly uncertain. What can be done is to use physical concentration techniques to produce fractions that are enriched with the chemical of interest. These include particle size fractionation (Pierzynski et al., 1990), density gradient separation, and magnetic separation (Essington & Mattigod, 1990). These methods are less invasive and destructive than chemical treatment and are often necessary for the study of trace constituents in soil. For example, it is very difficult to identify crystalline solids in soil with x-ray diffraction (XRD) at concentrations <1 % by weight, yet toxic trace element concentrations in soil can be <0.1 % by weight.

Solid phase speciation can be divided into three general components with very different analytical requirements and challenges: (i) sorbed species; (ii) inorganic solid phases; and (iii) organic solid phases.

Molecular information on sorption of ions, complexes, and organics onto soil solid phases is limited. Tools have included infrared (IR), electron spin resonance (ESR) and fluorescence spectroscopy. Modern fourier-transform IR (FTIR) and lifetime fluorescence spectrophotometry methods have increased the sensitivity and precision of surface species characterization, but defining the reacting surfaces is still difficult for real soil systems. Electron microscopy and AFM provide physical information on the nature of surface-sorbed species from which molecular knowledge can be deduced. As molecular level information becomes available for sorbed species, they can be modeled with chemical affinity computational methods like the surface complexation models (Sposito, 1984) that treat the solid phase sorbent as a chemically reactive phase that interacts with specific solutes. This approach provides molecular specificity that is not available with physically based models like the diffuse double layer model.

Identification of inorganic solid phases has traditionally required XRD for crystalline solids and thermal methods (TGA, DTA) for amorphous materials. Improvement in XRD detection and peak analysis in the last decade or so has permitted the identification of poorly crystalline solids like ferrihydrite and schwertmannite (Bigham et al., 1990, 1994), and advanced analytical electron microscopes permit XRD analysis of single grains in complex mixtures (Pierzynski et al., 1990). A discrete Pb solid, hydroxypyromorphite ($Pb_{10}(PO_4)_6(OH)_2$) was identified by XRD in soil, even though total Pb content was <4 % by weight and

Pb was found to exist in other crystalline forms including cerrusite ($PbCO_3$; Laperche et al., 1996). XRD detection was facilitated by Pb enrichment with particle size fractionation and density gradient separation.

Scanning transmission electron microscopy and electron microprobe analysis have been used to identify weathering products of trace element inorganic solids in soils and mine wastes (Davis et al., 1992; Pierzynski et al., 1990). Weathering products are often observed as coatings or rinds on the weathering particles.

An exciting new analytical technique for characterization of inorganic solids is x-ray absorption spectroscopy (XAS; Brown, 1990). In this method, coherent x-ray beams from synchrotron sources are used to probe soil samples. The resulting spectrum can be refined to provide information on next neighbor atomic coordination in solids. This is one of the only tools that can provide molecular information on trace element coprecipitation in solids (Manceau et al., 1996). The power of the technique is also its limitation in that it requires relatively homogeneous samples. Soils by their very nature are intimate physicochemical associations of solids, and soils have to be fractionated or otherwise cleaned up to produce reasonably homogeneous solid phases prior to analysis. The problems with chemical cleanup procedures have been discussed above, and only physical fractionation and enrichment techniques should be employed. Several researchers (Cotter-Howells et al., 1994; Laperche et al., 1996) have claimed to have identified pyromorphite in soil with XAS, following enrichment with density gradient separation.

Solid phase characterization of soil organic matter remains an intractable problem. Some advances have been made with solid state NMR, but molecular information on inorganic and organic contaminant interactions with solid phase organic matter still eludes us. Unresolved questions include understanding the nature of bound pesticide residues (sorbed, polymerized, physically protected; Alexander, 1995), and the long-term stability of organically bound trace elements (McBride, 1995).

In the following section, many of the approaches discussed above are illustrated by a body of work conducted by our group on Pb contaminated soil.

10–5 SOIL CONTAMINANT EXPOSURE ASSESSMENT AND REMEDIATION: AN EXAMPLE OF LEAD CONTAMINATED SOIL

Lead is a toxic element that can cause brain damage in small children that affects learning. The major sources are Pb paint and leaded gasoline. One of the most important pathways for child exposure is direct soil ingestion (Table 10–1). The exposure assessment questions are: (i) what is the chemical form of Pb in the Pb contaminated soil? and (ii) what is its bioavailability? The chemical form will tell us if Pb is in its most stable form, if it is acid labile, organically complexed and potentially mineralizable, or oxidizable. Bioavailability will provide the transfer coefficient that can be used in an exposure model such as the USEPA Pb biokinetic model.

Speciation also is important in assessing the effectiveness of remediation techniques. We have developed a method to immobilize Pb in contaminated soils and wastes with hydroxyapatite (HA) or phosphate rock (PR; Ma et al., 1995; Laperche et al., 1996). Studies with aqueous Pb and Pb contaminated soil showed that the product of Pb reaction with orthophosphate is pyromorphite (Ma et al., 1995). Geochemical modeling shows that pyromorphite is the most stable oxidized Pb solid phase. Since direct soil ingestion is the most important pathway of child exposure, effectiveness of remediation with HA or PR should demonstrate reduction in blood Pb when the contaminated soil is ingested. Toxicity studies were performed with rats and pigs (Logan et al., 1995), and with an in vitro assay that mimics stomach acidity (Macklis, 1996). Both showed significant reductions in Pb bioavailability with apatite addition. Pyromorphite was included in the toxicity studies to determine its inherent bioavailability relative to Pb acetate, a commonly used Pb control. Phytoavailability of pyromorphite was assessed in a greenhouse study (Laperche et al., 1997) in which Pb uptake was determined and energy dispersive x-ray spectroscopy (EDS) was used to determine the association of Pb and P at the root surface (Fig. 10–2). In an associated study, soil cont-

Fig. 10–2. Formation of a PbCa apatite on the surface of Sudax roots grown in Pb contaminated soil amended with hydroxyapatite (Laperche et al., 1997). SEM micrograph. Root diameter is 70 μm.

aminated with paint Pb was treated with HA and a density enriched fraction was examined with XRD and scanning electron microscopy (SEM; Laperche et al., 1996). XRD confirmed the presence of pyromorphite and SEM revealed the characteristic needle clusters of this mineral (Fig. 10–3). Direct proof of pyromorphite formation in situ is essential if predictions of long-term Pb stability in the treated soil are to be made.

Eventually, a field study will be conducted in which Pb contaminated soil will be characterized to determine Pb forms, the soil will be treated with PR, and the soil will be characterized periodically to determine the Pb reaction products. Treated and untreated soil will be fed to test animals (rats or pigs) and in vitro bioavailability will also be assessed. Ultimately, effects of the remediation on blood Pb concentrations in resident child populations will be determined. This process is shown schematically in Fig. 10–4, and the general approach can be used to assess any soil chemical contamination and remediation problem. Data on the chemical forms of the contaminant in the unremediated and remediated soil, and our ability to relate speciation to long-term stability and to bioavailability, will enhance exposure assessment and pre- and post remediation evaluation.

Cluster of pyromorphite crystals formed in Pb contaminated soil treated with hydroxyapatite

Fig. 10–3. Formation of pyromorphite, a Pb apatite, in Pb contaminated soil amended with hydroxyapatite (Laperche et al., 1996). SEM micrograph.

10–6 SOCIETAL FORCES AFFECTING ENVIRONMENTAL QUALITY AND THE ROLE OF SOIL CHEMISTS

Every generation argues that it is living in pivotal times. As we approach the new century, however, there are significant social, economic and technological forces at work that will determine our world view of environmental science and the role that soil chemists will play.

In the USA and other developed countries, the end of the Cold War has freed the focus of technology from weapons development to other enterprises. Sadly, this freed human and economic capital has not been redirected towards environmental protection, development of renewable natural resources, or increased food production. Nor do we have the commitment to global infrastructure investment that followed World War II. Technology development is centered on economic competitiveness in a global economy, and corporate profit and loss dominates national policies. Government investment in Research and Development has declined as a percentage of GNP, as has corporate Research and Development. Research is much more short-term and with focused objectives. Research support for environmental soil chemistry from USDA, USEPA and DOE has declined. Fewer soil chemists are finding employment in universities or government agencies. So, where will soil chemists be employed in the future? I believe that opportunities do exist in academia, in government, and in the private sector for soil chemists, but the jobs will be broadly described and will require experience or training in the environmental sciences: hydrogeology, water chemistry, toxicology and epidemiology, environmental engineering, environmental law and policy, geographic information systems (GIS), environmental monitoring

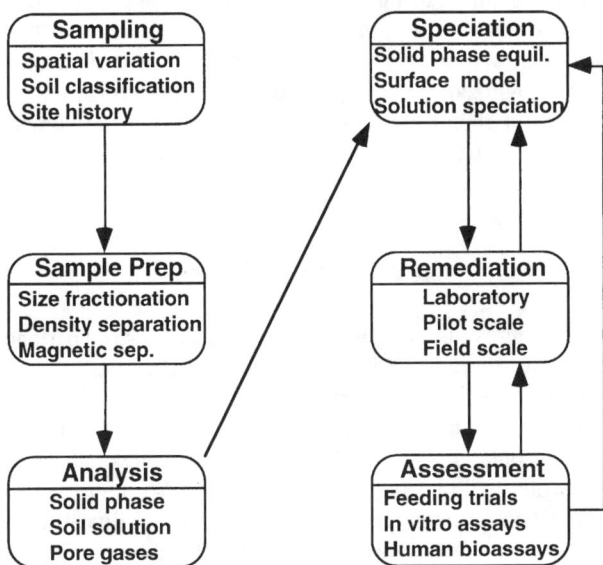

Fig. 10–4. A process for soil exposure assessment for a human health contaminant.

and modeling. In my institution, M.S. and Ph.D. soil chemists are increasingly taking their degrees in the Interdisciplinary Environmental Science Graduate Program that I helped to found in 1990. They develop their research projects in soil chemistry on an environmental application, but their academic course work is required to include the topics identified above.

It also is likely that soil chemists will work for consulting engineering companies, particularly if the companies have analytical testing divisions. Soil assessment for environmental remediation is an increasing activity in these companies. I also would suggest that many soil chemists will end up working for themselves as independent consultants. Outsourcing of work by companies and government agencies is an increasing trend, not just for unskilled labor but for more specialized positions. Independent consulting will be achieved over time after apprenticeships in government, academia, or industry. Independent consulting also will require life-long learning and professional development through short courses, and continuing education credits for certification. A basic knowledge of engineering problem solving methods is invaluable. Scientists rightfully place a high value on experimental proof of cause-and-effect, and are cautious in making representations of truth. This is frustrating and nonresponsive to the client who has an immediate environmental problem to solve. The engineer assesses the state of knowledge and proposes a solution that compensates for the degree of uncertainty. The soil chemist often has critical information that the engineer can use in the treatment design but the information is not expressed in a way that is readily used by the engineer.

Soil chemists also will have to broaden their professional activities to include societies other than SSSA. Other important groups are American Chemical Society (publisher of *Environmental Science and Technology*), Water Environment Federation, Society For Environmental Toxicology and Chemistry, and the International Union of Pure and Applied Chemistry (IUPAC). My participation in several of these groups has shown me that soil chemists have important roles to play in their activities. That we are not more visible, or that our profession is not well recognized by the larger environmental infrastructure is our fault. Contributions by soil chemists to these groups have been welcomed and valued where they have occurred.

10–7 CONCLUSIONS

We live in interesting and changing times. My own experiences and perspectives tell me that there is a bright future for soil chemists in the environmental sciences because of the dominant impact of soil and soil contamination on environmental quality. Our focus and our contributions will be in elucidating the chemistry of soil contaminants, predicting environmental fate of soil contaminants, and assessing the impacts of soil remediation. We must, however, prepare ourselves for these new challenges. We have a responsibility in Division S-2 of SSSA to expose the next generation of soil chemists to the realities they will face and to prepare them for the opportunities that will arise.

REFERENCES

Alexander, M. 1995. How toxic are toxic chemicals in soil? Environ. Sci. Technol. 29:2713–2717.

Allison, J.D., D.S. Brown, and K.J. Novo-Gradac. 1990. MINTEQA2/PRODEFA2: A geochemical assessment model for environmental systems. Version 3.00. USEPA-600/3-91-921. U.S. Environmental Protection Agency, Athens, GA.

Bigham, J.M., U. Schwertmann, L. Carlson, and E. Murad. 1990. A poorly crystallized oxyhydroxy-sulfate of iron formed by bacterial oxidation of Fe(II) in acid mine waters. Geochem. Cosmochim Acta. 54:2743–2758.

Bigham, J.M., L. Carlson, and E. Murad. 1994. Schwertmannite, a new iron oxyhydroxysulfate from Pyhasalmi, Finland, and other localities. Miner. Mag. 58:641–648.

Brown, G.E. 1990. Spectroscopic studies of chemisorption reaction mechanisms at oxide–water interfaces. p. 309–364. In M.F. Hochella and A.F. White (ed.) Mineral–water interface chemistry. Vol. 23. Reviews in Mineralogy. Mineralogical Soc. of Am., Washington, DC.

Code of Federal Regulations. 1986. Identification and listing of hazardous waste. Part 261. p. 359–408. U.S. Gov. Print. Office, Washington, DC.

Cotter-Howells, J.D., P.E. Champness, J.M. Charnock, and R.A.D. Patrick. 1994. Identification of pyromorphite in mine-waste contaminated soils by ATEM and EXAFS. Eur. J. Soil Sci. 45:393–402.

Davis, A., M.V. Ruby, and P.D. Bergstrom. 1992. Bioavailability of arsenic and lead in soils from the Butte, Montana, mining district. Environ. Sci. Tech. 26:461–468.

Essington, M.E., and S.V. Mattigod. 1990. Element partitioning in size- and density-fractionated sewage sludge and sludge-amended soil. Soil Sci. Soc. Am. J. 54:385–394.

Felmy, A.R., D.C. Girvin, and E.A. Jenne. 1983. MINTEQ, a computer program for calculating aqueous geochemical equilibria. Final Project Rep. Contract 68-03-3089. U.S. Environmental Protection Agency, Athens, GA.

Laperche, V., S.J. Traina, P. Gaddam, and T.J. Logan. 1996. Chemical and mineralogical characterizations of Pb in a contaminated soil: Reactions with synthetic apatite. Environ. Sci. Technol. 30:3321–3326.

Laperche, V., T.J. Logan, P. Gaddam, and S.J. Traina. 1997. Effect of apatite amendments on plant uptake of Pb from contaminated soil. Environ. Sci. Technol. 31:2745–2753.

Lindsay, W.L., and W.A. Norvell. 1978. Development of a DTPA soil test for zinc, iron, manganese, and copper. Soil Sci. Soc. Am. J. 42:421–428.

Logan, T.J., S.J. Traina, J. Heneghan, and J.A. Ryan. 1995. Effects of phosphate addition on bioavailability of lead in contaminated soil fed to rats and pigs. Contaminated Soils. In Third Int. Conf. on the Biogeochemistry of Trace Elements, Paris. 15–19 May.

Ma, Q.Y., T.J. Logan, and S.J. Traina. 1995. Lead immobilization from aqueous solutions and contaminated soils using phosphate rocks. Environ. Sci. Technol. 29:1118–1126.

Mackay, D., and S. Paterson. 1991. Evaluating the multimedia fate of organic chemicals: A level III fugacity model. Environ. Sci. Technol. 25:427–436.

Macklis, S.A. 1996. Gastrointestinal absorption of lead from soils and compost and associated health effects. M.S. thesis. Ohio State Univ., Columbus.

Manceau, A., M.C. Boisset, G. Sarret, J.L. Hazemann, M. Mench, P. Cambier, and R. Prost. 1996. Direct determination of lead speciation in contaminated soils by EXAFS spectroscopy. Environ. Sci. Technol. 30:1540–1552.

McBride, M.B. 1994. Environmental chemistry of soils. Oxford Univ. Press. New York.

McBride, M.B. 1995. Toxic metal accumulation from agricultural use of sewage sludge: Are USEPA regulations protective? J. Environ. Qual. 24:5–18.

Miller, G.T. 1991. Environmental science: Sustaining the earth. Wadsworth Publ. Co., Belmont, CA.

Miller, G.T. 1994. Living in the environment, principles, connections and solutions. 8th ed. Wadsworth Publ. Co., Belmont, CA.

O'Connor, G.A. 1988. Use and misuse of the DTPA soil test. J. Environ. Qual. 17:715–718.

Parker, D.R., L.W. Zelazny, and T.B. Kinraide. 1987. Improvements to the program GEOCHEM. Soil Sci. Soc. Am. J. 51:488–491.

Pierzynski, G.M., T.J. Logan, S.J. Traina, and J.M. Bigham. 1990. Phosphorus chemistry and mineralogy in excessively fertilized soil: Quantitative analysis of phosphorus-rich particles. Soil Sci. Soc. Am. J. 54:1576–1583.

Pierzynski, G.M., J.T. Sims, and G.F. Vance. 1994. Soils and environmental quality. Lewis Publ., Boca Raton, FL.

Sposito, G. 1984. The surface chemistry of soils. Oxford Univ. Press, New York.

Sposito, G. 1989. The chemistry of soils. Oxford Univ. Press, New York.

Sposito, G., and J. Coves. 1988. SOILCHEM: A computer program for the calculation of chemical speciation in soils. Kearney Foundation Soil Science, Univ. of California, Davis.

Sposito, G., and S.V. Mattigod. 1980. GEOCHEM: A computer program for the calculation of chemical equilibria in soil solutions and other natural water systems. Kearney Foundation of Soil Science, Univ. of California, Riverside.

U.S. Environmental Protection Agency. 1995. A guide to the biosolids risk assessments for the USEPA Part 503 rule. Office of Wastewater Management. USEPA832-B-93-005. USEPA, Washington, DC.

11 Innovations in Curricula and Teaching Methods in Graduate Education of Soil Chemists

Paul R. Bloom

University of Minnesota
St. Paul, Minnesota

Wayne P. Robarge

North Carolina State University
Raleigh, North Carolina

11–1 INTRODUCTION

The world that awaits M.S. and Ph.D. graduates of U.S. universities is quite different from what it was 10 yr ago. The post World War II expansion of colleges and universities has halted and budgetary constraints have resulted in the downsizing of many university departments. The era of retirement of the early post World War II generation soil scientists that accounted for a large fraction of faculty hiring in many soil science and agronomy departments in the past two decades is essentially complete. Budgetary constraints also have reduced hiring by government laboratories. This means that only a few of our best graduate students can follow the traditional path of employment in university or government laboratories and that the path to traditional employment may include a series of post doctoral appointments before a permanent position is found. On the other hand, during the last 25 yr soil chemists have become more recognized as environmental scientists, expanding job opportunities beyond work related to agricultural production. This shift in emphasis has opened some opportunities for our graduates in regulatory agencies and the private sector.

The changes in employment opportunities and the shift away from our traditional agricultural production roots have placed new demands on soil chemistry educators. We must structure our curricula so that our graduate students have the maximal opportunity for employment in jobs that will use the skills we have helped them develop. At the same time we must look at the courses we teach to see if we are meeting the current and future needs of our students. New developments in computer-based computational and communications technology have the potential to help soil chemists make changes in their teaching to meet these changing demands.

Copyright © 1998. Soil Science Society of America, 677 S. Segoe Rd., Madison, WI 53711, USA. *Future Prospects for Soil Chemistry.* SSSA Special Publication no. 55.

This chapter is divided into two parts. The first is concerned with changing employment options awaiting our graduates and the changes needed in our curricula to ensure that our M.S, and Ph.D. graduates qualify for these positions. Because more data are available concerning employment of Ph.D.s the first part of this chapter will primarily focus on Ph.D. curricula. The second part of this chapter focuses on new developments in soil chemistry instruction that can be employed to better prepare our students for competing in today's job market.

11–2 HOW SHOULD WE STRUCTURE GRADUATE PROGRAMS TO MAXIMIZE OPPORTUNITIES FOR EMPLOYMENT

11–2.1 Implications of the Ph.D. Glut in the USA

The current concern for reduced employment opportunities for Ph.D. graduates is not restricted to soil chemistry and appears to be more than a cyclical downturn. Across the scientific disciplines there is a discussion of the over supply of Ph.D.s, and the 1995 careers issue of *Science* entitled "Careers '95: The Future of the Ph.D." (Holden, 1995a) was devoted to this topic. In that issue, Robert McIntosh, former president of the American Society of Cell Biology, was quoted as referring to the current situation as a "Maltusian Crisis." In an effort to promote discussion about the Ph.D. glut and graduate education in science and engineering the publishers of *Science*, the American Association for the Advancement of Science (AAAS), established a World Wide Web site entitled "Science's Next Wave" (http://nextwave.sciencemag.org). In one of the articles on the "Next Wave" David Goodstein, a well known physicist and vice provost at the California Institute of Technology, contends that the exponential growth in science that began in the era of Newton and Bacon is over, which in turn has profound implications with respect to employment opportunities for scientists. Goodstein characterizes the era that extended from 1950 through 1970 as the golden age for science and that the end of that era marked the beginning of the end of the exponential growth in employment of scientists. The period from 1970 to 1990 marked the beginning of the employment crunch that, at the time, seemed to be a product of temporary trends rather than a historic trend. During this period expansion of science did not halt altogether but by 1990 it became obvious that, at least in the USA, the era of expansion was over.

In the early 1980s many scientists argued that the combination the baby boom echo and the retirement of the of college and university faculty hired during the rapid expansion in the 1960s and early 1970s would result in a strong demand for faculty positions in the 1990s (Walsh, 1980). It is now clear that these predictions were wrong and the assurance given many entering graduate students in the late 1980s and early 1990s about the coming availability of academic jobs was incorrect. Retirements did open positions but not in sufficient numbers to be commensurate with the pool of available applicants.

Post secondary education has reached a plateau; government laboratories are not expanding, and the remarkable era of post World War II growth in industrial research is over. Yet, each university professor continues to produce, on average, 15 Ph.D. graduates during his/her career with the hope that many of these graduates will in turn become university professors (Committee on Science and Engineering Public Policy, 1995). The resulting oversupply of Ph.D.s has led to a reduced interest among U.S. born students in pursuing graduate studies. To fulfill the needs for teaching and research assistants, a greater number of students from other countries have been accepted into graduate schools. Currently in the USA, about 50% of Ph.D. students in science are foreign students. Some of these students will return to their home countries, which helps in solving the Ph.D. glut, but historically only one-half of the foreign students return home (Committee on Science and Engineering Public Policy, 1995).

The rate of production of Ph.D.s has continued to grow in most disciplines. Data for Ph.D. output for science and engineering from 1959 to1993 show that with the exception of a dip in the physical sciences from 1970 to the early 1980s the production of Ph.D.s has continued to rise (Holden, 1995b). In the agricultural sciences, however, production of Ph.D.s since 1973 has been relatively constant (Holden, 1995b). With the growth in supply of Ph.D.s there also has been an increase in the number of post doctoral scientists in colleges and universities to the point that in most disciplines the number currently employed in post doctoral positions is more than double the annual number of Ph.D.s granted (Holden, 1995b). This has resulted in a large number of young scientists spending several years in post doctoral positions before they find a more permanent research position or give up and look for alternative employment. The frustration felt by some of these young scientists is illustrated by what some of them had to say in a 1992 article in *Science* entitled "Tales of Woe from the 'Invisible University'" (Travis, 1992). Two of the post doctorates interviewed said:

> " Post docs are rapidly becoming the hamburger flippers of science: they're cheap temporary highly trained laborers."

and

> "When you're 30, married, and considering having a family, temporary positions aren't that great."

The discussion on the "Next Wave" provides quick access to the nature of the national discussion concerning the employment of young scientists. In addition to original articles "Next Wave" includes articles from *Science* on employment and education of graduate students. These articles are meant to initiate discussion and an opportunity for response *via* electronic mail (e-mail) is provided. Among the titles of the articles written for the "Next Wave" are:

> "Learning to Survive in a Changing Landscape," by Floyd Bloom, neuropharmachologist and editor-in-chief of *Science*;
>
> "Burned by the Torch of Knowledge," by David Goodstein, vice provost California Institute of Technology;
>
> "Re-engineering the System," by James McGroddy, physicist and senior vice president for research at IBM; and
>
> "We need More Scientists—Not Fewer." Congressman Steven Schiff, New Mexico.

Despite the problems with Ph.D. graduates not being able to readily find permanent employment, unemployment is low. During the period from 1973 to

1991, within 4 yr of graduation, <1.5% of recent Ph.D.s were unemployed and looking for employment (Committee on Science and Engineering Public Policy, 1995). In 1991, however, 5 to 8 yr after completing their Ph.D., 3% were still in post doctoral positions (for biologists the fraction still in post doctoral positions was an astounding 9%). During the period from 1971 to 1993 the number of employed Ph.D.s in science and engineering increased from 220 000 to 437 000. While some of the growth has been in colleges, universities, and medical schools, the growth in nonacademic employment has been more rapid. The percentage in academe has decreased from 57 to 49% while the percentage in business and industry increased from 24 to 36%. The data indicate that at least through 1991, Ph.D.s were finding jobs; but it does not mean that all found work in their fields or found jobs that they expected or wanted. These data do not reflect the events since 1991 which include the recent pressure on public spending that has negatively affected employment in universities and government.

In physics the low availability of academic positions relative to non academic positions is such that some physicists contend that we should refer to academic employment, rather than nonacademic employment, as the alternative career path (Perkowitz, 1996). Despite this employment picture many faculty in physics departments continue to prepare graduate students for academic careers in *pure* research.

Data for M.S. graduates in science and engineering indicate that these graduates also have a low rate of unemployment (Committee on Science and Engineering Public Policy, 1995). Data obtained in 1990 for 1988 and 1989 graduates show only 1.8% where unemployed. Of these graduates, 23% went on to further graduate study. In 1990, however, more than one-third of science and engineering master's degree recipients were working outside the field of their degree, This included the 17% who were not employed in science and engineering occupations.

11–2.2 The Situation for Soil Chemists

No employment data are available for soil chemists (nor in any subdiscipline in soil science), but those of us who have been working in major universities during the last 20 yr have observed some obvious trends. The number of young scientists employed in post doctoral positions has greatly increased. At the same time, growth, in academic departments and government research laboratories has ceased, with some decrease in employment opportunities due to the closing of positions following retirements or resignations. Most of us know frustrated young scientists who have had to take a series of post doctoral positions in order to continue doing research in their chosen field of study. We have seen many Ph.D. soil scientists leaving research to take non traditional employment in regulatory agencies or as environmental consultants. The success of these scientists outside the mainstream research environment has demonstrated that Ph.D. soil chemists can have productive careers outside of research. Because of the breadth of opportunities for soil chemists, the Ph.D. oversupply problem is probably not as acute as in most of the other sciences, but we still have problems that should be addressed. The problem is serious and worthy of more attention by the discipline.

11–2.3 What Changes are Needed in Graduate Education?

Recently the National Academy of Science (NAS) published the results of a study on graduate education entitled *Reshaping the Graduate Education of Scientists and Engineers* (Committee on Science and Engineering Public Policy, 1995). The committee that produced the study consisted of leaders in science, engineering, academia, government, nonprofit scientific organizations, and business. Unfortunately, none of the panel members were from agricultural or earth sciences. The entire text of this report is available on the National Academy of Science World Wide Web site (http://www.nap.edu/readingroom) as well in a bound volume.

The major conclusion of the report was that the shift in job opportunities has resulted in considerable disappointment among young Ph.D.s who have prepared themselves for traditional research positions in academia, but often after several years in post doctoral positions, have been forced accept employment elsewhere. The report, however, states that the expertise of science and engineering Ph.D.s is very much needed as the country responds to ". . .expanded economic competition, urgent public health needs, environmental degradation, new national security challenges, and other pressing issues . . ." The NAS committee rejected suggestions for limits on enrollment or on the number of foreign graduate students. They suggested instead that graduate programs be made more flexible and more information be made available to students so they better prepare for employment in a more diverse job market.

The following changes in the education of Ph.D.s were recommended by the NAS committee.

1. More versatility. Students should take a broader array of classes and they should have a personal familiarity with several subfields. Students should have the opportunity to gain career skills beyond the classroom and laboratory, including the possibility of off-campus internships. Students interested in nontraditional careers should be given more flexibility in developing their programs within the bounds of being required to produce a body of high quality original research.
2. Better career information and guidance. Access to timely information on employment trends, careers, and sources of student support should be provided. National data bases with this information should be made available via the Internet.
3. Less time to the degree. Currently, for science and engineering the mean time from entering a U.S. graduate school to a Ph.D. degree is 6.7 years.
4. Education–training grants. Currently there is a heavy reliance on research assistantships that meet the needs of projects, rather than the needs of the students. Education–training grants are needed for local programs to help develop student oriented programs.

The NAS report did not make any recommendation for M.S. programs, but based on information in the report and our personal observations we conclude that M.S. programs may be more important in the future. In regulatory and consulting fields, the M.S. degree is often the preferred degree.

An important recommendation in the NAS report, which is not made explicit in the above list, is that our students need better communication skills. The committee concluded that in order for our Ph.D. graduates to work outside the traditional research fields better skills in oral and written communication are needed. In addition, graduates should be prepared to work in teams.

Implementation of the NAS recommendations will be difficult. The committee suggests adding to the standard Ph.D. program by broadening of the curriculum, including the possibility for internships, while not lowering the standards for the Ph.D. dissertation research, and at the same time shortening the time for the completion of a degree. The only way to solve this contradiction is to somehow make the learning the material taught in courses more efficient and perhaps requiring less in the quantity of research for the Ph.D. dissertation. Integration of course work material will be necessary. Integration of the teaching of communication skills into most courses can be done relatively easily. The more difficult task will be to integrate across scientific subdisciplines and disciplines.

11–2.4 How Does This Apply to Soil Chemists?

11–2.4.1 Are We Already Reducing the Supply of Graduate Students?

Cut backs in Experiment Station research support at many land grant universities during the last 20 yr has drastically reduced the number of graduate research assistantships supported by continuing state and federal funds. Because very few teaching assistantships are available in soil science, outside research grant support has become more important for support of graduate students. This results in less flexibility and more difficulties in planning the support of a Ph.D. student, who will likely take three or more years to complete his/her research while the duration of research grants today is usually 1 or 2 yr. An additional pressure on graduate programs is the recent steep rise in the cost for research assistantships due to an increase in the rate for fringe benefits, and federal requirements regarding funding of tuition subsidies. At the same time, the compensation for post doctoral scientists has increased much more slowly. The rising cost of graduate research assistantships, plus the possibility of hiring for the exact period of a grant, has resulted in more interest in the hiring of post doctoral scientists, or recent M.S. graduates on a full-time appointment, and less interest in employing graduate research assistants. Thus, many soil science graduate programs have experienced, or will soon experience, a very significant decline in the number of graduate students.

11–2.4.2 Do We Need To Broaden the Curriculum?

We need to continue to make sure our students have competency in a broad range of scientific disciplines, but we also must consider other competencies that our students may need for employment. In considering curricular changes we must not weaken the emphasis on chemistry including, analytical, physical, organic, and biochemistry. Competency in mathematics, including calculus and statistics also must be retained and we must make sure our students are knowledgeable in the subdisciplines of soil science including: soil physics, microbiology, pedology, and fertility. It is the understanding of the nature of soil materials and biological processes in soils that distinguishes our students from those in geology or environmental engineering programs. Continued competency of our students in these broad areas is critical to their employment in environmentally related positions.

Generally we have shown less concern for employment oriented courses that are not necessary to do thesis research or meet basic departmental requirements. Because our students are, to a great extent, funded to produce results that satisfy the objectives of a grant, their entire program is often tailored more to meet the needs of a project rather than to preparing them for their futures. This reduces the opportunity for course work or internships that would broaden competencies useful in obtaining employment outside of academia. Such course work might deal with the subject areas of toxicology, environmental law, hydrology, ecology, plant physiology, business, communication, environmental monitoring, and environmental remediation. Additional development of communication skills may not always require additional course work. As is discussed in the second section of this chapter, soil chemists are adding additional communication exercises to existing courses. More emphasis on problem solving skills also may be necessary. Our students gain considerable competence in problem solving doing research for completion of a thesis but adding real world problem solving exercises to our courses would be helpful for students. Obviously, the list of possible competencies for students is large and students have limited time for courses and for noncourse work experience. We need to make changes in curricula to maximize opportunities for students to gain competencies useful for employment and to work with our students to help them formulate reasonable employment goals. We must allow them the opportunity to develop these employment related competencies. Obviously, without changes that will integrate soil science courses, students will not be able to broaden their curriculum without extending the time for completion of a degree.

11–2.4.3 Do Our Students Need Opportunities for Internships?

Internships could be very valuable for both our M.S. and Ph.D. students. Internships also could be valuable for faculty advisors in establishing communication with employers; however, internships are incompatible with the almost total reliance on research grant funding for graduate student support. Education–training grants could allow for graduate programs to include internships. These grants also would allow for more flexibility in course work taken by students. It is hard to see, however, where funding could be obtained for training grants . The National Science Foundation does have a program for education–training grants but it is not funded at the level needed to meet the needs for a large numbers of graduate students. It will require a substantial effort at the college and university level to convince state legislators and Congress to provide adequate funding for a suitable number of education–training grants for graduate students.

11–2.4.4 How Can the Soil Science Society of America Work to Develop More Employment Opportunities for Our Students?

Our graduates have much to offer to a wide range of employers but often there is insufficient recognition of the breadth and depth of subject matter encompassed in the training of soil chemists. We have many examples of soil chemists who have performed well in a variety of public and private sector agricultural or environmentally related positions. It is still common for our graduates to encounter the view that their capabilities are limited to production agriculture. The Soil Science Society of America needs to do more to help project an image of soil scientists (and in particular soil chemist) having a wide range of capabilities. Participation in licensure programs that establish the professional competency of soil scientists as environmental scientists is important. In some states soil scientists are active in developing such licensure programs.

Soil Chemistry students and their academic advisors need more and better employment information. The Soil Science Society of America should be active in surveying graduates as to employment history. Also, information should be maintained on emerging employment needs. This information should be made available on the Soil Science Society of America World Wide Web site: (http://www.agronomy.org/sssa.html).

11–3 WHAT IS NEW IN THE TEACHING OF SOIL CHEMISTRY?

Note: *Much of the material in this section is based on information obtained from an e-mail survey of the members of Soil-Chem an e-mail discussion list established by the members of the Soil Chemistry Division (S-2) of the Soil Science Society of America. The list has more than 500 members in more than 20 countries (to subscribe send a one line e-mail message "subscribe soil-chem your name" to listproc@soils.umn.edu. The survey was structured to obtain information on: (i) curricular innovations added to soil chemistry classes in the last 10 years or contemplated in the near future, and (ii) opinions on changes needed to better prepare graduate students for the world of work. Twenty one responses were obtained, 15 were from teaching faculty, two were from people who assist in teaching soil chemistry classes and the remainder were from other professionals in the public and private sector. Only a few responses were obtained for items in Part ii; to few to draw any conclusions. The suggestions made by respondents to section ii were included in the first part of this chapter.*

11–3.1 Shift in Courses to Reflect More of an Environmental Focus

The most obvious shift in the teaching of soil chemistry in the last few years is the overt inclusion of an environmental focus. Previously, most soil chemistry teachers included examples of the use of soil chemistry for the solution of environmental problems in their courses, but the main focus was within agriculture. The shift to more of an environmental focus in soil chemistry classes coincides with the shift in much of soil chemistry research away from agricultural production to environmental concerns, and a shift in the interest of many students in careers directly related to environmental concerns. Soil chemistry presented as an environmental science is much more effective in drawing students from outside of soil science, than when soil chemistry is taught as a branch of agricultural science. This is important as we seek to maintain enrollment in our classes in the face of a decreasing population of graduate students and more emphasis by university administrators on student numbers.

The change in soil chemistry teaching is illustrated by the titles of four recent soil chemistry textbooks:

1. Soils and Environmental Quality (Pierzynski et al., 1993),
2. Environmental Chemistry of Soils (McBride, 1994),
3. Soil Solution Chemistry: Applications to Environmental Science and Agriculture (Wolt, 1994).
4. Environmental Soil Chemistry (Sparks, 1995),

These books cover the traditional topics of soil chemistry but rely on extensive environmental oriented examples to illustrate basic concepts as well as adding topics directly related to environmental pollution. In addition, soil chemists are teaching environmental courses in undergraduate environmental programs that do not necessarily cover all of the traditional topics taught in soil chemistry.

11–3.2 Innovations in the Soil Chemistry Classroom

11.3.2.1 Computer Models

In the last 10 yr computer equilibrium speciation programs have become popular in soil chemistry classes. These programs can readily solve complex solution equilibrium problems and many include subroutines for surface adsorption or ion exchange. The list of computer programs currently being used in soil chemistry classes includes all of the more popular programs plus some less well known programs including: MINTEQA2, SOILCHEM, MINEQL+, SPECIES, HYDRAQL, GEOCHEM-PC, FITEQL, PHREAQ, WATEQ, EQ3/6, MYRGT, AND LIBRA. The program most used in the classroom by the survey respondents is MINTEQA2. One soil chemist responded that his students seem to like MINTEQA2 and another commented that the dozen problems he assigns are "fun to solve" but a third soil chemistry teacher reported he prefers EQ3/6. Two respondents reported difficulty with not having enough time in their courses to get the students adequately prepared to use GEOCHEM or MINTEQA2. The teacher who had difficulty with MINTEQA2 stated that the program needs to be more a part of the course and not just added for one problem set.

The incorporation of speciation programs in soil chemistry courses not only gives the students the opportunity to solve complex real world problems, but also gives them hands-on training with a type of computer program they may use either in their thesis research or in future employment. These programs are readily available; while many are user friendly, it does take time to train students to use them correctly. It is best to incorporate them into the overall structure of a course and not limit student exposure to these computer models to one or two problem exercises.

Some soil chemists are beginning to use computer models for visualization of mineral or molecular structure. Two respondents use ATOMS, while another uses CLAYPLOT and another reported that she will soon be using BYOSIM. One respondent reported he is using *ab initio* and semi-empirical models for calculation of molecular interactions at surfaces in his surface chemistry course. The models used are HYPERCHEM and GAUSSIAN94. These programs can assist students to visualize the complex structures that we discuss in soil chemistry. Many students when introduced to complex structures, like clay structures, have a difficult time understanding the importance of the spatial relationships within these structures; computer graphics can be a valuable visual aid for students.

Process models that illustrate the role of chemical processes in the response of various natural systems to environmental perturbations are beginning to be used in the teaching of soil chemistry. One respondent uses SOILFILM to visualize chemical transport in agricultural soils and to illustrate the processes of ion exchange, plant uptake, and mineralization and others, while another uses LEACHM. One respondent plans to use MAGIC, a watershed acid rain model. These models can help students integrate the knowledge they have obtained concerning reactions in soils; however, considerable effort must be expended to incorporate process models into a course structure.

11–3.2.2 Communication and Group Skills

Soil chemists, like other teachers in colleges and universities, are becoming more concerned about the ability of students to communicate ideas and concepts. In response to this concern, exercises designed to help teach communication skills are being added to soil chemistry courses. Some of the changes are small, like the addition of peer evaluation of written assignments. Others are more major in scope, such as the use of case studies and group discussions. One teacher reported that inclusion of peer evaluation in the process of completion of written assignments resulted in an immediate improvement in the quality of the articles. One person who teaches a laboratory course requires an industry format for the laboratory reports so students can get accustomed to writing in a format that is used in the real world. Another requires that students, in groups of two, prepare reports using the *Soil Science Society of America Journal* format. Several respondents are using group processes including group discussions, group projects, and case studies.

11–3.2.3 Use of Communication Technology

Soil chemistry teachers are beginning to use electronic communications technology in their courses. This may be as simple as using e-mail to communicate with students, or as complex as using the World Wide Web for class exercises. The use of the capabilities of the World Wide Web is more advanced in some of the other scientific disciplines. Examples of some the ways the Web is being used by chemistry educators are available from a site maintained by University of Texas (http://www.utexas.edu/world/lecture/ch). Sites containing course work material in soil chemistry are at http://agrss.sherman.hawaii.edu/, and an example of the utilization of Web technology in teaching soil fertility is at http://bob.soils.wisc.edu/~barak/soilscience326. With Web technology, links can be included to the many sites across the world containing information related to soil chemistry. One of the sites being developed to assist soil chemists is the Soil Chemistry home page (http://www.soil.ncsu.edu/S2/ home.html) which is sponsored by the Soil Chemistry (S-2) division of the Soil Science Society of America.

In the future, we will see much more sharing of information related to the teaching of soil chemistry. The Soil Chemistry home page will help soil chemists share information, and will provide for links to other important sites related to soil chemistry. As more course material is put on Web sites, more sharing of problem sets etc., will be possible. It also will be possible for faculty at different universities to cooperate in the development of a course with class notes placed on the Web and class discussions conducted across several locations using the forthcoming broad band interactive communications technology. These technologies also create the possibility of easily offering courses to students remote from a university location without having to leave the main campus.

11–4 CONCLUSIONS

We conclude that in the near future only a minority of our M.S. and Ph.D. graduates will get jobs in academia or government research laboratories. This means that they must be prepared to work in regulatory agencies and the private sector, mostly in positions that do not directly involve being part of an active research group. Thus, our graduates will need a broad range of capabilities in scientific disciplines related to soil chemistry, as well as the ability to work in teams and communicate with a wide range of groups of differing scientific background. Academic advisors need to work with students to help them prepare for the realities of the world of work. Changes in the system of education so students can get more on the job experience would be useful. Internships could add greatly to the student experience, but the current system of funding assistantships does not allow for time off for internships. The Soil Science Society of America should be more active in gathering and disseminating data related to employment of our graduates. Currently there is no source of information on the employment of soil chemists and no projection of trends. Such information would be of great help to students and their advisors.

Soil chemistry teaching has evolved to include the incorporation of environmentally related subject material, computer based technology, and an emphasis on developing better communication skills. Environmental examples are being used to illustrate chemical processes in soils instead of the more traditional agricultural examples. Speciation programs are commonly used for calculation of chemical equilibria, and some teachers are using process models to illustrate the role of chemistry in complex soil systems. Computer graphics programs are being used for visualization of chemical structures. Communication technology is beginning to be used for teaching of soil chemistry. With the development of

World Wide Web sites for soil chemistry, more sharing of ideas, approaches and concepts among teachers of soil chemistry around the world is possible. In addition, soil chemists are including writing exercises, and exercises involving group processes, to help students improve communications skills.

REFERENCES

Committee on Science and Engineering Public Policy. 1995. Reshaping the graduate education of scientists and engineers. National Academy Press, Washington, DC.

Holden, C. 1995a. Careers'95: The future of the Ph.D. Science (Washington, DC) 270:121–122.

Holden, C. 1995b. Is it time to begin Ph.D. population control. Science (Washington, DC) 270:123–128.

McBride, M.B. 1994. Environmental chemistry of soils. Oxford Univ. Press, New York.

Perkowitz, S. 1996. Moving the goal posts. Am. Sci. 84:426–427.

Pierzynski, G.M., J.T. Sims, and G.F Vance. 1993. Soils and environmental quality. Lewis Publ., Boca Raton, FL.

Sparks, D.L. 1995. Environmental soil chemistry. Academic Press, New York.

Travis, J. 1992. Tales of woe from the "Invisible university." Science (Washington, DC) 257:1738–1740.

Walsh, J. 1980. Small colleges strong in science. Science (Washington, DC) 233:412–413.

Wolt, J.D. 1994. Soil solution chemistry: Applications to environmental science and agriculture. John Wily & Sons, New York.

12 Interdisciplinary Approaches to Environmental Research: Examples from the National Science Foundation[1]

Margaret A. Cavanaugh and Maryellen Cameron

National Science Foundation
Arlington, Virginia

12–1 INTRODUCTION

The National Science Foundation (NSF) is an independent agency of the U.S. federal government that supports basic science, mathematics, and engineering research and education programs at all levels (NSF, 1995). NSF does not conduct intramural research or operate laboratories, but rather supports projects primarily at academic or nonprofit institutions. The research sponsored is typically carried out by individuals, small university-based research groups, or large groups at facilities such as observatories and polar stations or at science and technology centers.

The research supported by NSF is largely disciplinary in nature, but the Foundation's science and engineering centers (NSF, 1992b) and several recent solicitations discussed below emphasize interdisciplinary research activities. In this report, we focus on interdisciplinary research, as we consider it to have a value-added component that results from synergism among the various disciplines involved. Additionally, we believe that a very important spin-off of interdisciplinary research is the benefit to graduate students and postgraduates engaged in such projects. We think that it promotes flexibility, intellectual openness, and a breadth of technical expertise that derives from exposure to scientists in a variety of disciplines. Such experience is particularly valuable in industrial settings. By stressing interdisciplinary work, we do not wish to diminish the importance or need for disciplinary research. The disciplines provide, and will continue to provide, the building blocks for science and engineering accomplishments, and in our opinion disciplinary research is likely to remain a principal focus at NSF.

---

[1] The topic of this publication was part of an oral presentation on NSF's environmental research programs that was delivered at the 1995 SSSA meeting. Although the authors are employees of the NSF, any opinions, findings, and conclusions or recommendations expressed in this publication are those of the authors and do not necessarily reflect the views of the National Science Foundation.

Copyright © 1998. Soil Science Society of America, 677 S. Segoe Rd., Madison, WI 53711, USA. *Future Prospects for Soil Chemistry.* SSSA Special Publication no. 55.

In this report, we present our perceptions regarding several of NSF's interdisciplinary research activities. Our discussion is not intended to be exhaustive. We emphasize only programs in relevant areas that have been functioning for at least three years. Thus, we do not include relatively recent activities, such as those on integrated assessment (Methods and Models for Integrated Assessment) or on the proposed interdisciplinary training of graduate students (Integrative Graduate Education and Research Training).

12–2 DISCIPLINE-BASED vs. INTERDISCIPLINARY APPROACHES IN RESEARCH

In recent years, there has been renewed discussion on the limits of discipline-based scholarship and on interdisciplinary or cross-disciplinary approaches in both research and educational endeavors. Selected comments from recent publications that provide the context for our discussion follow:

1. "In the physical sciences most of the opportunities for significant, easy-to-do monodisciplinary research have been exhausted. A major portion of the frontier lies in dealing with complex problems that require access to expensive equipment or participation in interdisciplinary programs" (Abelson, 1994).
2. "The idea that all scientific progress takes place within the boundaries of current disciplines is historically invalid and currently counterproductive. But the need for development of sustained collaboration by interdisciplinary groups of active research scientists does not arise only from the urgency of social problems; it is intrinsic to the scientific process itself . . ." (Kahn & Prager, 1994).
3. "The NSF [should] encourage interdisciplinary work and cooperation among sectors. Nature knows nothing about disciplinary boundaries." "... the NSF, and the science and engineering community, must come to grips with the reality that many fields not covered by traditional disciplines offer challenges for new knowledge and opportunities for creative, investigative research worthy of the most gifted scholar" (NSF, 1992a).
4. "Many of the most compelling and significant problems of science today are interdisciplinary in nature. Virtually all of the academic participants agreed on the value of multidisciplinary research for graduate education, although some questioned the wisdom of undergraduate multidisciplinary degree programs" (G-U-I, 1994).
5. "A world of work that has become more interdisciplinary, collaborative, and global requires that we produce young people who are adaptable and flexible, as well as technically proficient . . . Employers favor potential employees who can collaborate across disciplines . . . In some cases, multiple advanced degrees or multidisciplinary backgrounds will be useful" (NAS, 1995).

In this chapter, we use the term interdisciplinary to describe integrated research activities that develop at the margins of traditionally recognized disciplines. Such research typically draws ideas or techniques from disparate scientific fields, integrates and further develops them, and then utilizes the resulting new approaches in a novel manner to solve complex research problems. It varies in level of interdisciplinarity depending on whether the disparate fields involved share paradigms or similar approaches to research. An example of collaboration at the first (and lowest interdisciplinary) level would be one between an organic chemist and an inorganic chemist, who on a university campus may have offices down the hall from each other. Establishment of a working relationship at this level certainly requires flexibility and some adjustments, but, from a larger perspective, bridging from one such subdiscipline to the other is relatively easy. After all, language, conceptual models, problem-solving methods, and scale and location of work are commonly shared at this first level. At the next higher level of interdisciplinarity, the task of understanding another viewpoint and developing a common vocabulary or a new research approach becomes more formidable. This level could be exemplified by a collaboration between a geochemist and a microbiologist, who on many campuses are located in adjacent buildings. A third level—one in which the participants in this analogy are located perhaps across campus from each other—could be represented by a collaboration involving an ecologist, a hydrologist, and an economist. From one level to the next, the challenge of designing a joint research project and maintaining an integrated activity increases substantially. Collaborations of the third type are the least common in our NSF experience. At this level, melding of highly disparate disciplines to develop projects that are interdisciplinary, as per our definition, requires truly innovative and creative approaches. Even research alliances of the second type, in which the extent of interdisciplinarity is substantial and there are shared paradigms, are not common and must be encouraged in order to flourish.

12–3 INTERDISCIPLINARY ENVIRONMENTAL RESEARCH AT THE NATIONAL SCIENCE FOUNDATION

In the last few years, it has become clear that certain environmental problems are extraordinarily complex and that discipline-based approaches alone do not always provide adequate solutions. A growing recognition of the importance of interdisciplinary approaches to environmental research is reflected in the initiation of several interdisciplinary NSF programs, either alone or in partnership with other agencies. Most of these interdisciplinary activities are superimposed on existing disciplinary NSF programs and are matrix-managed with few additional personnel. Proposals submitted in response to these solicitations typically occur in one of two formats. One involves multiple investigators who are experts in different disciplines, but are collaborating on a single research problem. A second and much less common venue involves an investigator who has expertise in more than one field and whose research results are expected to have significant impact on multiple disciplines. The extent of synergy among various disciplinary components is commonly included as a review criterion, and thus the perceived level of interdisciplinarity is influential in funding decisions. Evaluation of these projects normally cannot be handled within a single disciplinary structure since it requires expertise from multiple scientific areas.

Interdisciplinary NSF programs that have been functioning for at least three or more years include: *Environmental Geochemistry and Biogeochemistry*, an NSF activity that targets research on chemical processes underlying the fate and transport of contaminants and nutrients; *Terrestrial Ecology and Global Change*, a joint NSF-DOE-NASA-USDA program that supports investigation of the susceptibility of ecosystems to change and quantification of processes regulating change; *the NSF/EPA Partnership for Environmental Research*, which supports work (i) in water and at watershed scale encompassing ecology, physical sciences such as hydrology and engineering, and the social sciences (Fig. 12–1), (ii) technology for a sustainable environment, and (iii) valuation and environmental policy; *Water & Energy: Atmospheric, Vegetative, and Earth Interactions*, a program that studies Earth's hydrologic and energy cycles sufficiently to allow assessment of the impact of human activities; and *Environmentally Benign Chemical Synthesis and Processing* that supports development of chemical, materials, and engineering methods that prevent pollution.

Following are several examples that further clarify the distinction between disciplinary and interdisciplinary research and among the different levels of interdisciplinarity. Environmental Geochemistry and Biogeochemistry (EGB), an interdisciplinary activity established in 1995, is based on a partnership that involves eight largely disciplinary NSF divisions: earth sciences, ocean sciences, atmospheric sciences, chemistry, chemical and transport systems engineering, environmental biology, molecular and cellular biosciences, and mathematical sciences. If researchers in disciplines represented by these divisions were to develop a *disciplinary* research proposal related to "soils", the following scenario

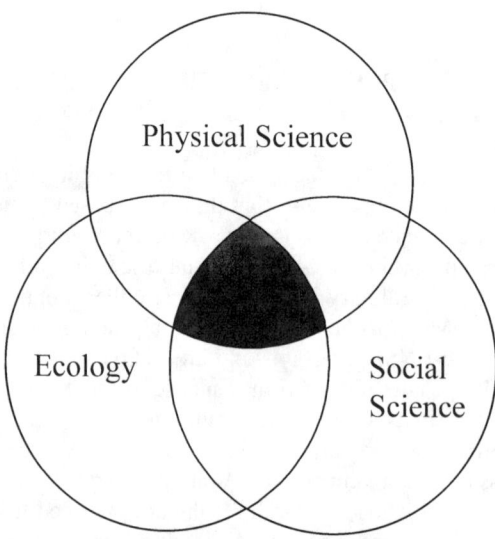

Fig. 12–1. Venn diagram illustrating the three scientific areas emphasized in the National Science Foundation–U.S. Environmental Protection Agency Water and Watersheds program. The shading denotes the only area in which proposals were accepted for the 1996 competition.

could be envisioned. The chemistry community might focus, for instance, on the role of surface structure on the behavior of nutrients or contaminants in soils, or on photochemical and photoelectrochemical transformations. In earth sciences, the emphasis could be on transport phenomena and chemical reactions between ground waters and natural soil components, or on the kinetics of dissolution/precipitation processes. Engineers might have interests in the mechanisms of contaminant transport and removal from soils and in factors affecting the geotechnical properties of soils. Biologists could study microbial diversity or the role of forest ecosystems or keystone species such as earthworms in nutrient and element cycling in soils. If these same scientists and engineers were to develop an *interdisciplinary* proposal to conduct soils research, the project might, for example, address the effect of biotic factors (which have been largely ignored until recently) on the surface structure of natural minerals and on inorganic dissolution/precipitation and transport processes in soils.

Actual examples of integrated level two projects are common in the 1995 and 1996 awards made by the EGB program. (For a list of awards, see the NSF Web Site at www.nsf.gov. Select crosscutting programs, then environment and global change, and then environmental research opportunities.) For example, one project that is examining weathering reactions at mineral–solution interfaces draws on the experience of a chemist, a geochemist, and a microbiologist. Another team is using concepts from microbiology, biogeochemistry, and mathematical modeling to address Hg and methylmercury dynamics in semi-arid environments. A third project involving hydrologists and microbiologists is using bioluminescent microbes to elucidate aquifer bio-geo-hydrochemistry.

Biogeochemical areas are not the only ones in which level two interdisciplinary environmental research is being conducted. Interest in supercritical carbon dioxide, a benign alternative to many volatile organic solvents, has brought chemists and process and electronic design engineers together to develop new approaches to chemical manufacturing. Biotechnology studies offer additional examples of focusing the disparate disciplinary methods of biology and engineering on development of remediation technologies. Projects such as these are considered in the technology component of the *NSF/EPA Partnership for Environmental Research* or in the *Environmentally Benign Chemical Synthesis and Processing* initiative.

Interdisciplinary research at the third level is supported in the NSF/EPA Water and Watersheds activity. For example, one project with social, geomorphological, and ecological components focuses on the premise that watershed management is socially-influenced. The research will elucidate the effect of new scientific knowledge on valuation of natural resources by local stakeholders and ultimately on their decisions determining watershed management policies.

12–4 SUMMARY AND CONCLUSIONS

The National Science Foundation supports both disciplinary and interdisciplinary environmental research. The disciplinary research is most often funded in permanent, established programs having traditional emphases. Highly interdisciplinary research is encouraged in specialized solicitations. These solicitations typically target specific research problems, and their associated programs have a limited lifetime. Proposals in the latter category have varying degrees of interdisciplinarity and varying research formats. They require additional attention during review. Environmental initiatives with an interdisciplinary emphasis include *Environmental Geochemistry and Biogeochemistry, NSF/EPA Partnership for Environmental Research, Environmentally Benign Chemical Synthesis and Processing*, and *Terrestrial Ecology and Global Change*.

Involvement in interdisciplinary scholarship is challenging both on an intellectual and on a personal level. For scientists and engineers, building teams that include participants from a variety of disciplines is time-consuming, and it requires adopting new perspectives, vocabulary, and procedures. For program managers at funding agencies and for the peer review community, there is a need to develop interdisciplinary expertise. Also largely lacking are innovative funding mechanisms and diverse funding sources that involve new types of partnerships. Most of these issues are not new: Schneider (1988) commented on many of them nearly 10 years ago. The new factor today may be that interdisciplinary activities are touching a broader segment of our communities, the majority of whom are still learning how to nurture, conduct, and manage such activities.

Investigators who wish to pursue fundamental research in interdisciplinary projects - whether in the laboratory, the field, or in modeling - will find that it is intellectually worth the effort to be involved in developing innovative approaches through collaborative efforts. We are confident that interdisciplinary environmental research will have a significant impact over the next decade. Combined with on-going disciplinary work, it will facilitate comprehensive solutions to important environmental problems.

REFERENCES

Abelson, P.H. 1994. Editorial: Multidisciplinary research. Science (Washington, DC) 266:951.

Government–University–Industry Research Roundtable. 1994. Stresses on research and education at colleges and universities: Institutional and sponsoring agency responses. Report of a Collaborative Inquiry Conducted Jointly by the National Science Board and the Government–University–Industry Research Roundtable, Washington, DC.

Kahn, R.L., and D.J. Prager. 1994. Interdisciplinary collaborations are a scientific and social imperative. *The Scientist*, 11 July 1994.

National Academy of Sciences. 1995. Reshaping the graduate education of scientists and engineers. NAS, Washington, DC.

National Science Foundation. 1992a. A foundation for the 21st Century: A progressive framework for the National Science Foundation. Rep. of the National Science Board Commission on the Future of the National Science Foundation. NSB 92-196. Natl. Sci. Foundation, Washington, DC.

National Science Foundation. 1992b. NSF Science and Technology Centers. NSF 95-104. Natl. Sci. Foundation, Washington, DC.

National Science Foundation. 1995. NSF in a changing world: The National Science Foundation's Strategic Plan. NSF 95-24. Natl. Sci. Foundation, Arlington, VA.

Schneider, S.H. 1988. The whole earth dialogue. Issues Sci. Technol. Spring:93–99.

13 Role of Soil Science in the International Council of Scientific Unions: Soil Chemistry

Winfried E. H. Blum

Institut für Bodenforschung
Universität für Bodenkultur
Vienna, Austria

P. M. Huang

Department of Soil Science
University of Saskatchewan
Saskatoon, Canada

13-1 INTRODUCTION

The International Council of Scientific Unions (ICSU) was created in 1931 to promote international scientific activity in the different branches of science and their applications for the benefit of humanity (Leonard, 1995). It pursues a policy of nondiscrimination, affirming the rights and freedom of scientists around the world.

ICSU is a nongovernmental organization with two categories of membership: scientific academies or research councils, which are national, multidisciplinary bodies (at the moment 94 members; Table 13–1) and scientific unions, which are international, disciplinary organizations (at the moment 23 members, including ISSS; Table 13–2). The combined effort of these two groups provides a wide spectrum of scientific expertise, enabling the members to address major international interdisciplinary issues, which none of them could handle alone. In addition, ICSU has 28 Scientific Associates. Administratively, ICSU is governed by the Executive Board, the General Committee and the General Assembly (Table 13–3).

The objective of this chapter was to address the function and new initiatives of ICSU, the role of Soil Chemistry in ICSU, and the proposed mechanisms to fulfill this role.

Copyright © 1998. Soil Science Society of America, 677 S. Segoe Rd., Madison, WI 53711, USA. *Future Prospects for Soil Chemistry.* SSSA Special Publication no. 55.

Table 13–1. National, multidisciplinary bodies of the International Council of Scientific Unions (Leonard, 1995).

Argentina	Lebanon
Armenia	Lithuania
Australia	Madagascar
Austria	Malaysia
Bangladesh	Mexico
Belarus	Moldova
Belgium	Monaco, Principauté de
Bolivia	Mongolia
Brazil	Morocco
Bulgaria	Nepal
Burkina Faso	Netherlands
Canada	New Zealand
Caribbean	Nigeria
Chile	Norway
China: CAST	Pakistan
Taipei	Panama
Colombia	Philippines
Côte d'Ivoire	Poland
Croatia	Portugal
Cuba	Romania
Czech Republic	Russia
Denmark	Saudi Arabia, Kingdom of
Egypt, Arab Republic	Senegal
Estonia	Seychelles
Finland	Singapore
France	Slovak Republic
Georgia	South Africa
Germany	Spain
Ghana	Sri Lanka
Greece	Sudan, Republic of
Guatemala	Swaziland
Hungary	Sweden
India	Switzerland
Indonesia	Thailand
Iran	Togo
Iraq	Tunisia
Ireland	Turkey
Israel	Uganda
Italy	Ukraine
Jamaica	United Kingdom
Japan	United States
Jordan	Uruguay
Kazakhstan	Uzbekistan
Kenya	Vatican City State
Korea, Dem. People's Republic of	Venezuela
Korea, Republic of	Vietnam Socialist Republic
Latvia	Zimbabwe

13–2 FUNCTION AND INITIATIVES OF THE INTERNATIONAL COUNCIL OF SCIENTIFIC UNIONS

ICSU acts in a number of ways. First, it initiates, designs and coordinates major international, interdisciplinary research programs, such as the International Geophysical Year (1957–1958), the International Biological Program

ROLE OF SOIL SCIENCE

Table 13-2. Scientific Union Members of the International Council of Scientific Unions (Leonard, 1995).

International Union of Anthropological and Ethnological Sciences (IUAES)
International Astronomical Union (IAU)
International Union of Biochemistry and Molecular Biology (IUBMB)
International Union of Biological Sciences (IUBS)
International Union of Pure and Applied Biophysics (IUPAB)
International Brain Research Organization (IBRO)
International Union of Pure and Applied Chemistry (IUPAC)
International Union of Crystallography (IUCr)
International Union of Geodesy and Geophysics (IUGG)
International Geographical Union (IGU)
International Union of Geological Sciences (IUGS)
International Union of the History and Philosophy of Science (IUHPS)
International Union of Immunological Societies (IUIS)
International Mathematical Union (IMU)
International Union of Theoretical and Applied Mechanics (IUTAM)
International Union of Microbiological Societies (IUMS)
International Union of Nutritional Sciences (IUNS)
International Union of Pharmacology (IUPHAR)
International Union of Pure and Applied Physics (IUPAP)
International Union of Physiological Sciences (IUPS)
International Union of Psychological Science (IUPsyS)
Union Radio Scientifique International (URSI)
International Society of Soil Science (ISSS)

Table 13-3. General Structure of the International Council of Scientific Unions (Leonard, 1995).

General assembly
Voting members:

Scientific Union Members   National Scientific Members

Representatives of Associates, Interdisciplinary ICSU bodies and cooperating Organizations may attend. Assembly Finance Committee, Assembly Nominating Committee, Assembly Resolutions Committee

Meets triennially

General committee
52 Members

Representatives of Scientific Union Members
(23, three of whom are Executive Board members)

Representatives of National Scientific Members
(23, three of whom are Executive Board members)

6 Officers

Meets annually

Executive Board

Officers:		Six Ordinary Members:
President	Past-President	Three from Scientific Unions
2 Vice-Presidents	Treasurer	Three from National Scientific Members
Secretary General		

Each member has one vote

Meets when necessary

Secretariat
51, Boulevard de Montmorency, 57016 Paris, France
Telephone: (331) 4525 0329, Telefax (331) 4288 9431, Telex 645554 F
Emails Omnet: ICSU.PARIS, Econet: icsu - Telecom
Gold: 10075:dbi0126, internet: icsu@paris7.jussieu.fr

Table 13-4. Joint initiatives of the International Council of Scientific Unions (Leonard, 1995).

Committee on Science and Technology in Developing Countries/ International Biosciences Networks (COSTED/IBN)
International Geological Correlation Programme (IGCP)
World Climate Research Programme (WCRP)
Global Climate Observing System (GCOS)
Global Ocean Observing System (GOOS)
Global Terrestrial Observing System (GTOS)
Joint Lectureship/Professorship Programme
UNESCO/ICSU Short-Term Fellowships Programme

(1964–1974), or the more recent International Geosphere–Biosphere Program (IGBP): A Study of Global Change (IGBP), which complements the joint World Meteorological Organization (WMO)–ICSU climate research program, and aims to describe and understand interactive physical, chemical, and biological aspects of the total earth system (Table 13–4).

Secondly, ICSU creates interdisciplinary bodies, which undertake activities and research programs of interest to several member bodies (Table 13–5). Examples of such activities include arctic, oceanic, space, and water research problems of the environment, genetic experimentation, solar-terrestrial physics, and biotechnology.

In addition to these programs and activities, which seek to break the barriers of specialization, several bodies set up within ICSU address matters of common concern to all scientists, such as capacity building in science, data, science, and technology in developing countries, ethics, and freedom in the conduct of science.

Moreover, ICSU acts as a focus for the exchange of ideas, the communication of scientific information, and the development of scientific standards. Scientific conferences, congresses, and symposia, in total about 600 per year, are organized around the world, and a wide range of newsletters, handbooks, learned journals, and proceedings of meetings are published. ICSU assists also in the creation of international and regional networks of scientists with similar interests and maintains close working relations with a number of intergovernmental and nongovernmental organizations, in particular with the United Nations Educational, Scientific, and Cultural Organization (UNESCO) and WMO. Through cooperation with these organizations a number of international programs have been launched and are currently in progress.

Finally, because ICSU is in contact through its membership with hundreds of thousands of scientists world-wide, it is being increasingly called upon to act as a spokesman for the world scientific community, and as an advisor in topics ranging from ethics to the environment.

13–3 ROLE OF SOIL CHEMISTRY IN THE INTERNATIONAL COUNCIL OF SCIENTIFIC UNIONS

The International Society of Soil Science (ISSS) is a learned international organization, founded in 1924, with about 7500 members in 155 countries. Since 1993 ISSS has been a full Union Member of ICSU, together with 94 national

Table 13–5. Interdisciplinary bodies of International Council of Scientific Unions (Leonard, 1995).

Ad hoc Group on Agriculture, Aquaculture and Forestry
Scientific Committee on Antarctic Research (SCAR)
Scientific Committee on Biotechnology (COBIOTECH)
Committee on Capacity Building in Science (CCBS)
Committee on Data for Science and Technology (CODATA)
Scientific Committee on Problems of the Environment (SCOPE)
Scientific Committee on Genetic Experimentation (COGENE)
Scientific Committee for the International Geosphere-Biosphere Programme (SC-IGBP)
Special Committee for the International Decade for Natural Disaster Reduction (SC-IDNDR)
Scientific Committee on Oceanic Research (SCOR)
Scientific Committee on Science in Central and Eastern Europe and the former Soviet Union (COMSCEE)
Scientific Committee on Solar-Terrestrial Physics (SCOSTEP)
Committee on Space Research (COSPAR)
Scientific Committee on Water Research (SCOWAR)
World Data Centre (WDC)
Federation of Astronomical and Geophysical Data Analysis Services (FAGS)
Inter-Union Commission on Frequency Allocations for Radio Astronomy and Space Science (IUCAF)
Inter-Union Commission on the Lithosphere (ICL)
Inter-Union Commission on Spectroscopy (IUCS)

academies or research councils and 22 other scientific unions (Leonard, 1995). This means that ISSS is now parallel with the International Union of Biological Sciences (IUBS), the International Union of Geological Sciences (IUGS), the International Geographical Union (IGU), the International Union of Geodesy and Geophysics (IUGG), the International Union of Pure and Applied Physics (IUPAP), and the International Union of Pure and Applied Chemistry (IUPAC), citing only some of those that might be of special importance in future cooperation.

For many years, ISSS has been closely cooperating with IUBS through its Sub-Commission D (Soil Zoology) and future cooperations with other organizations are planned in the near future. Of special interest for soil chemists is a future cooperation with the International Union of Pure and Applied Chemistry (IUPAC). In this respect some activities have already taken place. A few soil chemists (e.g., W.L. Earl, C.T. Johnston, G. Sposito, and W.H. van Riemsdijk) have been invited to contribute chapters in IUPAC Environmental Analytical and Physical Chemistry Series (Buffle & van Leeuwen, 1992, 1993). The first book on soil particles in the IUPAC book series has been published (Huang et al., 1998a). There are 12 chapters in this book. This book covers fundamentals of minerals–organics–microbes interactions in the soil environment, fractal approach in the evaluation of soil particle dimensions and aggregation, advanced instrumentation in analysis of soil particles, and reactions and processes at the soil particle–solution interface. Two out of the three editors of the book and the vast majority of authors of the book chapters are soil chemists invited to participate in this book project. Moreover, an "International Symposium on Surface and Colloid Science and its Relevance to Soil Pollution" in Madras, India, in 1994 was sponsored by IUPAC, cosponsored by ISSS, and in close cooperation with the Indian Soil Science Society and the Indian Society of Surface Science and Technology. Further, at the Annual Meeting of the Soil Science Society of

America (SSSA) in St. Louis, MO, in October 1995, a Workshop was held on "Soil Chemistry and Ecosystem Health," which was sponsored by SSSA and cosponsored by IUPAC. The papers presented at the Workshop have been published as a SSSA special publication (Huang et al., 1998b). This special publication begins with an overview of ecosystem health, followed by a description of molecular structure reactivity–toxicity relationships, to pave the way for addressing the issue of "Soil Chemistry and Ecosystem Health." The impact of fundamental chemical and biochemical reactions of metals, radionuclides, and organics of environmental concern on food chain contamination and ecosystem health, the ecotoxicological risk assessment and the development of remediation strategies are covered in this publication. This book is expected to build a linkage between soil chemistry and other disciplines pertaining to ecosystem health on a global scale. It also is hoped that this book will facilitate the interaction of soil chemists with IUPAC in fulfilling the role of Soil Chemistry in ICSU.

The actual scientific structure of ISSS and IUPAC as possible counterparts is shown in Tables 13–6 and 13–7 to facilitate discussion of possibilities of future cooperation, especially in the field of Soil chemistry. The structure of IUPAC was recently changed in the General Assembly, that took place in 1995. Of special interest for Soil Chemistry is the newly founded "Division of Chemistry and the Environment," which is subdivided into six commissions (Table 13–7). The "Commission of Fundamental Environmental Chemistry," and the "Commission of Soil and Water Chemistry" are of special interest for future cooperation with ISSS. The "Commission of Atmospheric Chemistry" and "Commission of Agrochemicals and the Environment" also are related to Soil Chemistry.

On the side of ISSS the following Commissions, Sub-Commissions and Working Groups could cooperate closely with the Commissions of Fundamental Environmental Chemistry and the Commission of Soil and Water Chemistry of IUPAC: Commission II, "Soil Chemistry," Commission VIII, "Soils and the Environment,"Sub-Commission A, "Salt Affected Soils," Sub-Commission G, "Soil Remediation," Working Group MO, "Interactions of Soil Minerals With Organic Components and Microorganisms," Working Group SG: "Soils and Geomedicine," Working Group AS, "Acid Sulphate Soils," Working Group FA, "Soil Organic Fertilizers and Amendments," Working Group RZ, "Rhizosphere," and Working Group SP "Soil and Groundwater Pollution."

This future cooperation could be developed on three levels:

1. The general exchange of information about ongoing activities in the respective organizations. ISSS is already sending its Bulletin to the respective chairpersons of Commissions VI.1 and VI.3 of IUPAC, in order to keep them informed about activities within soil science.
2. The co-organization of national and international scientific conferences, workshops, and meetings, such as the ones that have already taken place, for example in Madras, India, 1994 and in St. Louis, MO, in 1995, as discussed above.
3. The cooperation in concrete projects, either focusing on the development of new methods and methodological approaches, or looking into specific environmental problems, related to chemistry and soil science.

Table 13–6. Scientific structure of the International Society of Soil Science (ISSS; Blum, 1995).

Commissions		Sub-commissions		Working groups		Standing committees	
I	Soil Physics	A	Salt Affected Soils	AS	Acid Sulphate Soils	CSS	Statute and Structure
II	Soil Chemistry	B	Soil Micromorphology	CR	Cryosols	CIP	International Programmes
III	Soil Biology	C	Soil & Water Conservation	DE	Soil Resources of Desert Ecosystems	CST	Standardization
IV	Soil Fertility & Plant Nutrition	D	Soil Zoology	DM	World Soils and Terrain Digital Database	CBF	Budget and Finances
V	Soil Genesis, Classification & Cartography	E	Forest Soils	FA	Soil Orgainc Fertilizers and Amendments	CES	Education in Soil Science
		F	Land Evaluation	LI	Land Evaluation Information Systems	CHP	History, Philosophy and Sociology of Soil Science
VI	Soil Technology	G	Soil Remediation	MO	Interactions of Soil Minerals with Organic Components and Microorganisms		
VII	Soil Mineralogy			MV	Soil and Moisture Variability in Time and Space		
VIII	Soils and the Environment			PM	Pedometrics		
				PP	Paleopedology		
				PS	Paddy Soils Fertility		
				PT	Pedotechnique		
				RB	World Reference Base for Soil Resources		
				RS	Remote Sensing for Soil Survey		
				RZ	Rhizosphere		
				SG	Soils and Geomedicine		
				SP	Soil and Groundwater Pollution		
				US	Urban and Periurban Soils		

Table 13–7. Scientific structure of the International Union for Pure and Applied Chemistry (IUPAC; Leonard, 1996).

I Division of Physical Chemistry	II Division of Inorganic Chemistry	III Division of Organic Chemistry	IV Division of Macromolecular Chemistry	V Division of Analytical Chemistry	VI Division of Chemistry and the Environment	VII Division of Chemistry and Human Health
I.1 Physicochemical Symbols, Terminology, and Units	II.1 Atomic Weights and Isotopic Abundances	III.1 Nomenclature of Organic Chemistry	IV.1 Macromolecular Nomenclature	V.1 General Aspects of Analytical Chemistry	VI.1 Fundamental Environmental Chemistry	VII.C. Clinical Chemistry Section
I.2 Thermodynamics	II.2 Nomenclature of Inorganic Chemistry	III.2 Physical Organic Chemistry	IV.2 Polymer Characterization and Properties	V.2 Microchemical Techniques and Trace Analysis	VI.2 Atmospheric Chemistry	VII.C.1 Nomenclature, Properties, and Units
I.3 Electrochemistry	II.3 High Temperature and Solid State Chemistry	III.3 Photochemistry		V.3 Separation in Analytical Chemistry	VI.3 Soil and Water Chemistry	VII.C.2 Toxicology
I.4 Chemical Kinetics	II.4 Isotope Specific Measurements as References			V.4 Spectrochemical and Other Optical Procedures for Analysis	VI.4 Agrochemicals and the Environment	VII.C.3 Components of Quality Systems in the Clinical Laboratory
I.5 Molecular Structure and Spectroscopy				V.5 Electroanalytical Chemistry	VI.5 Food Chemistry	VII.M. Medicinal Chemistry Section
I.6 Colloid and Surface Chemistry including Catalysis				V.6 Equilibrium Data	VI.6 Oils, fats, and derivatives	VII.M.1 Nomenclature and Terminology
I.7 Biophysical Chemistry				V.7 Radiochemistry and Nuclear Techniques		VII.M.2 Training and Development
				V.8 Solubility Data		VII.M.3 New Technologies and Special Topics

Such a cooperation also could be achieved on national levels, and is not necessarily being coordinated through the international organizations. The Secretary General of ISSS would be very pleased to assist in any kind of cooperation between the organizations within ISSS, or organizations on a national level within soil science and with respective counterparts in IUPAC.

The acceptance of ISSS as a full Union Member of ICSU is a very important step forward to closer cooperation with IUPAC and other Unions within ICSU, and therefore a chance and a challenge for Soil Science in the future, especially for Soil Chemistry. The role of chemistry in increasing agricultural productivity and in improving the environment is of great concern in soil and environmental sciences. Chemical processes occurring in soils profoundly affect the transformation and fate of nutrients and pollutants in the terrestrial ecosystem and their subsequent transport to atmospheric and aquatic environments. Consequently, soil processes do have tremendous impacts on air and water quality and ecosystem health.

Approaching a more concrete stage of cooperation would raise the question: What role can Soil Chemistry play within ICSU?

There are several possibilities:

1. To build a bridge of communication between soil chemists and IUPAC, which strongly promotes the interaction of pure and applied chemists on the global scale;
2. To facilitate continuing communication of soil chemists with the progress made in pertinent fundamental chemical research;
3. To maximize application of research findings in pertinent pure and applied chemistry to improving the production of foodstuffs and fibers, and sustaining the integrity of the ecosystem, which is vital to human and animal health and the quality of life;
4. To provide valuable inputs of soil chemists to IUPAC Division VI "Chemistry and the Environment," which concerns itself with many problems relating to the quality of the environment as well as that of the foodstuff produced;
5. To provide pertinent expertise of soil chemists to IUPAC, which has been playing an increasingly important role of advising other international bodies concerned with chemical problems, including specialized agencies of the United Nations, such as UNESCO, the Food and Agriculture Organization (FAO), and World Health Organization (WHO); and
6. To cooperate with IUPAC to advance the frontiers of knowledge and to apply such knowledge to the solution of soil- and environment-related problems of chemical matters on a global scale.

The proposed mechanisms for fulfilling the role of Soil Chemistry in ICSU could be the following:

1. To appoint a representative of International Union of Soil Sciences [(IUSS) ISSS officially changed to IUSS in 1998] to IUPAC to promote

and facilitate the communication and interaction of soil chemists with IUPAC on the global scale;
2. To disseminate pertinent information of IUPAC's work (e.g., international nomenclature or terminology, symbols, units, standards of purity, analytical methods, biennial International Congresses, symposia, workshops, books, journals, and bulletin publications) to soil chemists;
3. To disseminate pertinent information on Soil Chemistry (e.g., books, journals, monographs, symposia, and workshops) to IUPAC;
4. To invite IUPAC to cosponsor symposia, workshops, and book projects on Soil Chemistry;
5. To cosponsor pertinent IUPAC symposia, workshops, and book projects;
6. To promote the appointment of soil chemists to pertinent commissions, divisions, advisory boards and review panels of IUPAC;
7. To invite the inputs of IUPAC to pertinent committees, advisory boards, and review panels pertaining to Soil Chemistry; and
8. To promote communication between IUSS and all national soil science societies to facilitate the interaction between soil chemists and IUPAC in fulfilling the role of Soil Chemistry in ICSU.

13-4 SUMMARY AND CONCLUSIONS

The International Council of Scientific Unions promotes international scientific activity in the different branches of science and the applications of scientific findings for the benefit of humanity. It initiates, designs, and coordinates major international, interdisciplinary research programs to seek to break the barriers of specialization. It addresses matters of common concern to all scientists and acts as a focus for the exchange of ideas, the communication of scientific information, and the development of scientific standards. It serves as spokesman for the world scientific community and as an advisor in topics ranging from ethics to the environment. Soil Chemistry plays a vital role in sustaining agricultural productivity and improving environmental quality. A bridge of communication between soil chemists and the newly created Division of Chemistry and the Environment of IUPAC, which is a member of the ICSU, should be constructed to strongly promote the interaction of pure and applied chemists on the global scale to improving the quality of foodstuffs and fibers produced and sustaining the integrity of the ecosystem, which is vital to human and animal health and the quality of life.

REFERENCES

Blum, W.E.H. (ed.) 1995. Bulletin of the International Society of Soil Science 87. ISSS, Vienna, Austria.
Buffle, J., and H.P. van Leeuwen (ed.). 1992. Environmental particles. Vol. 1. IUPAC Environmental Analytical and Physical Chemistry Series. Lewis Publ., Boca Raton, FL.
Buffle, J., and H.P. van Leeuwen. (ed.). 1993. Environmental particles. Vol. 2. IUPAC Environmental Analytical and Physical Chemistry Series. Lewis Publ., Boca Raton, FL.

Huang, P.M., N. Senesi, and J. Buffle. (ed.) 1998a. Structure and surface reactions of soil particles. Vol. 4. IUPAC Series on Analytical and Physical Chemistry of Environmental Systems. John Wiley & Sons, Chichester, England.

Huang, P.M., D.C. Adriano, T.J. Logan, and R.T. Checkai. (ed.) 1998b. Soil chemistry and ecosystem health. SSSA Spec. Publ. 52, SSSA, Madison, WI.

Leonard, C. (ed.) 1995. International Council of Scientific Unions Year Book. ICSU, Paris.

Leonard, C. (ed.) 1996. International Council of Scientific Unions Year Book. ICSU, Paris.